T0213433

Species Problems and Beyond

Species and Systematics

Phylogenetic Systematics: Haeckel To Hennig, *by Olivier Rieppel*

Evolution By Natural Selection: Confidence, Evidence and the Gap, *by Michaelis Michael*

The Evolution of Phylogenetic Systematics, *edited by Andrew Hamilton*

Molecular Panbiogeography on the Tropics, *by Michael Heads*

Beyond Cladistics, *edited by David M. Williams and Sandra Knapp*

Comparative Biogeography: Discovering and Classifying Bio-Geographical Patterns of a Dynamic Earth, *by Lynee R. Parenti and Malte C. Ebach*

Species: A History of the Idea, *by John S. Wilkins*

What Species Mean: Understanding the Units of Biodiversity, *by Julia Sigwart*

What, If Anything, Are Species? *by Brent D. Mishler*

Biological Systematics, *by Igor Ya. Pavlinov*

Species Problems and Beyond: Contemporary Issues in Philosophy and Practice, *by John S. Wilkins, Frank E. Zachos, Igor Ya. Pavlinov*

From Observations to Optimal Phylogenetic Trees: Phylogenetic Analysis of Morphological Data: Volume 1, *by Pablo Goloboff*

Refining Phylogenetic Analyses: Phylogenetic Analysis of Morphological Data: Volume 2, *by Pablo Goloboff*

For more information about this series, please visit: https://www.crcpress.com/Species-and-Systematics/book-series/CRCSPEANDSYS

Species Problems and Beyond

Contemporary Issues in Philosophy and Practice

Edited By
John S. Wilkins, Frank E. Zachos, and Igor Ya. Pavlinov

CRC Press
Taylor & Francis Group
Boca Raton London New York

CRC Press is an imprint of the
Taylor & Francis Group, an **informa** business

Cover art by Scott Partridge

First edition published 2022
by CRC Press
6000 Broken Sound Parkway NW, Suite 300, Boca Raton, FL 33487-2742

and by CRC Press
2 Park Square, Milton Park, Abingdon, Oxon, OX14 4RN

CRC Press is an imprint of Taylor & Francis Group, LLC

ISBN: [978-0-367-42537-1] (hbk)
ISBN: [978-1-032-22147-2] (pbk)
ISBN: [978-0-367-85560-4] (ebk)

DOI: [10.1201/9780367855604]

Typeset in Times
by Deanta Global Publishing Services, Chennai, India

Contents

SECTION 1   Concepts and Theories

SECTION 2   Practice and Methods

SECTION 3 Ranks and Trees and Names

SECTION 4 Metaphysics and Epistemologies

Afterword

Acknowledgement

We are greatly indebted to our contributors for their work in preparing these chapters. In particular, we thank Rob Wilson for his afterword and advice on editing this volume, although of course, he is not to be held responsible for our errors of judgement.

We also thank our Editor at CRC Press, Alice Oven, and her staff for prompt replies to our numerous queries during a difficult time of pandemic and overall assistance with preparation.

Editors' Biographies

John S. Wilkins

John Wilkins did his PhD at the University of Melbourne. He has since researched and taught at the University of Queensland, the University of Sydney, the University of New South Wales, and the University of Melbourne. He has published several books: *Species: A History of the Idea* (2009) and its successor *Species: The Evolution of the Idea* (2018), *Defining Species* (2009), and *The Nature of Classification* (2013, with Malte C. Ebach), and edited *Intelligent Design and Religion as a Natural Phenomenon* (2010). John is currently subject coordinator at the University of Melbourne School of Historical and Philosophical Studies, where he has previously been a research fellow. His interests include species conceptions (obviously), the history of biology, philosophy, and sociology of religion, phenomena, evolution, taxonomy, and Terry Pratchett's *oeuvre*. He has not published on the last listed.

Frank E. Zachos

Dr Zachos is head of the Mammal Collection at the Natural History Museum in Vienna, Austria, and an affiliated professor at the Department of Genetics, University of the Free State, Bloemfontein, South Africa. His interests include molecular and morphological approaches in microevolutionary studies on mammals, with a taxonomic focus on ungulates, particularly cervids, and birds of prey. He carries out research in the fields of population genetics, molecular (mostly intraspecific) systematics and phylogeography, developmental homeostasis and fluctuating asymmetry, as well as conservation genetics and the Quaternary distribution history of mammals and birds. His theoretical interests comprise species concepts, the foundations of taxonomy and systematics, and the history and philosophy of (evolutionary) biology.

Igor Ya. Pavlinov

Igor Ya. Pavlinov was, until his retirement in 2018, leading researcher and the chief of the Mammal Division at the Zoological Museum at the Lomonosov Moscow State University. His DrS dissertation was 'Cladistic approach in phylogenetics and taxonomy: theoretical foundations of evolutionary cladistics' (1997). He is still affiliated with the Zoological Museum, where he is a curator of mammals. His principal research interests are in theoretical taxonomy and phylogenetics, systematics of mammals (mainly rodents), and morphometrics. Igor is the author of several dozen books on natural history, including: *Cladistic Analysis: Methodological Aspects* (1990), *Evolution of Life* (2001), *Systematics of Mammals of the World* (2003), *Foundation of the Contemporary Phylogenetics* (2005), *Nomenclature in Systematics* (2015) and its successor *Taxonomic nomenclature — What's in a name: Theory and History* (2022), *Foundation of Biological Systematics* (2018) and its successor *Biological Systematics: History and Theory* (2021), *Mammals of Russia: A Reference and Guide* (2019), several books on particular orders and families of mammals, etc.

List of Contributors

Yuichi Amitani
University of Aizu
Aizuwakamatu
Fukushima, Japan

Matthew J. Barker
Concordia University
Montreal, Canada

Bryan C. Carstens
Department of Evolution, Ecology, and
    Organismal Biology
Ohio State University
Columbus, OH
and
Centre for Logic and Philosophy of
    Science
Leuven, Belgium

Les Christidis
Honorary Fellow
University of Melbourne
Melbourne, Australia

Stijn Conix
University of Padua
Padua, Italy

John Dupré
University of Exeter
Exeter, UK

Stephen T. Garnett
Charles Darwin University
Darwin, Australia

Catherine Kendig
Michigan State University
East Lansing, MI

Brent D. Mishler
Department of Integrative Biology
University of California, Berkeley
Berkeley, CA

Alessandro Minelli
University of Padua
Padua, Italy

Jay Odenbaugh
Department of Philosophy
Lewis and Clark College
Portland, OR

Igor Ya. Pavlinov
Zoological Museum of Lomonosov
Moscow State University
Moscow, Russia

Aleta Quinn
Department of Politics and Philosophy
University of Idaho
Moscow, ID

Thomas A.C. Reydon
Institute of Philosophy & Centre for
    Ethics and Law in the Life
    Sciences
Leibniz University Hannover
Hannover, Germany

Julia D. Sigwart
Senckenberg Research Institute and
    Museum
Frankfurt, Germany

John S. Wilkins
University of Melbourne
Melbourne, Australia

Robert A. Wilson
University of Western Australia
Perth, Australia

Frank E. Zachos
Natural History Museum Vienna
Vienna, Austria
and

Department of Evolutionary Biology
University of Vienna
Vienna, Austria
and
Department of Genetics
University of the Free State
Bloemfontein, South Africa

Introduction

SPECIES PROBLEMS AND BEYOND: AN INTRODUCTION

Species, or more exactly, 'the' Species Problem, is a topic in science, in the philosophy of science, and in general philosophy. It follows, therefore, that it is important that practitioners of all three fields address it. There is not one species problem, but many; hence the title of this volume. And, they need to be dealt with, if not so that science can proceed (since science proceeds independently of philosophical ruminations), then at least so that it can proceed with more clarity.

At the end of the last century, two books, *Species: New Interdisciplinary Essays* and *Species Concepts and Phylogenetic Theory* (Wilson 1999a; Wheeler and Meier 2000), were published that did precisely this: they presented arguments and views from these disciplines regarding species, and they were influential on the subsequent discussion.* However, this was over two decades ago, and both the philosophical debates and the scientific definitions, practices, and protocols have moved on.

What we have tried to do in this volume is to cover some of the main issues current in these fields. It has been divided into four sections: **Concepts and Theories**; **Practice and Methods**; **Ranks and Trees and Names**; and **Metaphysics and Epistemologies**. Our contributors include systematists, philosophers, historians, and working biologists, often in the same person.

To understand the need for this book, we need to see what has changed in the two decades since the volumes mentioned were published.

There has been a considerable shift in emphasis in philosophical approaches to the problem in this period. The work of the late David Hull focused primarily upon the metaphysics of species, and along with Michael Ghiselin, he argued that species were not kinds at all but historical individuals (Ghiselin 1974; Hull 1978). One of the implications of that view is that if species are spatiotemporally restricted entities, then each species is a species in its own way, as evolution tends not to have patterns that exactly repeat. Hull therefore rejected monistic views of what species were. As he once said to one of the editors, biology breaks all generalisations, including that one.

Another focus in the Wilson volume was essentialism. According to the Mayrian history of the topic, pre- and non-Darwinian views of species held that they had essences, which Darwin overturned. Wilkins (2009, 2018) argued instead that essentialism was a feature of *logical* notions of kinds, but not of *biological* notions, and that Mayr, Hull, and others had mistaken the logical tradition of post-Aristotelian logics for the naturalists' tradition of just putting like things into groups. In fact,

---

* A third volume at the same time was *Endless Forms: Species and Speciation* (Howard and Berlocher 1998). This was written as an update from the scientific battlefront by scientists and provides an excellent introduction to the issues within the various biological disciplines at the time. A fourth was *Species: The Units of Biodiversity* (Claridge *et al.* 1997), also mostly from a scientific perspective.

biological essences were mostly *post*-Darwin, and the Mayrian account has since been dubbed the Essentialism Story (see also Winsor 2003).

In the Wilson volume, a new essentialism was proposed by Rob Wilson (1999b) and Paul Griffiths (1999), based on the work of Richard Boyd (1999), that species are *homeostatic property clusters*, which is to say that they are defined by the mechanisms that cohere the lineages. Kevin de Queiroz, a vertebrate biologist at the Smithsonian, proposed a 'general lineage species concept' about the same time (de Queiroz 1998, 1999), in which species were lineages. These two views have become widely accepted within the philosophical and biological communities.

Less widely accepted is the 'new essentialism'. Apart from Wilson's view, a *historical essentialism* of species was proposed by Griffiths (1999), in which the essence of the species individual is that it is a series of lineages that have a distinct origin. And, Michael Devitt (2008, 2010, 2018) has proposed a 'traditional' logical form of biological essentialism. Neither of these seem to have been taken up by systematists or philosophers. Griffiths' view requires something akin to, if not entirely, monophyly for species, and this is too restrictive a constraint, as many 'good species' are non-monophyletic. Devitt's view, which seeks to marry logical and linguistic essences with biology, has not carried the day yet, either. John Dupré in this volume carries his process ontology of biology (Dupré 2017; Nicholson and Dupré 2018; Dupré 2020) further into the species problem, arguing that some species are processes, while others, including John Wilkins, argue instead (or as well) that they are epistemic items. Another, more 'biological', view of 'species essentialism' presumes that species, as a natural phenomenon, possesses a certain emergent property that is a part of the whole of natural history of organisms and can be called *specieshood* (Pavlinov this volume).

Several of our contributors have taken up these questions afresh. Matt Barker begins the volume by asking whether we are finally ready to begin dealing with the species problem, again, for the first time.

Monism appears to be effectively dead (Reydon, but see Pavlinov this volume), as much as philosophical topics can die, largely due to Ereshefsky (1992, 1998). Essentialism seems to be either benignly revived or where malignant in Mayr's sense, rejected outright. The homeostatic property clusters account is widely accepted, at least with respect to epistemology if not ontology. But either way, the problem, as with de Queiroz's view, is that the *rank* of species, or specieshood, is not specified by these accounts. Why is a lineage a species rather than a subspecific population or gene tree? Why are Devittian essences the essences of *species* and not of something else? This problem also applies to Griffiths' historical essence, as it also does with the so-called *phylogenetic* definition and de Queiroz's view. What shared origin or ancestral population makes something a species? It seems we are left with the evolutionary species concept as a fundamental notion, and even this doesn't help, either operationally or definitionally, for how can we know whether a single isolated lineage has its own evolutionary fate (see also Kendig this volume)? The rank problem remains.

One solution, or rather, dissolution, to the species problem is to deny that the species rank is in fact 'real' (for some interpretation of realism that satisfies biological questions, usually theoretical). Eliminativism of this kind for species concepts was proposed by Marc Ereshefsky (1992; see also Wilkins 2003), who suggested that there are *many* distinct theoretical notions of 'species' that are not compatible, and that they should be given specific and distinct categorical terms. Some critiqued this pluralist approach (Brigandt 2003; Conix and Chi 2020), and it is not yet widely adopted, although there have been several recent papers in favor of eliminativism and pluralism (Reydon 2005a, 2005b; Mishler and Wilkins 2018; Reydon and Kunz 2019). Some of these critics are merely anti-rankists (no species rank), while others are entirely anti-species (pure eliminativism of both category and taxa).

Pluralism has the difficulty of why the diverse and plural concepts are even to be considered as *species* concepts. One answer might be psychologistic – humans tend to have what Scott Atran once called 'folk species', or as Bulmer called them, 'speciemes' (Bulmer 1967; Atran 1995). But in effect, this merely puts us back to species as 'kinds' and in no way solves the rank problem.

In addition to these philosophical issues, there are some practical and disciplinary issues in the biosciences themselves. One topic that has arisen in greater detail and urgency is the so-called *delimitation problem* of finding criteria, usually molecular, for identifying whether an organism is to be included in a species or not (Hedrén 2004; de Queiroz 2007; Wiens 2007; Tobias *et al.* 2010; see also Carstens *et al.* 2013). Smith and Carstens in this volume discuss this topic.

With an explosion of delimitation methods and criteria comes what has been called *taxonomic inflation* (Isaac *et al.* 2004; Harris and Froufe 2005; Padial and De la Riva 2006; Zachos *et al.* 2013; Zachos 2015, 2016; Gippoliti and Groves 2012; Zachos and Lovari 2013) – covered by Smith and Carstens this volume – which is causing concern about identifying biodiversity in a consistent and manageable fashion, leading to questions of *taxonomic governance* and its tractability (Conix *et al.*, Sigwart this volume). Also, the sheer volume of data and data types now available with molecular sequencing techniques raises its own technical and conceptual difficulties (Quinn this volume).

The reason for the popularity among both biologists and philosophers of some sort of pluralism – or at least, skepticism towards monism – is, of course, the fact that despite all efforts, no single definition of species that is both universal and non-arbitrary has been achieved in over 300 years since John Ray first advanced a truly 'biological' definition. Species lists, therefore, are at least potentially (and most of the time, de facto) lists of apples and oranges, with serious consequences for biological practice – from quantification of biodiversity, through analyses of diversification and macroecological patterns, to conservation and management (Zachos this volume). Species delimitation-as-(grouping) is not the same as species delimitation-as-(ranking), and based upon the insight that there is no universal species level, Mishler (this volume) sketches the picture of a rankless biology in a 'post-species world' that aims to circumvent the incommensurability problem of the species category. Doubts about the reality of a single species level are also supported by Minelli's historical review (this volume) and the conclusions he draws from it.

On the other hand, Reydon, and Amitani, and Pavlinov (this volume), argue, from the viewpoints of philosophical and cognitive science perspectives, that the term *species* can still be useful conceptually. It is perhaps no coincidence that this view is defended more by philosophers (but see Wilkins' idea about 'species as a phenomenon' and Barker's discussion of that idea in this volume), whereas for biologists, the species problem relates more to practice when it comes to applied situations where particular species are to be recognized and explored. Both sides are well advised to listen to each other, as science is at its best when it is cognisant of underlying theoretical, including metaphysical, issues, and philosophy of science is best when it is also philosophy for science.

We have also included some historical research (Minelli, Pavlinov, Wilkins) following on from the interdisciplinarity of the Wilson (1999) book. The role of history, however, in the species problem is not to strengthen or weaken a particular vision of *species* today but to explain how a concept so central to so much science and philosophy can have arisen in a manner that seems haphazard and often not at all theoretically driven.

We are very grateful to the contributors for their efforts to make this volume as interesting as it might be to all kinds of theoreticians interested in the various aspects of the species problem, and we hope that the reader will find much to like, productively dislike, and consider in this book.

The editors
John S. Wilkins, Frank E. Zachos, and Igor Ya. Pavlinov

REFERENCES

Atran, S., 1995. Causal constraints on categories and categorical constraints on biological reasoning across cultures. *In*: D. Sperber, D. Premack, and A.J. Premack, eds. *Causal cognition: A multidisciplinary debate*. Oxford, UK and New York: Clarendon Press and Oxford University Press, 205–223.

Boyd, R., 1999. Homeostasis, species, and higher taxa. *In*: R. Wilson, ed. *Species, new interdisciplinary essays*. Cambridge, MA: Bradford/MIT Press, 141–186.

Brigandt, I., 2003. Species pluralism does not imply species eliminativism. *Philosophy of Science*, 70(5), 1305–1316.

Bulmer, R., 1967. Why is the cassowary not a bird? A problem among the Karam of the New Guinea highlands. *Journal of the Royal Anthropological Institute*, 2(1), 5–25.

Carstens, B.C., Pelletier, T.A., Reid, N.M., and Satler, J.D., 2013. How to fail at species delimitation. *Molecular Ecology*. 22(17): 4369–4383. https://doi.org/10.1111/mec.12413.

Claridge, M.F., Dawah, H.A., and Wilson, M.R., 1997. *Species: The units of biodiversity*. London and New York: Chapman & Hall.

Conix, S., and Chi, P.-S., 2020. Against natural kind eliminativism. *Synthese*.

de Queiroz, K., 1998. The general lineage concept of species, species criteria, and the process of speciation. *In*: D.J. Howard and S.H. Berlocher, eds. *Endless forms: Species and speciation*. New York: Oxford University Press, 57–75.

de Queiroz, K., 1999. The general lineage concept of species and the defining properties of the species category. *In*: R.A. Wilson, ed. *Species, new interdisciplinary essays*. Cambridge, MA: Bradford/MIT Press, 49–88.

de Queiroz, K., 2007. Species concepts and species delimitation. *Systematic Biology*, 56(6), 879–886.

Devitt, M., 2008. Resurrecting biological essentialism. *Philosophy of Science*, 75(3), 344–382.

Devitt, M., 2010. Species have (partly) intrinsic essences. *Philosophy of Science*, 77(5), 648–661.

Devitt, M., 2018. Individual essentialism in biology. *Biology & Philosophy*, 33(5), 39.

Dupré, J., 2017. The metaphysics of evolution. *Interface Focus*, 7, 20160148.

Dupré, J., 2020. Life as process. *Epistemology & Philosophy of Science*, 57(2), 96–113.

Ereshefsky, M., 1992. Eliminative pluralism. *Philosophy of Science*, 59, 671–690.

Ereshefsky, M., 1998. Species pluralism and anti-realism. *Philosophy of Science*, 65(1), 103–120.

Ghiselin, M.T., 1974. A radical solution to the species problem. *Systematic Zoology*, 23, 536–544.

Gippoliti, S., and Groves, C.P., 2012. 'Taxonomic inflation' in the historical context of mammalogy and conservation. *Hystrix, the Italian Journal of Mammalogy*, 23(2), 8–11.

Griffiths, P.E., 1999. Squaring the circle: Natural kinds with historical essences. *In*: R.A. Wilson, ed. *Species, new interdisciplinary essays*. Cambridge, MA: Bradford/MIT Press, 209–228.

Harris, D.J., and Froufe, E., 2005. Taxonomic inflation: Species concept or historical geopolitical bias? *Trends in Ecology & Evolution*, 20(1), 6–7.

Hedrén, M., 2004. Species delimitation and the partitioning of genetic diversity, an example from the Carex flava complex (Cyperaceae). *Biodiversity and Conservation*, 13(2), 293–316.

Howard, D.J., and Berlocher, S.H., 1998. *Endless forms: Species and speciation*. New York: Oxford University Press.

Hull, D.L., 1978. A matter of individuality. *Philosophy of Science*, 45, 335–360.

Isaac, N.J.B., Mallet, J., and Mace, G.M., 2004. Taxonomic inflation: Its influence on macroecology and conservation. *Trends in Ecology & Evolution*, 19(9), 464–469.

Mishler, B.D., and Wilkins, J.S., 2018. The hunting of the SNaRC: A snarky solution to the species problem. *Philosophy, Theory, and Practice in Biology*, 10(1), 1–18.

Nicholson, D.J., and Dupré, J., eds., 2018. *Everything flows: Towards a processual philosophy of biology*. Oxford, UK and New York: Oxford University Press.

Padial, J., and De la Riva, I., 2006. Taxonomic inflation and the stability of species lists: The perils of ostrich's behavior. *Systematic Biology*, 55(5), 859–867.

Reydon, T.A.C., 2005a. On the nature of the species problem and the four meanings of 'species'. *Studies in History and Philosophy of Science Part C: Studies in History and Philosophy of Biological and Biomedical Sciences*, 36(1), 135–158.

Reydon, T.A.C., 2005b. *Species as units of generalization in biological science: A philosophical analysis*. Rotterdam: Self-Published.

Reydon, T.A.C., and Kunz, W., 2019. Species as natural entities, instrumental units and ranked taxa: New perspectives on the grouping and ranking problems. *Biological Journal of the Linnean Society*, 126(4), 623–636.

Tobias, J.A., Seddon, N., Spottiswoode, C.N., Pilgrim, J.D., Fishpool, L.D.C., and Collar, N.J., 2010. Quantitative criteria for species delimitation. *Ibis*, 152(4), 724–746.

Wheeler, Q.D., and Meier, R., 2000. *Species concepts and phylogenetic theory: A debate*. New York: Columbia University Press.

Wiens, J.J., 2007. Species delimitation: New approaches for discovering diversity. *Systematic Biology*, 56(6), 875–878.

Wilkins, J.S., 2003. How to be a chaste species pluralist-realist: The origins of species modes and the synapomorphic species concept. *Biology and Philosophy*, 18, 621–638.

Wilkins, J.S., 2009. *Species: A history of the idea*. Berkeley, CA: University of California Press.

Wilkins, J.S., 2018. *Species: The evolution of the idea*. 2nd ed. Boca Raton, FL: CRC Press.

Wilson, R.A., 1999a. *Species: New interdisciplinary essays*. Cambridge, MA: MIT Press.

Wilson, R.A., 1999b. Realism, essence, and kind: Resuscitating species essentialism? *In*: R.A. Wilson, ed. *Species, new interdisciplinary essays*. Cambridge, MA: Bradford/MIT Press, 187–208.

Winsor, M.P., 2003. Non-essentialist methods in pre-Darwinian taxonomy. *Biology & Philosophy*, 18, 387–400.

Zachos, F.E., 2015. Taxonomic inflation, the phylogenetic species concept and lineages in the tree of life – A cautionary comment on species splitting. *Journal of Zoological Systematics and Evolutionary Research*, 53(2), 180–184.

Zachos, F.E., 2016. Tree thinking and species delimitation: Guidelines for taxonomy and phylogenetic terminology. *Mammalian Biology*, 81(2), 185–188.

Zachos, F.E., Apollonio, M., Bärmann, E.V., Festa-Bianchet, M., Göhlich, U., Habel, J.C., Haring, E., Kruckenhauser, L., Lovari, S., McDevitt, A.D., Pertoldi, C., Rössner, G.E., Sánchez-Villagra, M.R., Scandura, M., and Suchentrunk, F., 2013. Species inflation and taxonomic artefacts—A critical comment on recent trends in mammalian classification. *Mammalian Biology*, 78(1), 1–6.

Zachos, F.E., and Lovari, S., 2013. Taxonomic inflation and the poverty of the phylogenetic species concept – A reply to Gippoliti and Groves. *Hystrix, the Italian Journal of Mammalogy*, 24(2), 142–144.

Section 1

Concepts and Theories

1 We Are Nearly Ready to Begin the Species Problem

*Matthew J. Barker**

CONTENTS

1.1 INTRODUCTION

What are species? More exactly, what are separately evolving species lineages?

They might be *feedback systems*, displaying a distinctive degree or grade of metapopulation-level cohesion, which manifests and varies dynamically with the magnitude and frequency of feedback relations between several causal variables at population levels.

* Thank you to Igor Pavlinov, John Wilkins, Rob Wilson, and Frank Zachos for helpful comments on a previous draft of the chapter, to Kevin de Queiroz for a clarifying email correspondence some years ago, and to David Baum, Marc Ereshefsky, Mark Hershkovitz, Josine Lafontaine, Ivan Prates, Elliott Sober, and again Robert Wilson for helpful past discussions of some of the ideas in this chapter.

DOI: 10.1201/9780367855604-2

3

The next section briefly outlines that recent proposal (see Barker 2019a) – not to try to convince you it is true, but rather, to provide just one example of a new *kind* of species theory. This will set the stage for the chapter's subsequent sections and its main aims, which involve next isolating a hard, long-standing problem for providing a theory of the species category (Section 1.3) and sorting species concepts that address that problem (Section 1.4). The chapter then uncovers instructive flaws in several views that imply we have either already *solved* that hard species problem or *dissolved* it altogether – so-called We Are Done views. The flaws are found in cause-focused species concepts, such as the biological species concept (BSC) (Section 1.5), and in non-cause-focused species concepts, such as the general lineage species concept (GLSC) and evolutionary species concept (EvSC). Other We Are Done views that are challenged include those stemming from Ereshefsky's eliminative pluralism about the species category (Section 1.7), Mishler's pessimism about the category (Section 1.8), and Wilkins' phenomenalism about it (Section 1.9). More constructively, the chapter argues for a Revving Up view: rather than being done with the hard species problem, we are nearly ready to *begin* in earnest, as the feedback model will help exemplify.

1.2   EXEMPLIFYING A NEW KIND OF SPECIES THEORY: FEEDBACK SPECIES

In a feedback system, processes loop or cycle. The values taken by variables over one time period feed back into the system, influencing the values taken by those *same* variables at later times. Put differently, output conditions turn around and serve as input conditions in further iterations of the same or similar pattern, as exhaust air produced from combustion in a turbo engine is fed back into the system to help (via interactions with other variables) amplify further combustion, thus increasing exhaust, and on and on until there is intervention or a dampening feedback mechanism is engaged. How about in an evolving species lineage? It is widely agreed that any species is a group of populations, a metapopulation (de Queiroz 2005). But, authors argue about which among a variety of processes or variables are *most important* for connecting and 'holding together' the populations in the group. Nearly everyone acknowledges that many variables often *interact*. But, those variables deemed most important are privileged over those with which they interact and are then referenced in differing definitions of species concepts to the exclusion of the other variables. Many have privileged gene flow processes in this way (e.g., Mayr 1942; 1963; Morjan and Rieseberg 2004; Bobay and Ochman 2017), others the sharing of selection regimes and adaptive zones (e.g., Van Valen 1976; Cohan 2002), and so on. Table 1.1 provides a partial list of variables that have been discussed. But fortunately, in recent years – including for both sexual and asexual populations – some experts have taken steps away from the privileging of just this or that small set of variables, making interactions between them central to their accounts of the nature of species (e.g., Templeton 1989; Boyd 1999; R.A. Wilson, Barker, and Brigandt 2007; Barker and Wilson 2010; Ellstrand 2014; Shapiro and Polz 2015; Novick and Doolittle 2021). The feedback model incorporates this but goes further, suggesting

TABLE 1.1
Provisional Running List of Causal Variables Entering Feedback Relations

Variable label	Name of variable	Example works appealing to variables
g	Gene flow	(Mayr 1963; Brooks and Wiley 1988; Morjan and Rieseberg 2004; Bobay and Ochman 2017)
s	Shared selection regimes	(Ehrlich and Raven 1969; Van Valen 1976; Lande 1980; Mishler and Donoghue 1982; Templeton 1989; Cohan 2011; Shapiro and Polz 2015)
h	Homeostatic developmental systems	(Ehrlich and Raven 1969; Mayr 1970; Wiley 1981)
c	Colonisation	(Hellberg et al. 2002, 275–277)
m	Mutation	(Mayr 1970; Hellberg et al. 2002, 275–277; Morjan and Rieseberg 2004)
r	Genetic recombination	(Carson 1957; Mayr 1970)
t	Trait similarities	(Barker 2019a)

Source: Modified from Barker, M.J, *Philosophy, Theory, and Practice in Biology,* 11, 2019

a long view on which interacting variables over one time period influence the values of these same and other variables at later time periods, which influences still later values of these variables at later time periods, and on and on over many periods, forming an evolving species system as the cycling feedback relations set it apart from other metapopulations.

For instance, gene flow between populations in a group over some time period, $t1$, distributes alleles between populations. That results at $t2$ in the populations having population-level trait frequencies closer to each other than to out-group populations. Subsequently, this helps ensure that the populations participate, over $t3$, in selection regimes in ways more like each other than out-group populations (e.g., Morjan and Rieseberg 2004). And in turn, that may help cause further gene flow at $t4$, and trait sharing at $t5$, and selection regime sharing at $t6$, and …. The exact sequence needn't and surely won't continue on with the same degree of regularity as in a turbo engine; a suite of variables may feed back into the system in diverse and changing ways over time, with varying degrees of influence or importance across cases. This suggests a relatively inclusive account of the species category (e.g., Ellstrand 2014; Novick and Doolittle 2021), on which diverse variables and relationships between them can be recognised as important across species groups.

However, the feedback model needn't be *so* inclusive as to imply that anything goes, or that species groups can't be distinguished from others. Part of the proposal is that *relative to many non-species metapopulations*, species groups feature distinguishing kinds of complex feedback regularity, even if this regularity is loose and liberal when compared instead with turbo engines. And, the model doesn't purchase inclusivity at the cost of vagueness, as we'll later see some other views of species do when they discuss the cohesion produced by species-forming and species-maintaining processes without comprehensively detailing the nature of such cohesion (e.g., Mayr 1963; Hull 1976; de Queiroz 1998; Wiley and Mayden 2000a; see Barker and Wilson 2010). On the feedback model, evolutionary cohesion is something that not only species but also other types of metapopulations can exhibit in different degrees or ways, and it can be given exacting but flexible mathematical descriptions that enable testable predictions. A basic central idea is that as feedback relations play out at the level of populations (within a group of them), this manifests *metapopulation feedback cohesion*, or *M*, at the higher level of the whole metapopulation (Barker 2019a). The values taken by recurring causal variables in the metapopulation feedback system can vary in magnitude, frequency, or both. When magnitude and frequency are both high across many variables, that manifests a high value for *M* at the level of the whole metapopulation system, perhaps indicating a species or even a sub-species. Low magnitudes and frequencies across many variables manifest low *M*, perhaps a genus or family. Intermediate values are of course possible. So, *M* at the metapopulation level varies dynamically with the magnitude and frequency of feedback relations between causal variables at population levels; evolving metapopulation lineages of many sorts may just *be* these dynamic feedback systems; lineages of the *species* type in particular may be those within a species-distinguishing range of *M* values.

Now, regarding prediction, consider a traditional view that has been widespread since the Modern Synthesis and will be part of this chapter's focus. It proposes that within a variety of biological theories, the species category plays a more important or fundamental role than both more inclusive metapopulation categories (e.g., genus, family, or class) and less inclusive ones (e.g., sub-species, variety, or stirp) (see Wilkins 2018). As biologist Frank Zachos summarises it, 'Species are the fundamental unit in many branches of biology' (Zachos 2015, 180). In combination with the feedback model, this could be developed to predict that we'll find a metapopulation level – the species level, presumably – at which the involved metapopulations clump in feedback variable space in a more distinctive or patterned *and* theoretically important way than at other metapopulation levels. Call this variable space *M* space (Barker 2019a). In it, each axis depicts the intensity (some combination of feedback and magnitude) of a particular variable's feedback relations, a value whose measure is scaled between 0 and 1. For illustrative purposes, Figure 1.1 borrows from Peter-Godfrey Smith's (2009) way of depicting such variable spaces (in a different, non-species context where feedback isn't central) and uses just three example variables as axes: gene flow (*g*), trait similarities between populations (*t*), and sharing of selection regimes (*s*). Suppose we plot a wide variety of metapopulations, from those hypothesised to be at or around the species level of inclusivity to many of those thought to be much more inclusive metapopulations. (The example and figure set aside less inclusive metapopulations.) The traditional view would then predict that we get a distinctive clustering of a small proportion of the metapopulations in some non-zero patch of the variable space, one of distinctive importance to biological theories. That would be the 'species category patch', which may well have fuzzy boundaries but stand out to some quantifiable degree, with the involvement of some variables

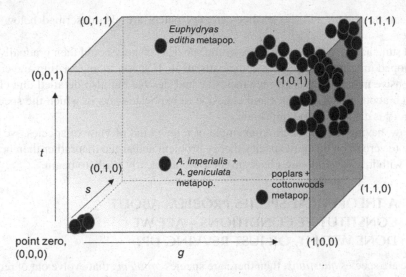

FIGURE 1.1 Plotting M values for metapopulations in M space. Black ovals represent metapopulations. On the prediction described in the text, many will form a dense cluster near point zero, while others form a species category patch in the back + top + right region. Outliers include the *Euphydryas editha* metapopulation, exemplifying the 1969 claim of Ehrlich and Raven that a California metapopulation within this butterfly group features very little gene flow between populations, but seemingly high t and s. The *A. imperialis* + *A. geniculata* example reflects the implication in Barbará et al. (22007, 1990) that sympatric groups from these two named *Alcantarea* species (bromeliads) feature low but non-zero g, s, and t. The poplars + cottonwoods example is from Templeton (1989), who claimed low levels of t and s for this inter-specific group, despite significant levels of gene flow between the involved poplars and cottonwoods. (Design aspects of this figure are taken directly from Figure 3.1 and Figure 5.1 in Godfrey-Smith, P., *Darwinian Populations and Natural Selection*, Oxford University Press, Oxford, 2009, and the figure here is reused from my paper: Barker, M.J, *Philosophy, Theory, and Practice in Biology*, 11, 2019.)

in feedback relations reinforcing the involvement of others in such relations. Along with this, the traditional view could predict that most of the remainder of the metapopulations (more inclusive ones, such as those associated with genera or families[*]) will feature significantly less intense feedback relations, perhaps 'layering out' in a portion of variable space between the species category patch and point zero, or even just clustering near point zero in M space. Figure 1.1 represents the latter of these, in the case where the species category patch is in the top + back + right (though a few other interesting metapopulations, cases scattered in regions of M space intermediate

---

[*] This will sound strange to biologists who think any metapopulation is *by definition* less inclusive than, say, a whole genus or family. But, the usual way that such inclusion is definitionally ruled out is by requiring that only members of the same *species* can form a metapopulation. We cannot accept that here, at least not without vicious circularity, because the very point is that we are seeking rather than presuming an account of the species category.

between point zero and the species category patch, are also described below the figure).

In sum, a theory about the nature of the species category could then gradually be developed in terms of species-specific patterns in M space, in an exacting and comprehensive manner that is at once inclusive and flexible but also detailed and clear enough about causal variables and cohesion to explicate ways in which the species category is distinctive and important.

Now, having in hand such an example of a new kind of view of species, we are ready to zero in on the hard species theory problem and argue that rather than being done with it, as several views have implied, we are nearly ready to begin.

1.3 A THEORETICAL SPECIES PROBLEM ABOUT CONSTITUTIVE CONDITIONS – ARE WE DONE WITH IT, OR JUST REVVING UP?

There are species *questions*, then there are species *problems* that evolve out of recalcitrant questions. Start with questions.

Earlier, we asked, 'what are species?'. But, we oughtn't be so simplistic. People often instead put the question in terms of definition, saying 'How should "species" be defined?' But, that is both simplistic *and* misleading, as it suggests the question is primarily about how we concisely summarise complex theories within pieces of language (definitions), when really, we are interested in the theories themselves – in discovering the very nature of a putatively important category in the living world.* So, 'what is the *nature* of the species category, understood as a category of separately evolving lineage?' is better. Yet, 'nature' is both vague and ambiguous. So, best to say 'what are the *constitutive conditions* of the species category?', where we're agreed on the minimal starting point that this is a category of separately evolving metapopulation lineage.

A category's constitutive conditions are those *in virtue of which* things belong to that category – conditions that *make* things (under usual circumstance) belong to the category, independently of our abilities to *tell* whether those things in fact satisfy the conditions and thereby belong to the category (Mayden 1997; de Queiroz 1998). That is a distinction between what a category's constitutive conditions *are* across cases, and when those conditions are *satisfied* in particular cases. For instance, according to the International Union of Pure and Applied Chemistry (IUPAC 2019), what makes something a chemical element is being a type of atom in which all instances have the same number of protons in the atomic nucleus. That theory about the chemical element category's constitutive conditions was developed and widely deemed accurate over a period of approximately 25 years (1910–1930s) (Scerri 2019), before our abilities to tell the proton counts of atoms were as sophisticated as now, and when practitioners would often tell elements apart from, say, compounds without looking for protons at all.

In seeking the species category's constitutive conditions, we aren't seeking the conditions that make certain organisms belong to certain species groups. Rather, we seek the

---

* See Cleland (2012) on this with respect to the category *life*.

conditions that set all species metapopulations apart from other categories such as the genus, paradivision, sub-species, and stirp categories. The theories that propose such conditions are found in so-called *species concepts*. Each of these is a construct that it is hoped will accurately correspond to the species category in nature. Although they are typically offered and interpreted as involving respective definitions of 'species', we are focusing on the constitutive conditions they propose. The (in)famous BSC, for instance, proposes that each species is a group of populations satisfying two constitutive conditions (Dobzhansky 1935; Mayr 1942; 1963; 2000b; Coyne and Orr 2004). One, the interbreeding condition, is there being gene flow between populations in the group. The other, the reproductive isolation condition, is there being a lack of gene flow between populations within the group and those outside it. Ernst Mayr urged these as constitutive because (or at least partly because) he thought each was *such an important cause* of conspecific populations becoming or remaining relatively homogenised in their traits – causes of holding the conspecific populations together, separated from other evolving lineages, and explaining why the living world is clumped into these lineages rather than, say, being 'continuous' (Mayr 2000a, 161; 1963).

The BSC also helps illustrate how our species question about constitutive conditions eventually grew into the hard theoretical species problem on which we're zeroing in, because it is widely agreed that the BSC's theory isn't *accurate in a general way* (Ereshefsky 2017). Many groups of populations appear to be species in virtue of satisfying conditions other than the two picked out as constitutive by the BSC. Indeed, these groups seem to be species despite failing to satisfy either the interbreeding condition (e.g., there isn't gene flow between all their populations) or the reproductive isolation condition (e.g., there is gene flow between some of their populations and others) or both. The BSC suffers these problems even though full elaborations of it seem to provide a relatively comprehensive and exacting theory that helps vindicate the traditional idea that the species category is of distinctive importance to several biological theories (e.g., Coyne and Orr 2004).

Many species experts (not all, as we'll see) now think that *every* species concept that has attempted to answer our species question has failed in one way or another. We can thus label and formulate the hard theoretical species problem that has emerged:

The species category theory problem: Despite laborious efforts, we lack a theory of the species category's constitutive conditions that meets, to high degrees, the following four desiderata:

(i)   Comprehensive;
(ii) Exacting;
(iii) Accurate in a quite general way;
(iv) Helps vindicate the traditional idea that the species category is more important in a variety of biological theories than other categories of evolving metapopulation lineages.

The We Are Done views and Revving Up view discussed in this chapter take very different stands on that problem.

According to one set of We Are Done views, we have already *solved* the problem via this or that particular species concept. This implies that I was (and many others are) mistaken to say we lack a theory of constitutive conditions meeting the four desiderata (e.g., Mayden 1997; de Queiroz 1998; 2005; 2007; Coyne and Orr 2004; Bond and Stockman 2008; Camargo and Sites 2013; Zachos 2016). These are *Done by Solution* variants of We Are Done views.

According to a second type of We Are Done view, we are done for a very different reason: even if each individual species group is a real biological entity, there is no general species category to which all and only they belong *and* which could do the presumed important work in biological theories (e.g., Ereshefsky 1992b; Mishler 1999; Pleijel 1999; Mishler and Wilkins 2018). These are *Done by Dissolution* varieties of We Are Done views.

I will argue that both sorts of We Are Done views are off the mark, each for different reasons. But I won't argue that *no* people working on the species problem should bother to further develop these views. We are fortunate that species problems are addressed by a large *community* of diverse researchers. At that more inclusive level, it is best, given our humble state of knowledge about species, that both types of We Are Done view, and my alternative Revving Up view, continue to be developed. So, my negative or critical thesis is this:

> *We Are Done views have not yet succeeded; currently, there is no We Are Done view that everyone should endorse.*

My positive or constructive thesis is:

> *Revving Up is plausible; some proportion of researchers working on species should endorse and further develop it while others work on We Are Done views.*

More specifically, the Revving Up view, which I will defend via examples and lessons drawn, says we have now learned enough from conceptual clarifications and past failures that we are well positioned to develop promising and testable new kinds of theories of the species category's constitutive conditions, theories that integrate strengths of prior ones while avoiding their weaknesses. The feedback model is an example of such a theory, and the time is ripe for developing other such theories as well. Nonetheless, I maintain that we haven't yet begun in earnest, and rather, are nearly ready to do so, because we haven't directly empirically tested such a theory, except via interpretations of past data. We are at a promising 'about to start' stage.

The next step in supporting my critical and constructive theses is to distinguish the types of species concepts that attempt to solve the species category theory problem.

1.4   SETTING ASIDE TAXONOMY TO FOCUS ON TWO KINDS OF FUNCTIONAL SPECIES CONCEPTS

We Are Done views must take firm stances on species concepts, because these are where we find proposed theories that attempt to solve the species category theory problem. One widely endorsed distinction will help us focus on the relevant species concepts, and a less familiar distinction will then help us differentiate between varieties of We Are Done views.

1.4.1 Taxonomic versus Functional Species Concepts

Many species experts have urged this distinction in one way or another (e.g., Cracraft 1989a; Endler 1989; Kimbel and Rak 1993; de Queiroz 1998; 2005; 2011; Ghiselin 2001; Zachos 2016). With minor tweaks, I will follow David Baum's (2009) way of characterising the distinction. What I am calling *taxonomic species concepts*, he calls species-as-taxa concepts and says that these 'are ones that emphasise the similarities between species and taxa at other ranks and mainly reflect a desire to guide taxonomists in the practice of assigning groups of organisms to species taxa' (2009, 74).

What I call functional species concepts, he calls species-as-functional-units concepts, which 'are ones that emphasise the functional cohesion or causal efficacy of species and generally emphasise the role of the term "species" in evolutionary and ecological theory' (2009, 74).

And about applying the distinction, he says:

> The clearest way to distinguish these 2 kinds of concepts is by asking the question: what is it about a group of organisms living at one moment that would make them one species as opposed [sic] to 2, or many, or a subset of a single species? If the answer is something about their functional integrity (e.g., interbreeding potential) or ecological cohesion, then the concept is 'functional'.... If the answer to the question stresses the same kinds of attributes that are used to delimit higher taxa, then the species concept is taxic in outlook.
>
> **(2009, 74)**

More generally, authors have characterised the functional concepts as understanding each denoted species to be an 'evolutionary unit' (Ereshefsky 1991), a dynamic or 'active' group of populations that partake in evolutionary processes connecting them (Ereshefsky 1992a, xiii; Eldredge and Cracraft 1980; Wiley 1981).

Clearly, it is functional species concepts that interest us here.* Those who attempt to solve the species category theory problem understand species to be metapopulations that *function* as active, separately evolving lineages (e.g., de Queiroz 1998). Their theories attempt to uncover the nature of these functional units by specifying the constitutive conditions of the species category. Taxonomic species concepts, on

---

* Without further refining Baum's distinction, some species concepts may count as both taxonomic and functional; e.g., Seifert's (2014) *pragmatic species concept* refers to both a) the lineage divergence and evolutionary explanations typical of species-as-functional-units concepts, and b) the operational diagnostic criteria typical of species-as-taxa concepts. So, let's understand the class of taxonomic species concepts as excluding any concepts that refer additionally to the functional features that Baum's other notion references. Then, the class of functional species concepts includes those that refer exclusively to functional considerations *and* those that refer to a mix of functional and taxonomic considerations.

the other hand, offer views about how to taxonomise species in relation to other types of groups within a taxonomic system. As such, they are typically more operational than functional species concepts.* This shouldn't be taken to imply that there are no relations between them that are very interesting and worth investigating in other contexts.

1.4.2   CAUSE-FOCUSED VERSUS NON-CAUSE-FOCUSED FUNCTIONAL SPECIES CONCEPTS

The less familiar distinction to apply is between functional concepts that are cause-focused and those that are non-cause-focused (Barker 2019a).[†] First, consider the following points, on which all functional species concepts seem to agree (Ereshefsky 1992b; de Queiroz 1998):

(A) *Metapopulations*: Each species is a *group* of populations, that is, a metapopulation.[‡]

(B) *Differentiating process participation*: The populations in each species group participate in the same or similar *evolutionary processes* – e.g., gene flow, niche sharing, selection regimes, and so on – and do so to degrees or in ways that are different from how they interact with populations outside the group.

(C) *Evolutionary cohesion*: Through their differentiating participation in evolutionary processes, the populations in a species jointly manifest and maintain a special kind of evolutionary unity or *cohesion*; i.e., the shared participation connects them into evolving lineages of the species type, diverged from other groups.

---

* In setting aside taxonomic species concepts, we set aside those that identify the species category exclusively in terms of genealogical exclusivity, or monophyly, or morphology, or diagnostic characters, or phenotype, or genotype, or some combination of those. See Baum (2009) for reasoning about this, and Zachos (2016) and Wilkins (2018) for species concepts that identify the species category by those criteria just listed.

† In the 2019a paper just cited, I referred to functional species concepts as 'ontological species concepts', which now seems to me a less informative and potentially misleading term.

‡ If a group of, say, three populations is reduced to one, is it still a group, still a metapopulation? I am not sure. But if not, then (A) will for some biologists be a merely contingent or typical feature of species, and for others a necessary one. The former type of biologist will allow that a species reduced to one population can still count as a species. The latter may be exemplified by Mayr. He was quite clear in his view that to be a species is partly to have a relational property – a group can only be a species if it is reproductively isolated from other groups, so that '[i]f only a single population existed in the entire world, it would be meaningless to call it a species' (1970, 14). He also insisted that to be a species is partly to be a *group* of *interbreeding* populations. If he were similarly strict about that property, then a species reduced from three populations to one would thereby cease to be a species, even if it were reproductively isolated from other groups. The feedback model helps bring out what is sensible in this view, as it highlights the importance, to evolutionary phenomena, of feedback relations *between* populations, relations that are lost when just one population – a mere population – remains.

So, (B) is thought to lead to or otherwise generate (C): the shared and differentiating participation in processes *results in* evolutionary cohesion that is distinctive of species. When outlining the feedback model earlier, I proposed further, more exacting specifications of these ideas. But, that was going beyond what is so far widely agreed upon. We can say that in the relationship between (B) and (C), experts are collectively picking out a *species category pattern*. However, among functional species concepts, there is much disagreement, and collectively much vagueness and ambiguity, about that pattern.

What is clear is that when focusing on this pattern, some functional species concepts pick out causal processes referenced in (B) as constitutive of the species category, e.g., gene flow, niche sharing, participation in selection regimes. Others abstract from causes and instead focus on the cohesion or lineagehood cited in (C). Hence, the distinction exemplified in Table 1.2 by two lists of functional species concepts. Bear in mind that the concepts listed as non-cause-focused do not imply that causes are irrelevant or unimportant to species. But, they deliberately refrain from stating any particular involved causes among the conditions deemed constitutive. Their theories of category constitution focus on (C) rather than (B), as I will further clarify later.

In now turning to Done by Solution variants of We Are Done views, we'll see there is a commonality among them. Each seizes on a particular functional species

---

TABLE 1.2

Cause-Focused versus Non-Cause-Focused Species Concepts, Both Types of Functional Species Concept

Cause-focused species concepts	Non-cause-focused species concepts
Biological species concept (BSC) (Dobzhansky 1935; Mayr 1942; 2000b, 200; Coyne and Orr 2004)	Evolutionary species concept (EvSC) (Simpson 1951; 1961; Wiley 1978; Mayden 1997; Wiley and Mayden 2000b)
Ecological species concept (EcSC) (Van Valen 1976)	General lineage species concept (GLSC)/Unified species concept (USC) (de Queiroz 1998; 2005; 2007)
Cohesion species concept (CSC) (Templeton 1989)	Internodal and/or Hennigian species concept (ISC/HSC)[1] (Willmann 1985; Ridley 1989; Kornet 1993; Meier and
Recognition species concept (RSC) (Paterson 1985)	Willmann 2000)

[1] To some, listing the ISC/HSC in the right-hand column may seem unintuitive. But consider Baum (2009, 74) on this issue: 'The internodal or Hennigian species concept argues that species are lineages (Ridley 1989). At first sight, this concept might seem to align with the species-as-taxa class because modern views of taxa assume that they are, like lineages, natural chunks of the tree of life. However, internodal concepts are usually species-as-functional-units concepts because the limits of a "lineage" at a moment in time are not governed by history, morphology, or similarity but by functional features. Indeed, internodal concepts are best viewed as versions of the evolutionary species concept (Simpson 1951, 1961; Wiley 1978) … the same can be said of the unified (or general lineage) species concept (de Queiroz 2005; 2007).'

concept and claims or implies that it meets the four desiderata identified in our statement of the species category theory problem. 'Mine wins!' has been a durable idea,
especially for the BSC, and for the EvSC and GLSC,* which probably continue to
have the most champions among functional species concepts despite their many
vocal critics. But, there are important differences too, between Done by Solution
views that build on cause-focused species concepts, and those instead building on
non-cause-focused counterparts. We start with the former.

1.5   DONE BY SOLUTION VIA CAUSE-FOCUSED SPECIES CONCEPTS?

Empirical counterexamples are among the most widely proclaimed criticisms of
cause-focused species concepts: for any one of these concepts, there are apparently
many kinds of living things said to violate its proposals about constitutive conditions.† Challenges to the BSC that were briefly introduced earlier best illustrate this,
as they have been representative, most thoroughly explored, and relentless. So, we
focus on them in more detail now.

Critics of the BSC have often interpreted it as though the interbreeding and reproductive isolation conditions it proposes as constitutive were also *each necessary* and
jointly sufficient for a group to be a species. Counterexamples then take one of three
forms:

- Challenging the necessity of the interbreeding condition by appealing to
  groups of populations that seem to be species even though not all the populations within them are connected by gene flow;
- Challenging the necessity of the reproductive isolation condition by documenting groups of populations that apparently are species despite gene flow
  processes stretching beyond them to include out-group populations;
- Challenging the sufficiency of the combination of the two conditions for
  specieshood by documenting groups that do *not* seem to be species even
  though they satisfy the interbreeding and reproductive isolation conditions
  that make something a species according to the BSC.

For counterexamples of the first two sorts, critics have often appealed to non-ani-
mals, including algae and fungi and, especially, plants and bacteria. Many botanists,
for instance, have argued that often there is little to no gene flow between populations *within* what certainly seems to be a species group – *Zea mays* (corn), or *Beta
vulgaris* (beet), or what have you – thus violating the necessity of the interbreeding
condition (e.g., Ehrlich and Raven 1969; Endler 1973; Levin and Kerster 1974; Lande
1980; Grant 1981; Futuyma 1998, 317). Conspecific populations often seem too geographically dispersed for processes of pollination (by wind or insects) to connect

---

* Both Naomi (2011) and Mayden (2013) discuss perceived similarities between the EvSC and GLSC.
† For reasons that can be inferred from our discussion earlier, we set aside criticisms based in operational considerations.

them. Botanists have also cited many putative cases of hybridisation and introgression *between* apparently distinct species groups, violating the necessity of the reproductive isolation condition (e.g., Anderson 1949; Eckenwalder 1984; Templeton 1989; Rieseberg and Soltis 1991).

Bacterial groups have seemed an even tougher case for the BSC's interbreeding condition, as their uniparental reproduction has suggested little gene flow between apparently conspecific populations – or at least, gene flow of a different sort than seems important to uniting sexual populations into species (Dupré 1993; Cohan 2002; Franklin 2007; Doolittle 2019).

And to make matters seem unbearable for the reproductive isolation condition, many authors have argued that disagreeable phenomena found in plants and bacteria are also much more common in *animals* than previously supposed (e.g., Chesser 1983; Templeton 1989; Dawley and Bogart 1989; Whittemore 1993; Mallet, Besansky, and Hahn 2016).

Finally, the most familiar proposed counterexamples to the sufficiency of the BSC's two constitutive conditions are cases of sympatric or clinal speciation, wherein a group splits into two or more species despite its populations being connected by gene flow and together reproductively isolated from out-groups. This has been urged, for instance, in some insects (e.g., Bush et al. 1989), birds (e.g., Sorenson, Sefc, and Payne 2003), fish (e.g., Barluenga et al. 2006), and palms (e.g., Savolainen et al. 2006)

While I too think the BSC, and indeed all extant cause-focused species concepts, have so far failed the general accuracy desideratum of the species category theory problem, I think the underlying reasons for this are often misunderstood in misleading ways. It is all too easy to overestimate the force of, or misinterpret, the sorts of counterexamples just summarised,[*] which sometimes motivates species experts to become unduly pessimistic about solving the species category theory problem at all. A more nuanced interpretation of the counterexamples yields lessons that should motivate some experts to see us as now well positioned, in light of them, to develop new kinds of theories about the species category's constitutive conditions.

First and most obviously, a biologist working on one of the sorts of group cited in the counterexamples wouldn't want to say, 'you see, this shows that the _____ [BSC, EcSC, whatever] is a bad choice for *detecting or delimiting the species boundaries of groups I study*'. This would be to put an operational spin on a challenge whose theoretical relevance, rather than operational relevance, we are assessing.

That distinction remains crucial when, for instance, interpreting Alan Whittemore's (1993) scathing presentation of plant counterexamples against the BSC. In it, he notes that when botanists are trying to detect species boundaries, they typically use taxonomic species concepts that employ similarity or character-based criteria rather than the gene-flow-based criteria of the BSC. But, his point isn't the much older and tangential one that the BSC is operationally inadequate for many plants. Rather, it is that once the taxonomic species concepts have uncovered species groups, then *further investigation* of those, including investigation of population

---

[*] As a graduate student, I sometimes tripped in these ways – see Barker (2007).

structure important to inferences about gene flow, suggests that those groups violate the BSC's proposed theory about constitutive conditions. Merely using a taxonomic species concept instead of the BSC, and seeming to uncover many 'good species' that way, is no problem in itself for the BSC; after all, it is possible in principle that the groups picked out by the taxonomic species concepts coincide with those the BSC *would* pick out. This is clearest when Whittemore notes that the most influential BSC champion – Ernst Mayr himself – used a taxonomic species concept for his own operational matters and then often argued that the groups thereby picked out agreed with the BSC. So, the counterexamples aren't about 'what works operationally for my kinds of groups'. They are about a theoretical problem: many good bets for species groups appear to violate the BSC's proposed constitutive conditions.

Once counterexamples are properly framed as theoretical, there are two general ways for a proponent of a cause-focused species concept to try defending against them.

One, *rejection*, is to give reason for rejecting some of the details of the supposed counterexample, e.g., to contest empirical claims or interpretations, implying that the conditions in question aren't violated after all.

The other, *revision*, is to grant the violation but then accordingly revise or clarify the initial proposal about constitutive conditions, so that going forward the violation is evaded.

There have been principled reasons for revising, in particular softening, the BSC. Both its proposed constitutive conditions are, of course, matters of degree. So, how much gene flow *within* a group does the interbreeding condition require? How little gene flow *out of or into* the group does the reproductive isolation condition require? What Coyne and Orr (2004) describe as 'hard line' versions of the BSC (Coyne and Orr 2004, 33–35) give an answer of *zero* to that second question. But, other BSC advocates, including themselves, have justified switching to softer, more flexible versions of the BSC upon growing appreciation that it is not just the amount of gene flow that matters but also the *kind*. In helpful historical work, Anya Plutynski (2019) documents how advances in molecular biology and genetics gradually overturned Mayr's view of 'the unity of the genotype', showing how species experts interpreted the continuous finding that not all genes in the genome are interdependent and co-adapted. Rather, key distinctions between species groups can sometimes depend on a relatively small number of genes. BSC softening could then reason that gene flow *into or out of* a candidate species needn't be zero for that group to retain its species candidacy, since small amounts of gene flow may not be enough to implicate the relevant genes (e.g., Coyne and Orr 2004). On the flip side, gene flow *within* a candidate group (between its populations) may be barely a trickle, even ten times less than the traditionally demanded equivalent of >1 migrants per generation,[*] yet still enough under the right conditions to spread the relevant genes for holding the group together (e.g., Morjan and Rieseberg 2004, 1341–1342).

With principled softenings of the BSC's proposed constitutive conditions, it has in turn been easier to deploy rejection responses to the supposed counterexamples.

---

[*] For the basis of this tradition, see Wright (1931).

BSC-friendly botanists have recently shown that many of the previous counterexamples were marshalled with what are now comparatively poor methods and outdated assumptions, and that newer and much more reliable and penetrating methods (especially genomic methods in the last ~15 years) suggest that the counterexamples are not nearly as widespread or clear-cut as claimed.[*] For the first time, Loren Rieseberg and colleagues quantitatively tested some of the assumptions packed into presentations of plant counterexamples and found much greater correspondence than supposed between the BSC and 'good species' picked out by taxonomic species concepts, e.g., '75% of phenotypic clusters in plants correspond to reproductively independent lineages … and thus represent biologically real entities. Contrary to conventional wisdom, plant species are more likely than animal species to represent reproductively independent lineages' (Rieseberg, Wood, and Baack 2006, 524).

Regarding uniparental groups, there has been a widely discussed resurgence in attempts to apply the BSC to bacteria, given growing appreciation of relationships between lateral gene flow, introgression, and homologous recombination (e.g., Dykhuizen and Green 1991; Fontaneto et al. 2007; Bobay and Ochman 2017). A study of 105 prokaryotic species found that 'only a minor fraction (<15%) undergoes too little gene flow to be assigned to species based on the BSC' (Bobay and Ochman 2017, 499–500).

And although there has long been healthy debate about how often sympatric and clinal speciation occur in nature, genomic methods have now afforded new appreciation of how complex these issues are and how simplistic associated past criticisms of the BSC were (e.g., Martin et al. 2015; E.J. Richards, Poelstra, and Martin 2018).

Such defenses often succeed in showing that many past criticisms were off the mark. But, they fall short of the loftier aim of showing that the BSC is generally accurate. A superficial reason is that although some of the cited percentages – 75% correspondence between BSC conditions and certain 'good' plant species, 85% in the case of some bacteria – are higher than past criticisms suggested, they are still far from establishing *general* accuracy (even short of having a precise percentage threshold to point to). A deeper reason, far more instructive, is that time and again, when gene flow has been important as the BSC would predict, this has been partly *because* of how it works in concert with *other* variables emphasised by apparently competing cause-focused species concepts (see Barker and Wilson 2010).

In his review that defends the relevance of gene flow to plant species, Ellstrand (2014) is up front about this, emphasising the importance of interactions between various types of gene flow and various types of selection. He notes that this importance of interaction 'simply was not recognised' from the 1960s through the 1990s (2014, 745).

In bacteria, Doolittle (2019) likewise points out that often when gene flow is important, it is partly because other variables such as selection are too, and vice versa. In more technical terms, it is rare in clear cases of bacterial species that there isn't distinctive interaction between:

---

[*] See Ellstrand (2014) for a comprehensive review.

- •. The frequency of clonal or periodic selection (labelled *s* by Shapiro and Polz [2015]);
- • The frequency of homologous recombination (labelled *r* by Shapiro and Polz [2015]);
- • The frequency of lateral gene transfer.

In the majority of cases, they interact importantly (Novick and Doolittle 2021).

The key empirical finding to clarify in the past 20 years is that often, *many or all* the disputed variables in given cases explain species coherence and distinctions by how they work *together* in complex ways.

But, old views die hard. Even when authors document the importance of variable interactions, they sometimes contort for ways of privileging their preferred variable. Morjan and Rieseberg (2004), for instance, uncover the variables of selective sweeps and small amounts of gene flow working together to help connect populations as evolutionary units, but then imply that the gene flow variables are more important than the selection variables for evolutionary coherence because the role of the gene flow variables is distinctly 'creative'. Barker and Wilson (2010) showed how such asymmetrical interpretations cannot be generally sustained, because often, when one attempts to privilege any one variable among interacting others in such conceptual ways, they ironically leave it open for the other variables to be similarly privileged.

However, it's not that all variables are important in the same ways at all times. Hardly. Sometimes, gene flow is especially important, perhaps in some creative sense and with the help of mutation, selection, a background of frequently shared related traits, and ...; in other cases, one or more of these other variables plays some critical role, and gene flow is among the mere enablers (e.g., Ellstrand 2014, 245–247, and Table 1.1). So, there are three lessons to draw from shortcomings of Done by Solution views that proceed via cause-focused species concepts.

First, we seek a theory of the species category's constitutive conditions – a functional species concept – that is more *inclusive* than these, i.e., that heeds the empirical finding of importance for many, not just a small set of, interacting variables. (See Table 1.1 again for example variables.)*

But second, we need the inclusivity to be *flexible*: sensitive to variation in the roles of variables across cases. This would be a theory that recognises distributed importance without demanding uniformity in importance. Proposing necessity for each causal variable is not going to work. We need flexible, even dynamic, inclusivity.

Third, whatever more flexible and inclusive approach we develop should be one that helps reconcile an ambiguity within these views about what the relevant *effects* of the causes are. A number of philosophers and biologists have discussed this in terms of 'species cohesion' or 'evolutionary cohesion', and these discussions deserve to be better known amongst species experts in general (e.g., Mishler and Brandon 1987; Ereshefsky 1991; 2001; de Queiroz 1998; J. Wilson 1999; Brooks and McLennan 2002; Barker and Wilson 2010). At least two notions of 'cohesion' have been uncovered in the literature. One is more akin to similarity or uniformity or

---

* Templeton's CSC was an early attempt at this, though not nearly as inclusive as is here being suggested.

homogenisation of traits within species groups – something Barker and Wilson 2010 elaborated as *response cohesion*. The other is a type of integration or organisation, which we explicated as *integrative cohesion*.* Proponents of cause-focused species concepts have collectively vacillated between both, as they see processes like gene flow as *homogenising* conspecific populations but also talk of these processes as *integrating* those populations into separately evolving lineages.

In sum, we can and should aim for a species concept that is more inclusive and flexible while adequately resolving confusion about and relationships between types of cohesion. The feedback model presented at the outset of this chapter was just one example of how this can be done. It is far more inclusive than the typical cause-focused species concepts, allowing that many important causal variables not only interact but do so in the unappreciated way of feedback relations, which help constitute species. This also provided flexibility, as the feedback relations needn't be of the exact same frequencies and magnitudes across cases; there are many ways in which the relations can result in an M value within a species category patch of M space. And, clarification of cohesion is proposed: the effects of cited causal variables are further implementations of those same variables, including similarity variables, and these feedback relations dynamically manifest M, a type of cohesion at the metapopulation level. Aside from the feedback model, there are numerous other ways one could develop a more inclusive and flexible concept that addresses cohesion considerations.†

1.6   DONE BY SOLUTION VIA NON-CAUSE-FOCUSED SPECIES CONCEPTS?

Turning now to attempts at Done by Solution via non-cause-focused species concepts, we find authors drawing much more radical lessons than I have advised from the failures of species concepts discussed earlier. They have agreed that each cause-focused species concept is too exclusive when it tries to elevate just this or that cause (or small set of causes) to the status of being constitutive. But, they have then banished all the usually cited causal conditions from having this status.

This is clearest from Kevin de Queiroz, who has repeatedly stressed with his GLSC/Unified species concept (USC) that *none* of the causal conditions deemed constitutive by others are constitutive for specieshood on his view (de Queiroz 1998; 1999; 2005; 2007). Keeping in mind that he doesn't free constitutive conditions from having to each be necessary, and that he then uses the term 'necessary property' instead of 'constitutive condition', he says: 'A unified species concept can be achieved by treating existence as a separately evolving metapopulation lineage as the only necessary property of species' (2007, 879). Causal conditions such as differentiating gene flow processes, sharing of selection regimes, adaptive zones, etc. are then not constitutive or necessary; rather, he says they have a 'secondary' status as 'lines of evidence (operational criteria)' (2007, 879).

---

* My more recent 2019a paper shows how I think we didn't fully explore and develop the consequences and applications of that distinction, nor supplement it in crucial ways, as the feedback model now does.
† For example, see Sterner (2019).

The EvSC of Simpson (1951, 1961) and that of Wiley and Mayden (2000) also expel the individual causal conditions from those that are constitutive, and they likewise appeal instead to how species lineages are separately evolving entities. Simpson elaborates that this involves the lineage having 'its own unitary role and tendencies' (1961, 153), while Wiley and Mayden underscore a long view of such tendencies – they extend from history to future fates (2000, 73).

I agree that these views avoid many of the counterexamples that haunt cause-focused species concepts. Counterexamples to necessity, in particular, are easier to avoid when you don't propose stringent necessity conditions in the first place.

But, there is more than one way to free us from the offending necessity claims, as the feedback model exemplified (also see Boyd 1999; Griffiths 1999; R.A. Wilson 1999; Pigliucci 2003; R.A. Wilson, Barker, and Brigandt 2007; Slater 2014). And, the way provided by these non-cause-focused species concepts is extreme, facing its own dilemma.

Either the banishing leaves us with so sparse a theory about the species category's constitutive conditions that it fails to be exacting and comprehensive, and we also aren't able to assess the extent of its accuracy; or there is enough theory to allow some assessment of its accuracy, but the assessment shows we so far have insufficient reason to accept it as accurate: its lone extremely inclusive constitutive condition recognises far more kinds of groups as species than we are given reason to do.

In other words, non-cause-focused species concepts either don't provide a clear theory of the species category's constitutive conditions, or they hint at a wildly unfamiliar one without the justification this demands.

The first horn of that dilemma amounts to *vagueness* about evolutionary cohesion rather than the cohesion *ambiguity* that plagued cause-focused species concepts. These species concepts are relatively clear that the lineage-identifying cohesion is of the integrative sort rather than similarity-based response cohesion. But, so little about this integrative cohesion is unpacked that the notion is nearly vapid. Criticisms by others have suggested that Simpson (1951; 1961) and Wiley and Mayden (2000) may be stuck on this horn of the dilemma, as it doesn't help much to be told that the evolutionary unity of a species consists in being a lineage with independent evolutionary fate and historical tendencies (e.g., Ghiselin 1987; Stamos 2003; cf. R.A. Richards 2010). This recalls Molière's quip, in *The Imaginary Invalid*, that we don't learn much about the 'dormitive virtue' of opium when a physician tells us this consists in 'the power to put one to sleep'.

I tend to think, however, that a more charitable read of non-cause-focused species concepts pushes them from the first to the second horn of the dilemma, where they at least get a little farther along. Proponents of these concepts stress the senses in which species are evolutionary *lineages* of a certain sort. They don't give elaborate theories about this, but enough to appreciate the dramatic consequences they then go on to infer from this.[*] What they don't give is sufficient principled reason to accept these consequences and thereby, the accuracy of their views.

---

[*] So, to be clear, we should of course be on board with the idea that species are lineages! It's the inferences they build from this that are questionable.

Again, this is clearest with de Queiroz:

> When I use the term lineage, I am not talking about a clade or a monophyletic group (see de Queiroz 1998, 1999), and thus, I am not advocating a version of the phylogenetic species concept. A lineage, in the sense that I am using the term … *is a line of direct ancestry and descent* … Lineages are formed by biological entities at several different levels of organization. For example, every person can trace his or her ancestry back along an organism lineage that passes through a series of ancestral organisms.

(de Queiroz 2005, 200, my emphasis).

Because he also specifies that he is talking about certain sorts of lineages – separately evolving metapopulation lineages – he *does* rule out some lineages from counting as species. Sensibly, a lone lineage of organisms, such as one tracing back from you to a parent, then a grandparent, then a great grandparent, and so on, will not count as a species because it is not a *metapopulation* lineage. Likewise, mere cell lineages, gene lineages, etc. will rightly be ruled out from specieshood. But, many other lineages are peculiarly let in because groups of populations (that is, metapopulations) are found at a staggering number of levels of biological organisation. Within many groups of populations that are widely deemed species, there are sub-groups, of just some of the populations, which are themselves lineages too; most readers may recognise these lineages as sub-species or varieties or otherwise mere samples of species, etc. But de Queiroz's view implies they are instead themselves also whole species, calling them 'species within species' (2005, 208).

He notes that we also find metapopulation lineages above the typically recognised species level, each of which comprises multiple species and would normally be deemed a higher taxon (more inclusive than species taxon). But on his view, these very inclusive groups will likewise themselves be single species (while also subsuming others). And, we cannot avoid recognising very inclusive groups as species by stepping back to the concept of a metapopulation and claiming that populations form a metapopulation in the first place only if they belong to the same species. In the present context, that would be viciously circular, helping ourselves to the very species category we are at pains to uncover.

We need to be clear about the sense in which the dramatic consequences are extreme enough to demand more justification. It is familiar by now to claim that there are evolutionary metapopulation lineages at many 'levels' of inclusivity, or that there are many degrees or grades of inclusivity. This is something for which I too have argued (Barker 2019a; Barker and Wilson 2010, 76). It has also been a basis for rank-free taxonomy proposals. So, that idea is well-known. What is instead radical is the further implication that *all* of these putatively many levels are *species* levels (or degrees or grades). That abandons the widespread idea that the more traditionally recognised species level is especially important or fundamental within biological theories and thus, abandons the fourth desideratum listed in our statement of the species category theory problem. Experts have thought that there are many reasons to believe that a level or grade of lineagehood corresponding to a more traditional conception of the species category is of more theoretical import than, say, the grade corresponding to our conception of stirps or paradivisions. The latter may well have some import – that, I think, is a valuable insight from appreciating that there are many levels or degrees or grades of evolutionary metapopulation lineages. But, it is a marked departure from our starting point to propose that such import is roughly *equal* to that of a level or degree or grade more traditionally associated with the species category. Is there really nothing of theoretical importance that distinguishes between a species lineage and a paradivision lineage? Even most rank-free taxonomy proponents would (whatever their misgivings about treating 'paradivisions' as a formal rank) demand further justification for this, as most have insisted on the distinct importance of lineages that are species in a more traditional and much less expansive sense than de Queiroz's (Mishler 1999, 311).

So, the sharp departure from prevailing views may well turn out to be correct, but we should think so only upon good reason. Thus far, we are short on this. De Queiroz points to analogous and now well-known departures from familiarity; e.g., some *taxonomic* species concepts (especially phylogenetic ones) also imply that what were once regarded as sub-species are instead species, and some concepts in *conservation* do something similar (de Queiroz 2005, 206–207). But, the reasons experts give for those very different departures from familiarity won't be of the sort needed when addressing the theoretical problem about functional species concepts on which we are working. In the taxonomic case, the reasons will be taxonomic, and in the conservation case, they will be based on the conservation value of various evolutionary groups. We are instead looking for a theory about units traditionally regarded as species that play distinguishing functional roles in evolution, roles *unlike* and more fundamental in many biological theories than those played by categories of variety, genus, stirp, paradivision, and the like. No surprise, then, that those departures from familiarity in more applied fields have already become familiar, but researchers have yet to depart nearly as much with respect to theories of the species category.

De Queiroz also stresses that all proponents of other functional species concepts agree that species are separately evolving metapopulations. Some of his sympathetic readers (e.g., Camargo and Sites 2013) then infer that if we stop our theory about the species category's constitutive conditions *right there*, we will preclude the disagreement that comes when those proponents move on to fill in their respective theories with more comprehensive and exacting details, e.g., about gene flow processes, niche sharing, selection regimes, etc.

We can agree that preclusion of disagreement would have some advantages. But, we would want such preclusion to be based at least partly on an accurate, comprehensive, exacting, and general theory that helps vindicate the importance of the species category, rather than on just the practical gains or relief we would get from less disagreement. Why think we now have such a theory, especially when all the usual causal processes studied and discussed have been banished from the status of being constitutive (relegated to lines of evidence)?

Notice how thinking through de Queiroz's view in this way can lead back to the first horn of the dilemma for non-cause-focused species concepts. At first, we moved these concepts away from that horn by appreciating some of their clear consequences as theories. But, consideration of those consequences raised further questions. Why accept a theory with such consequences? What are the good reasons to do so? It is very hard to imagine someone adequately answering those questions in favor of the non-cause-focused species concepts without further clarifying *what it is to be a separately evolving metapopulation lineage*. When the concepts have such dramatic consequences, we should not accept them as accurate until we hear more from their proponents about the proposed conditions in which *separate metapopulation evolution* consists – what *makes* a metapopulation exactly *one* separately evolving lineage instead of *two*, for instance? – so that we are positioned to tell whether these are plausible enough that we should overthrow deep and familiar views (for related concerns, see Haber 2013, 341; R.A. Richards 2010, 142; Pigliucci 2003, 598).

This helps address a view of Frank Zachos', which likely appeals to many biologists with pressing practical interests or species problem fatigue. On his view, the absent details, about what separate evolution consists in, are of operational rather than theoretical relevance (2016, 107–109). He cedes that we face hard operational challenges of *telling* when, for instance, a metapopulation is two or instead a single separately evolving lineage. But, he thinks these are not theoretical challenges – the theory is already comprehensive and exacting enough about what *makes* a metapopulation one, or instead two, or instead some other number, of separately evolving lineages. He also adds a nuance: what counts as comprehensive and exacting *enough* is relative to research discipline. Relative to philosophy standards, Zachos quickly grants that de Queiroz-type theories may in fact not be comprehensive and exacting enough. But relative to biology standards, he proposes they are.

I am sympathetic to ecumenical suggestions like Zachos'. And, what counts as comprehensive and exacting 'enough' is very probably relative to disciplines, projects, interests, resources, abilities, and other things. But, the dilemma presented earlier suggests that this doesn't help the non-cause-focused theories about the species category's constitutive conditions. It is hard to see how the consequences of what little theory we do get from these views are any less dramatic in *biological terms* than in *philosophical terms*. I am not exactly sure, however, how to apply the biology-versus-philosophy distinction here, so let me instead recount what seems to be the case.

There is a long-standing view *in biology* about the species level of organisation and about this having fundamental theoretical importance. This has prompted searches *in biology*, as in many analogous cases in the history of science, for an exacting, comprehensive, and generally accurate theory about the species category's constitutive conditions, like analogous theories that science has worked hard to attain for *chemical element, electron, tectonic plate, cancerous tumor, cell*, and so on. Then, some very thin non-cause-focused theories are proposed, ones that do not remotely approach how comprehensive and exacting our theories are for the other categories just mentioned. And, these thin theories imply that the long-standing *biological views* are wildly mistaken or off track, that the species category is vastly more inclusive than long presupposed, with correspondingly very different (or comparatively much less) theoretical import than countless biologists have thought. It seems that experts in biology should then allow that the thin theories *may be accurate*. But, it is *only after* being given much more detail about what it is to *be* a separately evolving lineage (not just details about how to detect or delimit lineages) that experts should also accept that the theories and their consequences *in fact are* accurate.

This does not mean that biologists should cease grappling with the delimitation challenges until they have sufficient reason to accept the thin theories. Of course, let many researchers push on with delimitation meanwhile. Indeed, research communities in biology are relatively massive – many can focus on *only* delimitation, with little regard to theoretical toil, while others continue as well with theories of the species category's constitutive conditions. Oncologists continued developing incredibly useful ways of detecting tumors for many decades in the late 19th and early 20th centuries, after having gained a first window into the nature of cancer but long before achieving the deep theoretical insights about it that came with molecular and genomic methods in the 1980s, 1990s, and 2000s. Likewise, chemists made striking progress from the 1770s through the 19th century in detecting separate elements (including remarkably accurate estimates of atomic weights for many elements), long before the nature of elements was uncovered in nuclear charge and corresponding relations between protons and electrons, from about 1910 to the 1930s. But, imagine where those disciplines would be now if, say in 1950, the community of cancer researchers had said, 'our theory of cancer's nature is now comprehensive and exacting enough for us, let's divert the remainder of our efforts to X-ray technology development'; or imagine the community of chemists and physicists had said, in 1890, 'the element is clear enough to us, let's now only sharpen element delimitation methods and applications'. Among the consequences, my mother would very probably not have survived her HR+, HER2– breast cancer, and you would not have a smartphone at hand. Cancer researchers in 1950 and chemists and physicists in 1890 could not possibly have foreseen such innumerable costs. Neither could biologists now see potential far-off costs of stopping work on the nature of the species category.

Or, worse (much, much worse?), go back to the biology-versus-philosophy distinction and imagine that these past scientific research communities had added, 'and let's leave the continued work on the nature of these categories to our fellow philosophers'. Shiver.

There are two lessons to draw from shortcomings presented in this section. First, it is now advised that at least some researchers seek a theory of the species category's constitutive conditions that is more *comprehensive and exacting about integrative cohesion* (or evolving separately or having distinct evolutionary roles) than those so far furnished by non-cause-species concepts. Second, we should investigate whether developments can *clarify and constrain* proposals about constitutive conditions in ways that vindicate familiar views about the species category being distinct, in fundamental theoretical ways, from other lineage categories or taxonomic categories.

As with the lessons drawn in the previous section, these further lessons are ones that new proposals can heed, as exemplified by the feedback model. Compared with cause-focused species concepts, it suggested a *more* inclusive and dynamic theory of the species category's constitutive conditions while addressing *ambiguity* about cohesion. But, it was also appropriately *less* inclusive than the different species concepts considered in this section and less *vague* about cohesion. Putting these innovations together resulted in a much clearer picture of the species category pattern that Section 1.4 showed to be accepted in more vague and ambiguous form by all functional species concepts.

1.7 DONE BY DISSOLUTION VIA ELIMINATIVE PLURALISM?

Aside from the above-examined attempts at solving the species category theory problem, the other main way of supporting a We Are Done view is to dissolve that problem. I'll consider what I think are three of the most interesting and influential ways of attempting this, starting with one of Marc Ereshefsky's most frequently cited papers (Ereshefsky 1992b). It argues that we should think there is no natural, general species category that is important to biological theories, and that we should eliminate the species concept from biology.[*] This implies that there can be no theory meeting the desiderata we have laid out here. That would dissolve the species category theory problem.

In a recent paper, I have argued that Ereshefsky's argument fails (Barker 2019b). If that is correct, dissolution does not follow from it. For the details, I refer readers to that paper; here, I will simply run main points from it through the lens of this chapter while also making new points helping to show that dissolution in general is difficult to defend.

Ereshefsky's focus is on what we are here calling cause-focused species concepts, and especially on the fact there are a plurality of these, each focusing on a different set of evolutionary causes.[†] Three premises are central to his argument. One (1), which I'll call *cross-cutting pluralism*, is that the multiple different sets of causes at issue (e.g., gene flow processes, niche sharing processes, etc.) each operate on and produce different types of 'basal lineages', with some of these different lineages cross-cutting each other. For example, according to Ereshefsky, there are interbreeding lineages, ecological lineages, and genealogical lineages, and often, there are cases like this: populations A and B belong to the same interbreeding lineage but to different ecological lineages and to no genealogical lineages. In a way, each cause-focused species concept is finer-grained than a general species concept, and no one of these is always compatible with any other.

A second central premise (2) is that we very probably will not discover a parameter common to all the different types of cross-cutting basal lineages created by the different evolutionary causes. Call that the *essence cynicism* premise: Ereshefsky doubts there is any underlying essence that could unify all the cross-cutting, finer-grained basal lineage concepts.[‡]

---

[*] In later work (e.g., Ereshefsky 2010), he sees practical value for the concept but maintains his old view about the category, leaving the old implication for us the same.

[†] Elsewhere (e.g., Ereshefsky 2010; 2017), he argues against some non-cause-focused species concepts.

[‡] Ereshefsky does not use the term 'essence' in his paper, though it seems suitable to his claims.

Finally, the following premise (3) is needed to move from the other two to the conclusion: *if* cross-cutting pluralism and essence cynicism are true (as is the separate presumption that no non-cause-focused species concepts succeed in general), *then* we should think there is no general functional species category that is of distinctive importance to biological theories.

A first problem for the argument is the appeal to basal lineages. Premise (1) assumes that each type of fine-grained group recognised by Ereshefsky shares in being such a lineage. But, premise (2) suggests there is no 'common parameter' or essence underwriting that category. It seems by the argument's own lights that (1) and (2) cannot be true at the same time, in which case it cannot issue its conclusion (Barker 2019b).

A second set of problems for the appeal to basal lineage in premise (1) arises even when setting aside its relation to premise (2). What should we take a basal lineage to be? Ereshefsky implies it is to be an evolutionarily cohesive thing that can be produced in a variety of ways – he cites interbreeding causes, ecological causes, and genealogical causes. But, this faces the first horn of the dilemma raised in Section 1.6 for non-cause-focused species concepts: what is it that these causes are producing and which we associate with being a basal lineage? Moreover, why think interbreeding groups count as basal lineages? Same for ecological groups, and genealogical ones. They may each be biologically interesting groups. But, what makes them *lineages*, and what makes them lineages of a *basal* sort? These aren't rhetorical questions. Rather, it seems that if there are answers to them for each proposed type of basal lineage, then we have the makings of a theory about a general species category, contrary to the argument's conclusion.

A third problem for the argument afflicts premise (3). We should not think that cross-cutting pluralism and essence cynicism would, if true, together suffice to take us to the conclusion. Neither of those would help rule out an alternative general theory – or just another theory among the plurality – of the species category, one proposing that each species lineage consists in dynamic interactions of multiple variables, with variations in these interactions across different examples of species. To see this, suppose cross-cutting pluralism is true, so that populations A and B belong to one 'interbreeding lineage' while also belonging to distinct 'ecological lineages'. (Perhaps Templeton's (1989) cottonwoods and balsam poplars are such a case.) This wouldn't rule out there being a third interactionist type of lineage, with 'feedback lineage' being just one example, and investigating how the two populations relate to it. Already, Ereshefsky's view implies that multiple kinds of lineages can overlap or have some of their member populations be the same. There is no reason to think a further kind of lineage couldn't also partake in such complexity. We can also model these types of lineage separately or jointly; e.g., the same metapopulations can be plotted in M space, or in a BSC space, or in an EvSC space, etc., and their relations studied.

Now, suppose essence cynicism is true, so that an essentialist standard is not met for the basal lineage category or a general species category. This shouldn't matter, because we have already seen that such a standard is neither necessary nor advised. We have learned that prodigious variation, change, and complexity in the living world challenge such kinds of essentialism, and so we have developed thorough alternatives to accommodate this, including cluster and family resemblance views of categories (e.g., Boyd 1999; Pigliucci 2003; Wilson et al. 2007; Slater 2014). On such views, each of the typically important causal variables can count as one of the constitutive conditions of the species category without the condition being necessary. Such views are implicitly routine in many other branches of biology; e.g., there is a particular mutation (i.e., ΔF508) at work in the majority of cases of cystic fibrosis, but not in all, and indeed, any one of over 1500 different mutations can play the role (Bobadilla et al. 2002). It does not follow that there is no general cystic fibrosis category.

That example also illustrates the point made earlier about cross-cutting pluralism not doing the work implied. Ereshefsky's fine-grained 'interbreeding lineages' and 'ecological lineages' are akin to more fine-grained cystic fibrosis categories distinguished by mutation type, say, 'ΔF508-cystic fibrosis' and 'G542X-cystic fibrosis'. Each of these four fine-grained categories can have their purposes: there are different treatments for the different fine-grained cystic fibrosis categories; different conservation measures may be advised for the different fine-grained lineages. Nonetheless, there can also be both a more general cystic fibrosis category and a more general species category, whether that be a feedback species category or some other. And, these more general categories can have importance additional to that of just the four fine-grained categories listed.

1.8   DONE BY DISSOLUTION VIA PESSIMISM?

Brent Mishler is well-known for skepticism about the species category. His argument (1999, 308–309) from failures of past attempts at solution within species concepts can be summarised in this way:

(1) 'When anyone has looked closely for an empirical criterion to distinguish the species rank uniquely and universally from all others, the attempt has failed' (1999, 308);
(2) 'We are unlikely to have any criterion for distinguishing species' (1999, 309);
(3) If (1) and (2) are true, then we should think there is no general species category that is of distinctive importance to biological theories. ('Species are not special', 1999, 309);
(4) Therefore, we should think there is no general species category that is of distinctive importance to biological theories.

Premise (1) is about *past* work, and our examination of Done by Solution views suggests we should agree with it, *if* by 'fail' we mean: did not succeed in providing the theory we seek. But, we should be less sanguine if 'fail' implies more scathingly that we have not even seen instructive defenses of and revisions to functional species concepts, or learned anything on which we have decent prospects of building. We *have* seen such things, as discussed above.

Mishler's premise (2) is a probability claim about the *future* and is thus difficult to assess. However, note that in (1), he is talking about respective searches for 'a' criterion. As with Ereshefsky, that sounds like the assumption of an essentialist standard – that we are seeking a single, necessary and sufficient, constitutive condition for the species category. So, there are two problems with (2) and its relation to the other premises. If (2) is pronouncing a low probability for satisfying an essentialist condition, that seems a reasonable inference from (1), but then it will cast doubt on (3). This is because, as we have seen, failed essentialist attempts have taught us better ways to proceed rather than closing off hope altogether. So, those better ways suggest a more optimistic conclusion rather than the pessimistic one that (3) says is in the offing.

If, instead, (2) is pronouncing poor prospects for un-essentialist attempts as well, then (2) itself seems dubious. Again, we have drawn guidance and hope from the failures; that doesn't *lower* the probability of future success as (2) presupposes.

Last, regarding (3), the point about communities of researchers surfaces again. Suppose Mishler responds with a strong case for thinking that past failures suggest poor prospects for even more sophisticated, un-essentialistic attempts in the future. Would those poor prospects be sufficient to conclude once and for all that 'species are not special' (p. 309), recommending that zero people in the communities should work on the species category theory problem? I don't think so. Given the lessons we've learned and new kinds of views that heed them, such as the feedback model, it would remain wise to have some portion of the research community revving up.

I shouldn't suggest more difference between Mishler's views and my own personal views than there is. I, too, often feel pessimistic about the species category theory problem. But, constructing an argument for such pessimism proves difficult.

1.9   DONE BY DISSOLUTION VIA PHENOMENALISM?

John Wilkins has recently and provocatively argued that species are not 'theoretical objects' and instead are 'phenomenal objects' (2018, ch. 14; see Wilkins in this volume for more on his views of species). What does this mean?

One function of scientific theories is explanation. The theory of evolution by natural selection, for example, is supposed to offer explanations for many things, including many changes in population trait frequencies. As such, the mechanisms of natural selection count as theoretical objects. Theoretical objects are things that do some explaining according to theories about them. Being 'theoretical' is not pejorative here and doesn't imply 'less real'. It is famously quite the opposite, according to W.V.O. Quine. Roughly, the more scientific confirmation a theory enjoys, the more assurance we have that the objects to which it appeals for the explanation are real (Quine 1948).

Wilkins thinks that whenever biological theories make reference to a general species category, it is *not* in order to explain anything. A general species category or features associated with it *get* explained, but, he thinks, *do* no explaining. This does *not* imply the category's unreality. He thinks phenomena that get explained – patterns, cohesive objects, effects, and so on – can also have reality, and he thinks this is the case for the species category. So, he thinks there is a real species category but that it is like the *mountain* category in geology – difficult to define and fuzzy at the edges, but a well explained phenomenon nonetheless (Wilkins 2018, 343). Other categories in geology, like *tectonic plate*, are different – they are theoretical objects that do explaining.

How might this threaten to dissolve the species category theory problem? The general species category we are trying to further uncover by solving that problem is supposed to be of fundamental importance to biological theories. If having such status entails being a category that helps do important *explaining* according to biological theories, but Wilkins is correct to *deny* such an explanatory role to any general species category, then there is no category *of the type sought*. (Again, he will insist that there indeed is a real species category but of a type different than the one sought by the species category theory problem articulated in this chapter.) Here is the argument:

(1) If we should think there is a general species category that is of distinctive importance to biological theories, then some general species concept does important explaining across the living world;

(2) But, it's not the case that some general species concept does important explaining across the living world;

(3) Therefore, it's not the case that we should think there is a general species category that is of distinctive importance to biological theories.

Start with (2). Wilkins provides two interesting reasons in its favor. First, when biological theories seem to appeal to a proposed species category as an explainer, they have actually (so far anyway) been appealing to categories and associated processes at *lower* levels of biological explanation, such as the flow of genes (much lower level) between less inclusive populations (also a lower level). This is a reductionist claim. As he puts it, 'the theories used, the explanations, are not theories of species; they are theories of gene exchange, reproduction, fitness, adaptation, and so on' (Wilkins 2018, 342; also see Cracraft 1989b).

This seems incorrect in some familiar cases, including with the BSC and other concepts he discusses under the label 'reproductive isolation species concepts' (RISCs). One of the central properties that BSCers think does a lot of explaining is reproductive isolation. That is at least often a property of *groups* of populations (Wilson 1996), and on the BSCer's view, it is a property of groups they recognise *as species*. I don't see that we are bound to fully reduce this explanatory property to lower-level entities and processes any more than we are bound to talk about the allele frequency in a population without at least some non-reductive deference, implicit or otherwise, to the whole population (Sober 1980). Populations are what have population allele frequencies and the dynamics that explain changes in these. By analogy, groups of populations (which BSCers happen to recognise as species) have the property of reproductive isolation.

Indeed, there are many explanatory properties like this, which aren't fully reduced. Trait frequencies across groups of populations, not just within populations, can also help explain how one group responds differently than another group of populations to selection regimes, and some views of species appeal to this (e.g., Barker and Wilson

2010; Barker 2019a). Some putatively explanatory models of speciation also have such non-reductionist aspects (Coyne and Orr 2004), as do attempted explanations of 'rules' such as Haldane's rule (e.g., Turelli and Orr 1995), Bateson-Dobzhansky-Muller (BDM) incompatibilities (Cutter 2012), the Island rule (e.g., Lomolino 2005), and Cope's rule (e.g., Hone and Benton 2005).

Then, there is what many have proposed as the big-ticket item: why is life clumpy rather than continuous? Or as Mayr put it, why 'has nature, and more precisely natural selection, favored the discontinuities' between some groups of populations but not others? (Mayr 2000a, 161). All cause-focused species concepts attempt to explain that big-ticket item. In doing so, they treat clumpiness as an *explanandum* (thing to explain) but include only the *explanantia* (things that do the explaining) among the conditions deemed constitutive of the species category. Mayr, for example, does this when insisting that we exclude phenotypic uniformity from the definition of 'species' while also urging that we explain that uniformity by the processes he *does* reference in the definition: gene flow processes between populations. This is analogous to how chemists treat the chemical element category. They take different sorts of clumpiness as the things to explain, e.g., clumpiness rather than continuity in chemical phenomena such as boiling points, densities, malleability, and so on, across the chemical world. They then explain such clumpiness by appeal to differences in nuclear charge (and associated electron configurations). And when next isolating the conditions that are constitutive of the chemical element category, they appeal only to the explanatory properties and processes (nuclear charge and associated electron configurations), not to the clumpiness they have explained. (As seen earlier, there is no mention of boiling points, densities, etc. in the IUPAC definition of 'chemical element'.) There is at least one apparent difference, though. Whereas explanatory and constitutive nuclear charges are intrinsic to their respective atoms, the explanatory and constitutive processes in the case of species are not all intrinsic to populations or species. At least some are unreduced relationships between or across populations and species, according to cause-focused species concepts. Thereby, proponents of those concepts imply that reference to species does a lot of *unreduced* explaining.

On more *general* reductionist grounds, one might try to fully reduce *all* those sorts of explanations. But although in other work I have defended some reductionist views against hasty anti-reductionist ones (Barker 2013), the prospects of ambitious general reductionist views also seem bleak, given the demise over the past 30 years of such views in biology and philosophy of biology (Brigandt and Love 2017).

If the reductionist point is not available in support of (2), then Wilkins' second point in its favor becomes more important. It is that none of the species concepts we might regard as explanatory (the functional ones) have succeeded in being accurate *in general*. If we opt for the BSC, for instance, then 'the vast bulk of life would not be in species' (2018, 344).

That is a much more promising defense of (2), as our discussion of Done by Solution views would suggest. But, it throws a troubling light on (1). It indicates that (1) demands that some existing species concept has *already* succeeded with its explanations in general terms if we are to think there exists a general and important species category of the sort we are discussing. This is even more demanding about past work on functional species concepts than is Mishler. Mishler demanded *more* from past work if we were to see promise for the future; we countered that past work has already provided enough for promise. Wilkins instead demands that past work *already have solved the problem.* So, our counter to Mishler suggests that the more demanding view of Wilkins' is even further from justified.

1.10 CONCLUSION

We began with the feedback model of species, which exemplified a new kind of approach we are positioned to develop. Then, in Section 1.3, we zeroed in on the hard theoretical species problem, which this new kind of approach might help with. This species category theory problem says that we lack a theory of the species category's constitutive conditions that meets, to high degrees, the desiderata of being exacting, comprehensive, accurate, and helpful in vindicating the traditional idea that the species category is especially important to many biological theories. We met opposing views about this problem. Revving Up says we are well positioned to begin work on the new approaches to species thanks to a host of conceptual and empirical lessons learned over recent decades. We Are Done views instead imply that we are finished with the species category theory problem because we have either already solved it or dissolved it altogether. Section 1.4 clarified the relevant extant species concepts that would help us assess Revving Up against some We Are Done views. These are functional species concepts of the cause-focused and non-cause-focused sorts. In Section 1.5, we saw how the cause-focused sorts don't support We Are Done views that think the problem at issue has been solved; in Section 1.6, we reached an analogous conclusion for non-cause-focused species concepts. In both cases, we drew specific lessons from the challenges to both types of species concept and exemplified via the feedback model that we are well positioned to build on these, thereby supporting Revving Up. Sections 1.7–1.9 then examined the other type of We Are Done views that oppose Revving Up – those that say we have dissolved the species category theory problem. These were found problematic. The overall critical conclusion is that We Are Done views have not yet succeeded, and there is currently no such view that everyone should endorse. More constructively, Revving Up is plausible, and some proportion of researchers working on species should endorse and further develop it while others work on We Are Done views.

REFERENCES

Anderson, Edgar. 1949. *Introgressive Hybridization*. New York: J. Wiley.
Barbará, T., G. Martinelli, M. F. Fay, S. J. Mayo, and C. Lexer. 2007. 'Population Differentiation and Species Cohesion in Two Closely Related Plants Adapted to Neotropical High-Altitude 'Inselbergs', Alcantarea Imperialis and Alcantarea Geniculata (Bromeliaceae).' *Molecular Ecology* 16 (10): 1981–92. https://doi.org/10.1111/j.1365-294X.2007.03272.x.
Barker, Matthew J. 2007. 'The Empirical Inadequacy of Species Cohesion by Gene Flow.' *Philosophy of Science* 74 (5): 654–65. https://doi.org/10.1086/525611.
———. 2010. 'Specious Intrinsicalism.' *Philosophy of Science* 77 (1): 73–91.

————. 2013. 'Biological Explanations, Realism, Ontology, and Categories.' *Studies in History and Philosophy of Science Part C: Studies in History and Philosophy of Biological and Biomedical Sciences* 44 (4): 617–22.

————. 2019a. 'Species and Other Evolving Lineages as Feedback Systems.' *Philosophy, Theory, and Practice in Biology* 11. https://doi.org/10/gg44bg.

————. 2019b. 'Eliminative Pluralism and Integrative Alternatives: The Case of Species.' *The British Journal for the Philosophy of Science* 70 (3): 657–81. https://doi.org/10/gjcmt9.

Barker, Matthew J., and Robert A. Wilson. 2010. 'Cohesion, Gene Flow, and the Nature of Species.' *The Journal of Philosophy* 107 (2): 59–77.

Barluenga, Marta, Kai N. Stölting, Walter Salzburger, Moritz Muschick, and Axel Meyer. 2006. 'Sympatric Speciation in Nicaraguan Crater Lake Cichlid Fish.' *Nature* 439 (7077): 719–23. https://doi.org/10/b5b9md.

Baum, David A. 2009. 'Species as Ranked Taxa.' *Systematic Zoology* 58 (1): 74–86. https://doi.org/10.1093/sysbio/syp011.

Bobadilla, Joseph L., Milan Macek, Jason P. Fine, and Philip M. Farrell. 2002. 'Cystic Fibrosis: A Worldwide Analysis of CFTR Mutations – Correlation with Incidence Data and Application to Screening.' *Human Mutation* 19 (6): 575–606. https://doi.org/10/dkcn24.

Bobay, Louis-Marie, and Howard Ochman. 2017. 'Biological Species Are Universal across Life's Domains.' *Genome Biology and Evolution* 9 (3): 491–501. https://doi.org/10/f98rt7.

Bond, Jason E., and Amy K. Stockman. 2008. 'An Integrative Method for Delimiting Cohesion Species: Finding the Population-Species Interface in a Group of Californian Trapdoor Spiders with Extreme Genetic Divergence and Geographic Structuring.' *Systematic Biology* 57 (4): 628–46. https://doi.org/10.1080/10635150802302443.

Boyd, Richard. 1999. 'Homeostasis, Species, and Higher Taxa.' In *Species: New Interdisciplinary Essays*, edited by Robert A. Wilson, 141–85. Cambridge, MA: MIT Press.

Brigandt, Ingo, and Alan Love. 2017. 'Reductionism in Biology.' In *The Stanford Encyclopedia of Philosophy*, edited by Edward N. Zalta. Metaphysics Research Lab, Stanford University. https://plato.stanford.edu/archives/spr2017/entries/reduction-biology/.

Brooks, Daniel R., and Deborah A. McLennan. 2002. *The Nature of Diversity: An Evolutionary Voyage of Discovery*. Chicago, IL: University of Chicago Press.

Brooks, Daniel R., and E. O. Wiley. 1988. *Evolution as Entropy: Towards a Unified Theory of Biology*. 2nd ed. Chicago, IL: University of Chicago Press.

Bush, Guy L., Jeffrey L. Feder, Stewart H. Berlocher, Bruce A. McPheron, D. Courtney Smith, and Charley A. Chilcote. 1989. 'Sympatric Origins of R. Pomonella.' *Nature* 339 (6223): 346. https://doi.org/10/ddrvdx.

Camargo, Arley, and Jack Jr Sites. 2013. 'Species Delimitation: A Decade after the Renaissance.' In *The Species Problem: Ongoing Issues*, edited by Igor Pavlinov. InTech. http://www.intechopen.com/books/the-species-problem-ongoing-issues/species-delimitation-a-decade-after-the-renaissance.

Carson, H. 1957. 'The Species as a Field for Gene Recombination.' In *The Species Problem*, edited by Ernst Mayr, 23–38. Washington, DC: American Association for the Advancement of Science.

Chesser, R. K. 1983. 'Genetic Variability within and among Populations of the Black-Tailed Prairie Dog.' *Evolution* 37: 320–31.

Cleland, Carol E. 2012. 'Life without Definitions.' *Synthese* 185 (1): 125–44.

Cohan, Frederick M. 2002. 'What Are Bacterial Species?' *Annual Review of Microbiology* 56: 457–87. https://doi.org/10.1146/annurev.micro.56.012302.160634.

————. 2011. 'Are Species Cohesive? – A View from Bacteriology.' In *Population Genetics of Bacteria: A Tribute to Thomas S. Whittam*, edited by S. T. Walk and P. C. H. Feng, 43–65. Washington, DC: ASM Press. http://www.asmscience.org/content/book/10 .1128/9781555817114.ch05.

Coyne, Jerry A., and H. A. Orr. 2004. *Speciation*. 1st ed. Sunderland, MA: Sinauer Associates, Inc.

Cracraft, Joel. 1989a. 'Speciation and Its Ontology: The Empirical Consequences of Alternative Species Concepts for Understanding Patterns and Processes of Differentiation.' In *Speciation and Its Consequences*, edited by Daniel Otte and John A. Endler, 28–59. Sunderland, MA: Sinauer Associates Inc. http://ci.nii.ac.jp/naid/10014820123/.

————. 1989b. 'Species as Entities of Biological Theory.' In *What the Philosophy of Biology Is, 31–52*. Nijhoff International Philosophy Series. Springer, Dordrecht. https://doi.org /10.1007/978-94-009-1169-7_3.

Cutter, Asher D. 2012. 'The Polymorphic Prelude to Bateson-Dobzhansky-Muller Incompatibilities.' *Trends in Ecology & Evolution* 27 (4): 209–18. https://doi.org/10/ dpq5xq.

Dawley, Robert M., and James P. Bogart. 1989. *Evolution and Ecology of Unisexual Vertebrates*. Albany, NY: New York State Museum.

Devitt, M. 2008. 'Resurrecting Biological Essentialism.' *Philosophy of Science* 75 (3): 344– 82. https://doi.org/10.1086/593566.

Dobzhansky, T. 1935. 'A Critique of the Species Concept in Biology.' *Philosophy of Science* 2 (3): 344–55. https://doi.org/10/d8zr4c.

Doolittle, W. Ford. 2019. 'Speciation without Species: A Final Word.' *Philosophy, Theory, and Practice in Biology* 11. https://doi.org/10/gk8ht3.

Dupré, John. 1993. *The Disorder of Things: Metaphysical Foundations of the Disunity of Science*. Cambridge, MA: Harvard University Press.

Dykhuizen, D. E., and L. Green. 1991. 'Recombination in *Escherichia coli* and the Definition of Biological Species.' *Journal of Bacteriology* 173 (22): 7257–68. https://doi.org/10/ gfkjbh.

Eckenwalder, James E. 1984. 'Natural Intersectional Hybridization between North American Species of Populus (Salicaceae) in Sections Aigeiros and Tacamahaca. II. Taxonomy.' *Canadian Journal of Botany* 62 (2): 325–35. https://doi.org/10/bznscd.

Ehrlich, P. R., and P. H. Raven. 1969. 'Differentiation of Populations.' *Science* 165 (899): 1228–32.

Eldredge, Niles, and Joel Cracraft. 1980. *Phylogenetic Patterns and the Evolutionary Process: Method and Theory in Comparative Biology*. New York: Columbia University Press.

Ellstrand, Norman C. 2014. 'Is Gene Flow the Most Important Evolutionary Force in Plants?' *American Journal of Botany* 101 (5): 737–53. https://doi.org/10.3732/ajb.1400024.

Endler, John A. 1973. 'Gene Flow and Population Differentiation: Studies of Clines Suggest that Differentiation along Environmental Gradients May Be Independent of Gene Flow.' *Science* 179 (4070): 243–50. https://doi.org/10/cr6fjv.

————. 1989. 'Conceptual and Other Problems in Speciation.' In *Speciation and Its Consequences*, edited by Daniel Otte and John A. Endler, 625–48. Sunderland, MA: Sinauer Associates.

Ereshefsky, Marc. 1991. 'Species, Higher Taxa, and the Units of Evolution.' *Philosophy of Science* 58 (1): 84–101.

————, ed. 1992a. *The Units of Evolution: Essays on the Nature of Species*. Cambridge, MA: The MIT Press.

————. 1992b. 'Eliminative Pluralism.' *Philosophy of Science* 59 (4): 671–90.

————. 2001. *The Poverty of the Linnaean Hierarchy: A Philosophical Study of Biological Taxonomy.* 1st ed. Cambridge, MA: Cambridge University Press.

————. 2010. 'Darwin's Solution to the Species Problem.' *Synthese* 175 (3): 405–25. https://doi.org/10.1007/s11229-009-9538-4.

————. 2017. 'Species.' In *The Stanford Encyclopedia of Philosophy*, edited by Edward N. Zalta, Fall 2017. Metaphysics Research Lab, Stanford University. https://plato.stanford.edu/archives/fall2017/entries/species/.

Fontaneto, Diego, Elisabeth A. Herniou, Chiara Boschetti, Manuela Caprioli, Giulio Melone, Claudia Ricci, and Timothy G. Barraclough. 2007. 'Independently Evolving Species in Asexual Bdelloid Rotifers.' *PLOS Biology* 5 (4): e87. https://doi.org/10/cbhmkj.

Franklin, Laura R. 2007. 'Bacteria, Sex, and Systematics.' *Philosophy of Science* 74 (1): 69–95.

Futuyma, Douglas J. 1998. *Evolutionary Biology.* 3rd ed. Sunderland, MA: Sinauer Associates.

Ghiselin, Michael T. 1987. 'Species Concepts, Individuality, and Objectivity.' *Biology and Philosophy* 2 (2): 127–43. https://doi.org/10.1007/BF00057958.

————. 2001. 'Species Concepts.' In *Encyclopedia of Life Sciences.* New York: Wiley. www.els.net.

Godfrey-Smith, Peter. 2009. *Darwinian Populations and Natural Selection.* Oxford: Oxford University Press.

Grant, Verne. 1981. *Plant Speciation.* Columbia University Press. https://www.degruyter.com/document/doi/10.7312/gran92318/html.

Griffiths, Paul. 1999. 'Squaring the Circle: Natural Kinds with Historical Essences.' In *Species: New Interdisciplinary Essays*, edited by Robert A. Wilson, 209–28. Cambridge, MA: MIT Press.

Haber, Matthew H. 2013. 'Species Problems.' *Metascience* 22: 333–42.

Hellberg, Michael E., Ronald S. Burton, Joseph E. Neigel, and Stephen R. Palumbi. 2002. 'Genetic Assessment of Connectivity among Marine Populations.' *Bulletin of Marine Science* 70 (Supplement): 273–90.

Hey, Jody. 2001. *Genes, Categories, and Species: The Evolutionary and Cognitive Cause of the Species Problem.* Oxford, New York: Oxford University Press.

Hone, David W. E., and Michael J. Benton. 2005. 'The Evolution of Large Size: How Does Cope's Rule Work?' *Trends in Ecology & Evolution* 20 (1): 4–6. https://doi.org/10/d799br.

Hull, David L. 1976. 'Are Species Really Individuals?' *Systematic Biology* 25 (2): 174–91. https://doi.org/10.2307/2412744.

IUPAC. 2019. 'Chemical Element.' *IUPAC Gold Book.* http://goldbook.iupac.org/terms/view/C01022.

Kimbel, William H., and Yoel Rak. 1993. 'The Importance of Species Taxa in Paleoanthropology and an Argument for the Phylogenetic Concept of the Species Category.' In *Species, Species Concepts and Primate Evolution*, edited by William H. Kimbel and Lawrence B. Martin, 461–84. New York: Springer.

Kornet, D. J. 1993. 'Permanent Splits as Speciation Events: A Formal Reconstruction of the Internodal Species Concept.' *Journal of Theoretical Biology* 164 (4): 407–35. https://doi.org/10.1006/jtbi.1993.1164.

Lande, Russell. 1980. 'Genetic Variation and Phenotypic Evolution during Allopatric Speciation.' *The American Naturalist* 116 (4): 463–79.

Levin, D. A., and H. W. Kerster. 1974. 'Gene Flow in Seed Plants.' *Evolutionary Biology* 7: 139–220.

Lomolino, Mark V. 2005. 'Body Size Evolution in Insular Vertebrates: Generality of the Island Rule.' *Journal of Biogeography* 32 (10): 1683–99. https://doi.org/10.1111/j.1365-2699.2005.01314.x.

Mallet, J., N. Besansky, and M. W. Hahn. 2016. 'How Reticulated Are Species?' *BioEssays* 38 (2): 140–49. https://doi.org/10/f3kg4h.

Martin, Christopher H., Joseph S. Cutler, John P. Friel, Cyrille Dening Touokong, Graham Coop, and Peter C. Wainwright. 2015. 'Complex Histories of Repeated Gene Flow in Cameroon Crater Lake Cichlids Cast Doubt on One of the Clearest Examples of Sympatric Speciation.' *Evolution* 69 (6): 1406–22. https://doi.org/10/f7gr3h.

Mayden, R L. 1997. 'A Hierarchy of Species Concepts: The Denouement in the Saga of the Species Problem.' In *Species: The Units of Biodiversity*, edited by Michael F. Claridge, H. A. Dawah, and M. R. Wilson 54:381–424. Boca Raton, FL: Chapman & Hall.

———. 2013. 'Species, Trees, Characters, and Concepts: Ongoing Issues, Diverse Ideologies, and a Time for Reflection and Change.' In *The Species Problem: Ongoing Issues*, edited by Igor Pavlinov. InTech. http://www.intechopen.com/books/the-species-problem-ongoing-issues/species-trees-characters-and-concepts-ongoing-issues-diverse-ideologies-and-a-time-for-reflection-an.

Mayr, Ernst. 1942. *Systematics and the Origin of Species from the Viewpoint of a Zoologist*. New York: Columbia University Press.

———. 1963. *Animal Species and Evolution*. 1st ed. Belknap Press.

———. 1970. *Populations, Species, and Evolution*. Abridged. Belknap Press of Harvard University Press.

———. 2000a. 'A Defense of the Biological Species Concept.' In *Species Concepts and Phylogenetic Theory: A Debate*, edited by Quentin Wheeler and Rudolf Meier, 161–66. New York: Columbia University Press.

———. 2000b. 'The Biological Species Concept.' In *Species Concepts and Phylogenetic Theory: A Debate*, edited by Quentin Wheeler and Rudolf Meier, 17–29. New York: Columbia University Press.

Meier, Rudolf, and Reimer Willmann. 2000. 'The Hennigian Species Concept.' In *Species Concepts and Phylogenetic Theory: A Debate*, edited by Quentin D. Wheeler and Rudolf Meier, 30–43. New York: Columbia University Press.

Mishler, Brent. 1999. 'Getting Rid of Species?' In *Species: New Interdisciplinary Essays*, edited by Robert A. Wilson, 307–15. Cambridge, MA: MIT Press.

Mishler, Brent, and R. N. Brandon. 1987. 'Individuality, Pluralism, and the Phylogenetic Species Concept.' *Biology & Philosophy* 2 (4): 397–414. https://doi.org/10.1007/BF00127698.

Mishler, Brent, and M. J. Donoghue. 1982. 'Species Concepts: A Case for Pluralism.' *Systematic Zoology* 31: 491–503.

Mishler, Brent, and John Wilkins. 2018. 'The Hunting of the SNaRC: A Snarky Solution to the Species Problem.' *Philosophy, Theory, and Practice in Biology* 10. https://doi.org/10.3998/ptpbio.16039257.0010.001.

Morjan, Carrie L., and Loren H. Rieseberg. 2004. 'How Species Evolve Collectively: Implications of Gene Flow and Selection for the Spread of Advantageous Alleles.' *Molecular Ecology* 13 (6): 1341–56. https://doi.org/10.1111/j.1365-294X.2004.02164.x.

Naomi, Shun-Ichiro. 2011. 'On the Integrated Frameworks of Species Concepts: Mayden's Hierarchy of Species Concepts and de Queiroz's Unified Concept of Species.' *Journal of Zoological Systematics and Evolutionary Research* 49 (3): 177–84. https://doi.org/10.1111/j.1439-0469.2011.00618.x.

Novick, Aaron, and W. Ford Doolittle. 2021. '"Species" without Species.' *Studies in History and Philosophy of Science Part A* 87 (June): 72–80. https://doi.org/10/gjnq6h.

Paterson, Hugh. 1985. 'The Recognition Concept of Species.' In *The Units of Evolution: Essays on the Nature of Species*, edited by Marc Ereshefsky, 139–58. Cambridge, MA: MIT Press.

Pigliucci, Massimo. 2003. 'Species as Family Resemblance Concepts: The (Dis-)Solution of the Species Problem?' *BioEssays* 25 (6): 596–602. https://doi.org/10.1002/bies.10284.

Pleijel, Fredrik. 1999. 'Phylogenetic Taxonomy, a Farewell to Species, and a Revision of Heteropodarke (Hesionidae, Polychaeta, Annelida).' *Systematic Biology* 48 (4): 755–89. https://doi.org/10/fq8tng.

Plutynski, Anya. 2019. 'Speciation Post Synthesis: 1960 – 2000.' *Journal of the History of Biology* 52 (4): 569–596.

Queiroz, Kevin de. 1998. 'The General Lineage Concept of Species, Species Criteria, and the Process of Speciation: A Conceptual Unification and Terminological Recommendations.' In *Endless Forms: Species and Speciation*, edited by D. J. Howard and S. H. Berlocher, 57–75. New York: Oxford University Press.

———. 1999. 'The General Lineage Concept of Species and the Defining Properties of the Species Category.' In *Species: New Interdisciplinary Essays*, edited by Robert A. Wilson, 48–89. Cambridge, MA: MIT Press.

———. 2005. 'A Unified Concept of Species and Its Consequences for the Future of Taxonomy.' *Proceedings of the California Academy of Sciences* 56: 196–215.

———. 2007. 'Species Concepts and Species Delimitation.' *Systematic Biology* 56 (6): 879–86. https://doi.org/10.1080/10635150701701083.

———. 2011. 'Branches in the Lines of Descent: Charles Darwin and the Evolution of the Species Concept.' *Biological Journal of the Linnean Society* 103 (1): 19–35. https://doi.org/10.1111/j.1095-8312.2011.01634.x.

Quine, Willard V. 1948. 'On What There Is.' *The Review of Metaphysics* 2 (5): 21–38.

Richards, Emilie J., Jelmer W. Poelstra, and Christopher H. Martin. 2018. 'Don't Throw Out the Sympatric Speciation with the Crater Lake Water: Fine-Scale Investigation of Introgression Provides Equivocal Support for Causal Role of Secondary Gene Flow in One of the Clearest Examples of Sympatric Speciation.' *Evolution Letters* 2 (5): 524–40. https://doi.org/10/gk8ht2.

Richards, Richard A. 2010. *The Species Problem: A Philosophical Analysis*. 1st ed. Cambridge: Cambridge University Press.

Ridley, Mark. 1989. 'The Cladistic Solution to the Species Problem.' *Biology and Philosophy* 4 (1): 1–16. https://doi.org/10.1007/BF00144036.

Rieseberg, Loren H., and D. E. Soltis. 1991. 'Phylogenetic Consequences of Cytoplasmic Gene Flow in Plants.' *Evolutionary Trends in Plants* 5 (1): 65–84.

Rieseberg, Loren H., Troy E. Wood, and Eric J. Baack. 2006. 'The Nature of Plant Species.' *Nature* 440 (7083): 524–27. https://doi.org/10.1038/nature04402.

Savolainen, Vincent, Marie-Charlotte Anstett, Christian Lexer, Ian Hutton, James J. Clarkson, Maria V. Norup, Martyn P. Powell, David Springate, Nicolas Salamin, and William J. Baker. 2006. 'Sympatric Speciation in Palms on an Oceanic Island.' *Nature* 441 (7090): 210–13. https://doi.org/10/b9nt7v.

Scerri, Eric. 2019. *The Periodic Table: Its Story and Its Significance*. 2nd ed. Oxford and New York: Oxford University Press.

Seifert, Bernhard. 2014. 'A Pragmatic Species Concept Applicable to All Eukaryotic Organisms Independent from Their Mode of Reproduction or Evolutionary History.' *Soil Organisms* 86: 85–93.

Shapiro, B. Jesse, and Martin F. Polz. 2015. 'Microbial Speciation.' *Cold Spring Harbor Perspectives in Biology* 7 (10): a018143. https://doi.org/10.1101/cshperspect.a018143.

Simpson, George Gaylord. 1951. 'The Species Concept.' *Evolution* 5 (4): 285–98. https://doi.org/10/gg4394.

———. 1961. *Principles of Animal Taxonomy*. 1st ed. New York: Columbia University Press.

Slater, Matthew H. 2014. 'Natural Kindness.' *The British Journal for the Philosophy of Science*, April, axt033. https://doi.org/10.1093/bjps/axt033.

Sober, Elliott. 1980. 'Evolution, Population Thinking, and Essentialism.' *Philosophy of Science* 47 (3): 350–83.

Sorenson, Michael D., Kristina M. Sefc, and Robert B. Payne. 2003. 'Speciation by Host Switch in Brood Parasitic Indigobirds.' *Nature* 424 (6951): 928–31. https://doi.org/10/b7nzw3.

Stamos, David N. 2003. *The Species Problem, Biological Species, Ontology, and the Metaphysics of Biology.* Lanham, MD: Lexington Books.

Sterner, Beckett. 2019. 'Evolutionary Species in Light of Population Genomics.' *Philosophy of Science* 86 (5): 1087–98. https://doi.org/10/gk8htz.

Templeton, Alan. 1989. 'The Meaning of Species and Speciation: A Genetic Perspective.' In *The Units of Evolution: Essays on the Nature of Species*, edited by Marc Ereshefsky, 159–83. Cambridge, MA: MIT Press.

Turelli, M., and H. A. Orr. 1995. 'The Dominance Theory of Haldane's Rule.' *Genetics* 140 (1): 389–402.

Van Valen, Leigh. 1976. 'Ecological Species, Multispecies, and Oaks.' *Taxon* 25 (2/3): 233–39.

Whittemore, Alan T. 1993. 'Species Concepts: A Reply to Ernst Mayr.' *Taxon* 42 (3): 573–83. https://doi.org/10/ddj596.

Wiley, E. O. 1978. 'The Evolutionary Species Concept Reconsidered.' In *The Units of Evolution: Essays on the Nature of Species*, edited by Marc Ereshefsky, 79–92. Cambridge, MA: The MIT Press.

———. 1981. *Phylogenetics: The Theory and Practice of Phylogenetic Systematics.* 1st ed. Wiley-Liss.

Wiley, E. O., and Richard Mayden. 2000a. 'The Evolutionary Species Concept.' In *Species Concepts and Phylogenetic Theory: A Debate*, edited by Quentin Wheeler and Rudolf Meier, 70–89. New York: Columbia University Press.

———. 2000b. 'The Evolutionary Species Concept.' In *Species Concepts and Phylogenetic Theory: A Debate*, edited by Quentin D. Wheeler and Rudolf Meier. Columbia University Press.

Wilkins, John. 2018. *Species: A History of the Idea.* 2nd ed. Berkeley, CA: University of California Press.

Willmann, Reimer. 1985 *Die Art in Raum Und Zeit. Das Artkonzept in Der Biologie Und Paläontologie.* Berlin: Paul Parey.

Wilson, Jack. 1999. *Biological Individuality: The Identity and Persistence of Living Entities.* Cambridge: Cambridge University Press.

Wilson, Robert A. 1996. 'Promiscuous Realism.' *The British Journal for the Philosophy of Science* 47 (2): 303–16.

———. 1999. 'Realism, Essence, and Kind: Resuscitating Species Essentialism.' In *Species: New Interdisciplinary Essays*, edited by Robert A. Wilson, 187–207. Cambridge, MA: MIT Press.

Wilson, Robert A., Matthew J. Barker, and Ingo Brigandt. 2007. 'When Traditional Essentialism Fails: Biological Natural Kinds.' *Philosophical Topics* 35 (1–2): 189–215.

Zachos, Frank E. 2015. 'Taxonomic Inflation, the Phylogenetic Species Concept and Lineages in the Tree of Life – A Cautionary Comment on Species Splitting.' *Journal of Zoological Systematics and Evolutionary Research* 53 (2). https://doi.org/10/ggbq8h.

———. 2016. *Species Concepts in Biology: Historical Development, Theoretical Foundations and Practical Relevance.* Cham, Switzerland: Springer.

2 Is the Species Problem *That* Important?

Yuichi Amitani

CONTENTS

2.1 INTRODUCTION

The species problem is a longstanding puzzle concerning the nature of the species category and how to define the notion of species. It is often described as difficult, or even impossible, to solve. This impression comes from the fact that biologists and taxonomists have proposed a number of definitions over a long period of time, but there is little agreement among biologists on the right definition. For example, Frank Zachos (2016) notes that there are 32 definitions of species proposed thus far, and that number increases when a new definition has been proposed (Seifert 2020).

DOI: 10.1201/9780367855604-3

The species problem is also described as a very important issue. There are *prima facie* reasons to believe that this problem is important for biologists and philosophers of biology. Species is a basic unit of any classification system. Thus, the lack of a universally agreed-upon definition of species may cause a serious communication breakdown among taxonomists. Species is a basic unit of evolution. Thus, the lack of proper understanding of the nature of species may represent our lack of understanding of a crucial part of the biological world and evolution. Additionally, species is a basic unit of biodiversity, and the number of species in a protected area is a good indicator of that area's biodiversity. Thus, the lack of a universally agreed-upon definition may mean that we are not able to compare biodiversity across areas.

However, there is a conflicting observation in regard to these assumptions. If the species problem is both important and difficult to solve, then one would expect that biologists would have deep trouble communicating their ideas regarding species and have made little progress in the research fields related to species and speciation. This does not appear to be a correct description of the current state of affairs. For example, research on speciation has made significant progress in spite of the lack of an agreed-upon definition of species (Schluter 2000; Coyne and Orr 2004; Price 2008; Nosil 2012). There also does not seem to be a serious communication breakdown regarding the meaning of species in many research contexts.

How are these three conflicting observations resolved? In this chapter, we will focus on the second point of the trilemma and argue that the species problem is not as important as commonly thought. As we have seen, there are three types of reasons cited for the importance of the species problem: concern regarding communication breakdown (without a definition agreed upon at the community level, we will not communicate with each other effectively), the theoretical significance of the concept of species (the concept of species will not be helpful in biologists' conduction of their research without a precise and agreed-upon definition), and the concern regarding practical implications (for example, without a precise definition of species, we cannot measure the degree of biodiversity in a given protected area).

If you look at how biologists deploy the concept of species in their research, however, those concerns can be significantly alleviated. For example, the stability of individual species names (e.g., *Homo sapiens* and *Drosophila melanogaster*, not species definitions) contributes to effective communication among naturalists and biologists when finding a serious disagreement on the nature of species. In addition, biologists do not have to have any precise definition of species to advance their research even if they study species and speciation. For example, if one studies a so-called *good species* (a taxon that is expected to be treated as a species among biologists, for instance, by satisfying most or all of the major species definitions), it *is* a species regardless of the definition one happens to use. In this case, biologists can conduct their research when they have disagreements on the nature of species. These observations describe how biologists and taxonomists manage to carry out their research despite a lack of agreement on the right definition of species. This means that the species problem is not as important as species theorists (biologists and philosophers) have assumed.

The remainder of this chapter is organised as follows. In the next section, I provide an overview of the arguments given for the importance of the species problem. Each argument corresponds to the concerns outlined earlier. In the third section, I address the concern regarding possible communication breakdown and argue that biologists can effectively communicate without a precise definition of species. Then, I argue that the concept of species can play significant theoretical roles in biology without a precise definition. In the following section, I address the concern regarding biodiversity and identification and point out that the practical interests mentioned here can be addressed without a precise and universally accepted definition of species. Finally, I address some of the possible objections and reply to them.

2.2 WHY IS THE SPECIES PROBLEM IMPORTANT?

In this section, I describe the reasons cited by species theorists to claim that the species problem is important. There are three types of reasons cited for the importance of the species problem: communication, theoretical significance, and practical consequences.

2.2.1 COMMUNICATION

One reason why the species problem is said to be important is that we may not ensure effective communication among biologists if we do not have an agreed-upon definition of species. In the history of the species controversy, this point has traditionally been made as an argument against species pluralism (Holsinger 1987; Hull 1999). Under pluralism, more than one definition of species can be legitimate. Consequently, supporters of different definitions of species, such as the biological species concept, the phylogenetic species concepts, and the ecological species concept, refer to different things when they use the word 'species' because different definitions of species divide the biological world differently. If different people use the same word to refer to different things, communication between them can break down. Think of the case in which one party uses 'bank' to refer to a financial institution, and the other uses the same word to refer to sloping land along a river. While it is relatively straightforward to spot and clarify the confusion in cases like this, one can imagine similar but trickier scenarios in the usage of 'species'.

2.2.2 THEORETICAL SIGNIFICANCE

A great number of prominent biologists and taxonomists have thought that species is a real and basic unit of the living world. Many of them believe that species is *the* real unit of biodiversity; categories under and above species, such as genus and subspecies, are artificial rather than real (Simpson 1961; Mayr 1969; 1970; Eldredge and Cracraft 1980; Hull 1997; Stamos 2003). A recent survey shows that most rank-and-file biologists hold that the concept of species is the fundamental one in biology (Pušić, Gregorić, and Franjević 2017). Biologists also use this concept in studies on the history of organisms (phylogeny), their spatial distribution (biogeography), and their evolutionary process (e.g., speciation).

Most biologists and philosophers seem to believe that since the notion of species plays an important role in research, the concept of species will not be helpful in biologists' conduction of their research if it does not have a precise and agreed-upon definition.* I do not find many biologists stating this explicitly, but there is reason to assume this. It is generally believed that no important concept will help its users proceed with their business without a precise and agreed -upon definition, whether or not it is in a scientific context.

One example of this attitude toward defining species can be seen in Ernst Mayr's position on the relationship between defining species and the study of speciation. The question is whether defining species or studying speciation has research priority. Mayr (1942; 1957; 1970) asserted that it is necessary to have the right definition of species *before* we begin to study speciation.

> A concise definition of species is, for [a student of speciation], a necessity, because his interpretation of the speciation process depends largely on what he considers to be the final stage of the process, the species.

> **(Mayr 1942, 114)**

In other words, since speciation is a production process of species after all, biologists cannot study it in a meaningful way unless they have a good understanding of what is supposed to be produced in that process.

Comparison of species is another point that researchers have made for the importance of the species problem (Cracraft 2000). If one area or field of biology uses one definition of species, while another uses another definition, 'species' refers to different things between those areas or fields. We would then compare apples and oranges when we make a comparison relating to 'species' between those areas. For example, if we use different definitions of species in the study of birds and the study of crustaceans, and a bird clade A has x species and a crustacean clade B has $y(>x)$ species, we still cannot conclude that the species richness of B is greater than A because the species in A are not equivalent to those in B.

2.2.3 PRACTICAL IMPORTANCE

The species problem is of practical importance. Biologists and philosophers have pointed out that the choice of species definition can have practical implications for the conservation and identification of species.

The species problem affects the conservation of biodiversity in many different ways, because a substantial portion of the current conservation efforts is based on the assumption that species is a unit of biodiversity (Claridge, Dawah, and Wilson 1997).† Controversy over species realism, for example, could undermine conservationists'

---

* As Neto (2020) suggests, at least for philosophers, this assumption may come from the logical positivist tradition.
† For example, many legal frameworks for conservation, such as the *Endangered Species Act* in the United States, focus on species rather than other units.

efforts to protect endangered species (Hull 1997); if the species category turns out to be an artifact like constellations, there may be little biological meaning in the protection of endangered species (see also Reydon 2019). An agreement on the right definition of species would help the conservation movements because we would not know exactly what entities to protect under the rubric of 'species' if we do not have a precise and agreed-upon definition of this concept.

Another practical implication of the species problem is the identification of species. If we confuse two forms with different properties as one and the same species, it may lead to practical problems in many areas. Sigwart (2018) and Cracraft (2000) offer many examples of this. Some species of snakes have similar morphologies, but their venoms have different chemical compounds. As antivenoms are most effective when applied to the right venom, misidentification of snake species can have grave consequences. Another example is mosquitoes as a vector of infectious diseases. If one of two similar-looking mosquitoes is a vector for malaria, and the other is not, identification of these species is crucial for the efforts to eradicate the disease.

2.2.4 MEANING OF THE IMPORTANCE OF THE SPECIES PROBLEM

We have seen that biologists and philosophers have cited various reasons for the importance of the species problem. Because these reasons indicate the conditions under which the species problem is taken to be solved, we can reconstruct their idea of an 'ideal' species definition from those reasons. Based on this idea, we can succinctly summarise the reasons why the species problem is thought to be important, because what they say amounts to the idea that it is important to have a species definition that includes those qualities.

From the observations earlier, we can infer that an ideal species definition is a *precise*, *non-arbitrary*, and *agreed-upon* definition. The 'precision' requirement is that an ideal species definition should not be overly vague (its boundaries are not blurred), ambiguous (it does not mean more than one thing), or under-specified (its description is not too broad to capture its reference). This is important because an overly ambiguous or under-specified conception of species will not be useful in various research and practical contexts. For example, an ambiguous or under-specified definition of species may not help us identify what is to be protected in conservation efforts. It may also prevent effective communication by leaving what is really meant by 'species' or a species name unclear. If ensuring effective communication and protecting biodiversity are important goals for biologists, then, all else being equal, they would prefer a precise definition.

The 'non-arbitrariness' requirement demands that an ideal species definition enable us to 'carve nature at its joints' so that we can make further inductions about it. For example, dividing the entire category of vertebrates into two groups based on body length, say, whether a specimen is longer than 2 cm, is an arbitrary classification and enables us to make only a few further inductions about organisms in the categories. We need this condition because biologists and philosophers have maintained that the concept of species plays a significant role in their research in diverse areas such as phylogeny, biogeography, and evolution.

Finally, an ideal definition of species should be agreed upon or accepted among biologists. 'Agreed upon' here means that the definition is accepted within the biologist community, not merely by a single biologist. An ideal situation that satisfies this requirement is that all the biologists have one and the same definition of species at their disposal. Needless to say, this is only an ideal. Biologists and philosophers will not have any unanimously agreed-upon definition of species in the near future (in fact, many biologists think that monism may not be even a desired position; Pušić, Gregorić, and Franjević (2017)). On the other hand, a definition held by a single biologist or a very small group will not be a solution to the problem. The right balance is likely between those two extremes. Nonetheless, it is sensible to require that an ideal definition of species be accepted by a group of biologists of an appropriate size.[*]

2.3 WHY NOT *THAT* IMPORTANT – COMMUNICATION WITHOUT DEFINITION

We have seen various reasons why the species problem is said to be important. We argue that the problem is not *that* important, that is, that some of the reasons do not have the strength they are claimed to have. In particular, we will discuss the precision requirement of a species definition and show that biologists can often do their work well, if not perfectly well, even when they have only a somewhat agreed-upon but imprecise conception of species.

I begin by discussing the problem of communication. Biologists and philosophers have expressed concern that researchers will have trouble exchanging their ideas efficiently without a precise and agreed-upon definition of species. However, a number of options are, in fact, available to biologists to avoid breakdowns in communication when encountering disagreement on the content of 'species'. There are several historical cases to support this. In those cases, biologists and taxonomists, despite holding conflicting notions of 'species', managed to keep communication between scholars effective. One example is a proposal for taxonomic nomenclature drafted by a nineteenth-century English naturalist, Hugh Strickland (McOuat 1996).

2.3.1 HUGH STRICKLAND'S CASE

Hugh Strickland drafted his proposal for biological nomenclature in the 1830s, when there were two conflicting views in England on naming taxonomic groups and the Linnaean hierarchical system of classification. The first camp, the conservatives, supported the Linnaean system and held that the taxonomic groups should be named following the Lockean view of naming. According to this view, a name is merely a

---

[*] Note that those requirements largely correspond to the criteria proposed by David Hull to evaluate different definitions of species (Hull 1997). His criteria are universality, applicability, and theoretical significance. Precision largely corresponds to applicability, non-arbitrariness to theoretical significance, and agreement to universality.

'tag' to its reference, and it does not have to tell us anything about its properties. For example, a species referred to as *Homo sapiens* does not have to be wise despite the fact that *sapiens* means wise, nor does humans' wisdom have to be a differentiating trait between human beings and other *Homo* species: even if our ancestors such as *Homo habilis* and *Homo erectus* were as wise as we are, *Homo sapiens* would be still a legitimate name for our species.

The second camp, called the reformists, disagreed with the views of the conservatives on both issues. This group called for reform of the Linnaean system. They also argued that a taxon's name should convey some information about the properties of its reference that are of use in discriminating that reference from its sister groups. To use a recent example, the name of an American moth species, *Neopalpa donaldtrumpi*, would be a good name according to the criteria of the reformists, based on the characteristics that its scales resemble Donald Trump's hair, provided that its hair helps discriminate the species from its sister groups.

Although he admitted the need for reform of the Linnaean system of classification, Strickland was a conservative naturalist and in 1837, drafted taxonomic nomenclature embodying the conservative naming philosophy ('Rules for Zoological Nomenclature', Strickland 1837). He then requested that the British Association for the Advancement of Science (BAAS) form a special committee to discuss his proposal. His goal was to convince the committee to recommend that the Association adopt his proposal as the official zoological naming rules. However, Strickland was met with radical disagreements within the committee on the nature of species. Far from reaching consensus on how to define species, they could not even agree that species was a tangible object; T.H. Bunbary maintained that species is merely an abstraction, and only individual organisms are tangible objects. Facing strong objections such as this, Strickland entirely gave up defining species in the final report he submitted to BAAS (Strickland 1843). Although the final report was rejected due to fierce opposition from the reformists at the BAAS meeting, Strickland managed to print the rules in the official BAAS report. This made the rules look as if they were an official document of the BAAS, which helped the acceptance of similar rules in the United States and Italy.

Strickland's effort as the chair of the special committee is significant to this study for two reasons. First, it reveals how a competent naturalist could successfully avoid communication breakdown when he found fundamental disagreements among his colleagues on the nature of species. By stripping much of the ontological and philosophical content from the notion of species, Strickland found common ground among naturalists with fundamentally different views. The second point is that this 'common ground' did not have any definition as its component. Rather, it comprised exemplars: taxa that most competent naturalists would classify as species. 'For the Rules, species were just what competent (read: institutional, published, gentlemanly, conservative) naturalists said they were' (McOuat 1996, 512). Indeed, the members of Strickland's committee, including Charles Darwin, shared many exemplars of species with each other (McOuat, personal communication). This suggests that taxonomists can go a long way without a precise definition of species – if they have an agreement on concrete species classification.

2.3.2 COYNE, ORR, AND BUSH

Another case in which biologists can effectively communicate despite disagreements on species is debates between Jerry Coyne and H. Allen Orr, in one camp, and Guy Bush, in the other (Amitani 2013). This example is striking in the sense that they have engaged in two debates on species at the same time. The first debate concerns the fundamental issues involving the nature of species and the study of speciation. The second regards a possible case of sympatric speciation of a maggot fly (*Rhagoletis pomonella*). Since both camps disagree on the fundamental and methodological issues regarding species and speciation, one might expect that the disagreement on this issue would make it hard for each camp to understand what the other camp said about the second issue. However, this is not what happened. Although Coyne and Orr disagree with Bush regarding whether *Rhagoletis pomonella* undergoes sympatric speciation, they were able to make themselves understood to their opponent in the second debate. In the following, we will explore how this was possible.

The first disagreement Coyne and Orr have with Bush concerns the research priority between defining species and studying speciation. The key question here is whether or not choosing their definition of species is helpful for students of speciation when they begin their studies. Coyne and Orr assert that defining species before commencing the study of speciation will advance it (Coyne and Orr 2004). This is because many species definitions reflect particular research interests. For example, one version of the phylogenetic species concept (de Queiroz and Donoghue 1988) reflects the systematists' interest in the phylogenetic patterns of species. This interest relativity, they argue, makes choosing a species definition before one begins their study desirable, because the choice of a definition directs their attention to particular aspects of species and speciation, thereby helping to design their research. Thus, Coyne and Orr (2004) chose the biological species concept as suitable for their research interest, which is to find the causal origin of the phenotypic and genetic discontinuity among different species.

In contrast, Guy Bush believes that the choice of a species definition is not necessary or beneficial to the subsequent study of speciation (Bush 1993; 1994; 1995; Bush and Smith 1998; Bush and Butlin 2004). He begins his argument with his criticism of the biological species concept. Bush first points out that there are many 'good' species[*] that are still not reproductively isolated from each other and that the establishment of reproductive isolation often does not coincide with the two groups evolutionarily parting from each other. These cases suggest that factors other than gene flow may play an important role in speciation and that one species could become two with different degrees of genetic exchange – different species have different 'thresholds' to become a distinct species.

This plurality of causes and speciation thresholds combined with the above-mentioned interest relativity of species definitions would have implications for the issue of priority between defining species and studying speciation. As Bush argues, if

---

[*] A good species here is a taxon that is deemed a legitimate species by many or most taxonomists or species criteria (Amitani 2015). I will discuss this concept later in the chapter.

different research contexts urge researchers to focus on different aspects of species, and if different species have different speciation thresholds, then it is always possible that the causal factor implied by one's choice of definition plays no essential role in the case at hand and that their research is thereby misguided. If there continues to be gene flow between populations while they are evolutionarily parting from each other, adopting the biological species concept may prevent researchers from finding true driving forces of the speciation process. Thus,

> Trying to explain speciation within the context of a preconceived species concept places the cart before the horse. It is an understanding of the factors that result in the reduction and eventual elimination of gene flow between sister populations – the very process of speciation itself – that is necessary before a clear species definition is possible.
>
> **(Bush 1994, 286)**[*]

The second disagreement Coyne and Orr have with Bush regards whether two host races of the apple maggot fly, *Rhagoletis pomonella*, are in the process of sympatric speciation. The two host races live on hawthorn and apple trees in parts of the United States where the two trees coexist. The hawthorn race also lives in other parts of the United States where no apple trees are found, and an isolated group lives on hawthorns in Mexico. After apple trees were introduced into the United States, *R. pomonella* was found to inhabit them in the mid-nineteenth century, and it later spread all over the country. Since the apple race is monophyletic, their shift to the new host occurred only once. The apple and hawthorn races breed on the apple and hawthorn trees, respectively, so genetic exchange between them is limited.

Where did the apple race come from? Bush and his colleagues (Bush 1969; 1992; Bush et al. 1989) believe that the two races are in the process of sympatric speciation. Some individuals of the hawthorn race found a new habitat in the apple trees in the regions where both trees coexist. Due to natural selection and the flies' tendency to find mates on their host tree, they have been evolutionarily and genetically diverging. Since the two host trees coexist, this would be an instance of sympatric speciation.

Coyne and Orr (2004) find Bush's conclusion too hasty. Here, we discuss one of their alternative interpretations of the evolution of the apple race. According to this interpretation, which was originally proposed by Carson (1989), the apple race did not come from the hawthorn race in the areas where their hosts coexist but rather, from another hawthorn race living outside those areas. Carson, Coyne, and Orr direct our attention to the fact that the hawthorn genus (*Crataegus*) has as many as 1,000 'species' named in the United States and Canada and suggest that the maggot flies may have diversified as the hawthorn trees did. Thus, '[b]efore the introduction of the apple, there may have been two or more genetically

---

[*] Bush may not be always right in this regard. I will return to this point later when I discuss how Peter and Rosemary Grant deal with the issue of defining species.

distinct races of *R. pomonella*, adapted to different hawthorn species or possibly to an endemic native crabapple' (Carson 1989, 304), and one of them may have turned into the apple race. In other words, the apple race may have not been from the sympatric hawthorn race but from another hawthorn race in a non-coexisting area.

The purpose of describing the disagreements on the two issues Coyne and Orr have with Bush at length is not to take a side on each issue but to see how they can communicate effectively in the second debate despite their disagreements in the first. Since the two camps disagree on the fundamental issues of species and speciation, one might expect that those disagreements would prevent them from communicating effectively on the second issue: they may well not understand what those in the other camp say in the second debate due to the disagreements in the first. However, this is not what has happened. In the second debate, disagreements are largely restricted to empirical issues, such as the origin of the apple host race. Although Coyne (1994) notes that the apple race is not a full species even if it arose sympatrically, this has not been Coyne and Orr's main argument against Bush.

How can this be possible? One reason, among others, is that those in both camps shared the reference of the species name of the apple maggot flies, *Rhagoletis pomonella* (Amitani 2013). Here it is important to understand that selection of a species definition can restrict the subsequent study of species and speciation in two ways: by restricting which population or species to study, and by restricting the aspects of species or the speciation process under study. If, for example, a species is defined by a property F, then a researcher who adopts the definition may not pay much attention to a population without that property. Likewise, if a species is defined by F, a researcher who adopts the definition may pay attention to the process in which a population acquires F but not to that in which it acquires another property, G, provided that G is not correlated to F. Either kind of restriction can cause miscommunication, because researchers with different definitions in mind may think that they are talking about the same species or process of speciation when they are not.

In the *R. pomonella* case, neither of the ways of restricting a researcher's focus affects Coyne and Orr or Bush. Both camps largely refer to the same population(s), that is, the apple race of *R. pomonella*. Thus, the first type of restriction did not lead to miscommunication. Since both camps take the establishment of reproductive isolation as an important event in the speciation process whether or not it defines species, the second type of restriction does not cause a breakdown in communication. This is partly because Bush still believes that the establishment of reproductive isolation occurs at the end of the speciation process, even though he does not take reproductive isolation as *the* defining feature of species. The Bush camp and the Coyne and Orr camp both look, in their study of speciation, at the processes in which a population acquires reproductive isolation.

This shows that even the lack of a universally accepted definition of species due to fundamental and methodological disagreements on the conception of species and the study of speciation may not lead to miscommunication if biologists may share the reference of a species name in question (*R. pomonella*). This lesson is the same as in the case of Strickland, because sharing the exemplars of species among naturalists saved them from miscommunication sans any definition of species.*

2.3.3 RHEINBERGER AND KELLER: THE AMBIGUITY OF 'GENE' HELPED BIOLOGISTS

A lesson from the last section is that the imprecise conception of species (i.e., understanding species without any definition in mind) may not be a barrier to effective communication. The following arguments suggest that adopting the 'broad' conception of species, rather than seeking a precise definition, may be *beneficial* for communication across different research contexts.

Hans-Jörg Rheinberger (2000) and Evelyn Fox Keller (2000) claim that this is the case for the concept of genes in biology.† Both claim that the ambiguity inherent in the concept of a gene has helped biologists make substantial progress in different fields. In his paper, Rheinberger (2000) points out that the concept of a gene plays different roles in different fields of biology. In molecular biology, for example, genes are parts of a chromosome that produce proteins, transfer RNA, and other gene products. In developmental biology, genes regulate development by turning on and off 'switches' of the developmental programs that produce parts of organs and tissues. Thus, 'gene' has not had any unified definition across different fields, but this has been beneficial to biology, he argues, because those areas of biology have made remarkable progress for decades without a precise and agreed-upon definition of what a gene is.

Rheinberger (2000) says that the ambiguity in the concept of a gene helped biologists but says little about how it did so. This is the gap Keller (2000) fills. She draws our attention to the fact that the concept of a gene refers to different things in various experimental contexts from transcription to translation. In one context, it may mean just an open reading frame, a sequence of nucleotides that is sandwiched between start and stop codons. In another context, it may mean RNA molecules ready for translation to produce proteins. Thus, researchers can use the concept of a gene in a way specific to a concrete context and eliminate ambiguity: '"gene" refers to an open reading frame with length *x* here'. This helps them sharpen their focus and clarify what they discuss, but it may also prevent them from applying the concept to other contexts, such as translation. That is where leaving the concept of gene as an ambiguous one could help researchers from different experimental contexts communicate effectively. If the concept of a gene is left somewhat ambiguous, it may enable

---

* Some might wonder whether this would hold if both camps had even more radically different views on speciation; for example, what if Bush and others held that the establishment of reproductive isolation is largely uncorrelated with the evolutionary parting of two populations? My reply is that disagreements may lead to miscommunication in this case. In other words, there is a certain limit to how far the taxonomic custom can keep disagreeing parties on the same page. However, my point is that there are a number of cases in which biologists can communicate effectively despite their (sometimes fundamental) disagreements on the nature of species.
† See Neto (2020) for other attempts to argue that imprecise scientific concepts have benefits, especially in integrating various fields in biology.

researchers to talk about roughly the same thing, even though it may not be exactly the same. In this way, she argues, it is the ambiguity inherent in the concept of a gene that actually facilitates effective communication among biologists.

We have seen how the ambiguity of the concept of a gene facilitated effective communication in biology. If this is true of *genes*, it may be true of *species* as well. Indeed, since the concept of species is used in different theoretical and practical contexts, including classification, the study of evolutionary processes (such as speciation), and conservation efforts, the concept of species, without any precise definition, may play a similar role in biology as does the concept of a gene.

2.4 WHY NOT *THAT* IMPORTANT – THEORETICAL SIGNIFICANCE

In the last section, we have seen that the concern over possible communication breakdown from the lack of a precise and agreed-upon definition of species is exaggerated. However, biologists have cited another reason why the species problem is important: its theoretical significance. In this section, we argue that this point is also exaggerated.

Their argument for the importance of the species problem was that since the concept of species plays important roles in various fields of biology, and no concept would help its users proceed with their research without a precise and agreed-upon definition, it is important to have such a definition for the concept of species. Our argument is that biologists can often carry out their research fruitfully without such a definition. We will begin by discussing what is referred to as the 'general' concept of species and its components (Amitani 2017). The general concept of species is what biologists conceive when they talk about species without any definition in mind. This is important for our purpose because species theorists have not paid much attention to the fact that there are many research contexts in which biologists use the concept of species without any individual definition in mind, and that suffices for them. The significance of finding a precise and agreed-upon definition of species is reduced by the extent to which using the notion of species without any definition in mind is helpful for biologists' research.

2.4.1 The General Concept of Species

The general concept of species cannot be expressed in the form of a definition. At the same time, it is not entirely nebulous, and it has several components (see Amitani 2017 for details). Some of these components are derived from what most or all of the biologists agree on with regard to the properties of species. For example, most biologists would agree that each species has distinctive (morphological, behavioral, and ecological) characteristics, and that there is some causal basis for the differences, even if we have little idea as to what that basis is (i.e., it is a placeholder in the concept of species). They also believe that a species is a lineage at the population level, because all of the major definitions of species imply this (de Queiroz 1999; 2005a; 2005b; 2007). Other components are things biologists have appealed to in order to avoid consequences from serious disagreements on the nature of species. We have seen that Strickland and other naturalists relied on the fact that they agreed on exemplars of species when they overcame such disagreements on the nature of species. Good species, a prototype of the concept of species, is something biologists can rely on when encountering disagreements, because a good species is a taxon that is deemed to be a species by most biologists or species criteria. These components are not powerful enough to discriminate species from non-species completely, but they convey some ideas about what a species is.

2.4.1.1   Good Species

Before discussing how the general concept of species helps biologists carry out their research fruitfully, let us describe what a good species is, as we will focus on this component in describing the functions of the general concept of species. 'Good species' is an unofficial term used in biology and systematics in general. This term has several usages, but here, we address two of them (Amitani 2015; see also Wilkins, this volume).* In the first usage, a 'good species' refers to a species that is distinctive or well defined by multiple species criteria, such as reproductive isolation, occupying a distinctive niche, and being monophyletic. When authors use 'good species' in this way, they often cite some of those criteria to suggest that a taxon in question is a good species.

> Polytene chromosomes of four members of the *Simulium perflavum* species group in Brazil are described. … Chromosomal, morphological and ecological evidence indicates that *S. maroniense* Floch & Abonnenc, previously considered synonymous with *S. rorotaense*, is a good species.
>
> **(Hamada and Adler 1999)**

In this quote, the authors cite evidence from multiple areas and conclude that *Simulium maroniense* is a good species.

In the second usage, a good species is a taxon generally recognised as a legitimate species by biologists. In contrast to the first usage, biologists do not cite any particular criterion when they note that a taxon is a good species. James Mallet explicitly states that he uses the term in this manner.

> I used the term 'good' species several times meaning that people generally agree that 'the blue whale' and 'the fin whale,' for example, are species, …. Unless taxonomists are mad, there is something reasonable about such species.
>
> **(Mallet 1996, 174)**

As different as these usages are, they point in similar directions. If a taxon is a good species in the first usage, it satisfies most or all of the major species criteria. This means that it will be deemed to be a species by most biologists because it *is* a species regardless of which definition one happens to adopt. It also implies that there is

---

* Other usages of this term can be found in Amitani (2015). Those usages come from a survey of scientific papers and an online mailing list.

no need to consider the right definition of species when discussing a good species, because again, a good species is a species regardless of which definition you happen to use.

2.4.2 EPISTEMIC ROLES PLAYED BY THE GENERAL CONCEPT OF SPECIES

The general concept of species plays positive epistemic roles in fostering research (Amitani 2017). The first role is agenda-setting: the general concept of species sets a research agenda for biologists by providing an *explanandum*. We have seen that a component of this concept is phenotypic differences among species with a 'causal placeholder'. In other words, according to this concept, species are phenomenologically different from each other, and there are causal factor(s) behind this, even if we are not aware of what they are. This indicates a future direction of research, that is, determining the causal factors responsible for phenomenological differences. We can think of several research projects conducted along this line of thought. For example, several studies of speciation (Mayr 1963; Coyne and Orr 2004) can be read as attempts to determine the origin of these differences in gene flow and reproductive isolation. The ecological species concept (Van Valen 1976) partially explains the differences from the standpoint of natural selection.[*]

This function of the general concept of species has been reported by several authors. Ingo Brigandt (2003) argues that species pluralism, the idea that there is more than one legitimate definition of species and that biologists can use different definitions from one research context to another, does not lead to the elimination of the concept of species as a whole. One of his reasons is that the research agenda it originally set (the causal explanation of phenotypic discontinuities) is still viable even after we have multiple definitions of species. In a similar vein, John Wilkins (2018) asserts that a species is not a theoretical object but a phenomenon. His starting point is the observation that there is no theory that needs species for explanation and that in a theoretical talk the concept of species can often be replaced with other concepts. For example, when medical and biological researchers refer to a particular species (such as *Mus musculus*, the common mouse) as a model organism, what they actually refer to is a particular strain, not the species as a whole, that is bred to exhibit a specific effect in response to a particular treatment. Thus, Wilkins argues that species is not an *explanans*. However, this does not necessarily mean that species are not real. Instead, Wilkins argues that species are real as a phenomenon, something calling for explanation. From our perspective, this means that the notion of species sets up a

---

[*] The late James Flynn, a scholar who discovered the Flynn Effect in intelligence studies, makes a similar point regarding the notion of intelligence (in Kaufman 2018). He says that the under-specificity of the notion of intelligence can help determine a research agenda, and giving it an overly precise definition will deprive it of that function.

research agenda for which finding causal factors behind the 'species phenomena', such as phenotypic discontinuities, is an explanatory goal.*

Another epistemic role of the general concept of species is to serve as a reference point. A "good species" is, in a nutshell, a species taxon regardless of which definition of species one happens to hold. Thus, the properties of a good species will give us a clue as to the properties of a species in general. Biologists use their findings on a good species in this manner.

> It is commonly assumed that 'good' species are sufficiently isolated genetically that gene genealogies represent accurate phylogenies. However, it is increasingly clear that good species may continue to exchange genetic material through hybridization (introgression).

> **(Chan and Levin 2005)**

In this quote, the authors imply that distinct species in general may genetically interact with each other from the observation that good species do so. The logic behind their reasoning runs that if a prototypical, or 'good', member of a category has a property F, then other members, including 'bad' ones, may have F too. If this expectation is satisfied with the category, then the concept of species without any particular definition in mind can have an epistemic virtue. Thus, an under-specified notion of species (the general concept of species and good species) can foster biological research.

We have seen cases in which the use of the general concept of species is epistemologically beneficial to us. In addition, there are a number of research contexts in which the general concept of species is good enough, if not positively beneficial, for our research purposes (Amitani 2017). This is because a precise definition of species is sometimes not needed considering the communication cost that it could impose. To understand why this is the case, let us imagine two biologists discussing the predation relationship between two species, with one saying, 'Species 1 is a predator of Species 2'. Without a precise definition of species being given, we cannot specify what this researcher actually means by 'species'; she might mean a reproductively isolated population, the smallest monophyletic group, or others.† In other words, the above statement is a 'rough and ready' generalization of the relationship between the two species. However, this does not mean that we should always avoid this kind of generalisation, because it would take time and mental effort to clarify which definition of species she follows here. Which definition she adopts matters in some contexts but not in others, and the use of the general concept of species will suffice

---

* In this sense, the general concept of species plays a role similar to that which what Richard Boyd calls 'theory-constituting metaphors' play in science (Boyd 1979). His examples include computer-related metaphors used in cognitive science, such as information processing and subroutines. These metaphors, Boyd notes, do not constitute a theory by themselves but rather, direct scientists' attention in particular directions by pointing out similarities between computers and human minds. These similarities build up an expectation that computers and minds may be similar in other respects, which helps researchers set up a new research program.

† Thus, this is an instance of what David Lewis calls *semantic indecision* (Lewis 1993).

when it does not. After all, it is one of the basic lessons from pragmatics that spelling out every little detail of a matter in conversation does not facilitate communication (Grice 1975). If both of them have reason to believe that they do not need to clarify their definition, avoiding doing so and simply using the general concept of species instead would save time and effort.*

2.4.3 BYPASSING SPECIES

In the previous section, we discussed how the general concept of species, being located 'above' individual definitions of species, is sufficient or beneficial to proceed with biological research in many contexts. We can also discount the importance of seeking a unified and precise definition of species from the opposite direction. Population-level phylogeny, being located 'below' the species level in the ontological hierarchy, can help biologists proceed with their research without any unified and precise definition of species.

This is what Joseph Felsenstein (2004) suggests when he discusses the utility of classification. His point is based on the observation that despite the advertised benefit of classification, biologists are often more interested in the study of phylogeny, and they do not need classification when they trace the evolution of a particular trait.

> Almost all systematists have considered taxonomy, the naming of organisms and their placement in an hierarchical classification, to be the basic task of systematics. ... And yet ... attending the annual meeting of a contemporary systematic society, such as the Society of Systematic Biology, will reveal that few of the speakers are concerned with classification. They spend their time making estimates of the phylogeny and using them to draw conclusions about the evolution of interesting characters. They use phylogenies a great deal. But, having an estimate of the phylogeny in hand, they do not make use of the classification.
>
> (Felsenstein 2004, 145)

Here, Felsenstein argues that phylogeny, not classification in general, including species classification, would be of great use to biologists whose interest is in the evolution of particular characters. It is true in most cases that biologists need some classification to determine which group is the terminal unit of a phylogenetic tree. Once they build a phylogenetic tree, however, they do not use classification to infer how a particular trait has evolved.

What this means to our project is as follows. With the general concept of species, we may have only an unclear picture of the biological world and evolution at the species level, as the general concept of species only allows us to have 'rough and ready' generalisations about the species in question. If we introduce a more precise definition of species, we can figure out what 'species' means in each case and thereby understand the biological world more accurately. However, this is not the only way to gain a clear

---

* Note that this role is to save communication cost, not to help us avoid miscommunication. This is why we discuss it in this section.

picture of the living world. We may bypass the species level and go directly to phylogeny; if we have population-level phylogenies, our understanding of the biological world may be accurate enough to track the evolution of a trait in question.* This is not to say that the problem of determining a fundamental biological unit would disappear if we focused on phylogenetic trees instead of species. In fact, the nature of lineages is no less serious an issue than species for biologists and phylogeny-minded philosophers (Neto 2019). However, if we can leave the problem of finding a fundamental unit of the biological world to the level of population lineages, we do not have to solve this problem at the species level, and the species problem is not important to that extent.

2.5 ALLEVIATING THE CONCERNS FOR BIODIVERSITY AND IDENTIFICATION

Another reason why the species problem is thought to be important regards concerns over biodiversity and identification, that is, the practical importance of the problem. We argue that these concerns can be alleviated if not eliminated.

2.5.1 BIODIVERSITY

Let us begin with biodiversity. It has been argued that a precise, non-arbitrary, and unified definition of species would be desirable for estimating and conserving biodiversity. We agree. With such a definition, measuring and comparing the biodiversity of various areas would be more straightforward and accurate. However, this does not mean that the lack of such a definition renders conservation efforts impossible. In fact, this concern will be partially alleviated if the general concept of species can be used as a unit of biodiversity. One piece of evidence for this comes from the fact that conservation biologists do not seem to pay as much attention to the minute differences between species definitions as species theorists often assume. For example, Richard Primack (2012), a renowned conservation biologist, briefly mentions only two definitions of species – a 'morphological' definition and a 'biological' definition – in his textbook. Although he is aware that we occasionally encounter difficulty in classifying species (p. 22), he does not seem to be particularly bothered by the fact thereafter; he takes for granted that the number of species is a good, if not perfect, measure of biodiversity. If the differences among definitions were important to him, then he would have extensively discussed it. This implies the possibility that these differences may not matter much to the students of biodiversity.

This case suggests that for conservation biologists, the concept of species as a whole may be a sufficient measure of biodiversity in many cases. If we look at a component of the general concept of species, we can advance an argument along similar lines. The concept of good species is a case in point. Because good species are species taxa that satisfy most or all of the major species criteria, they are mutually comparable regardless of which definition one happens to hold. Thus, the higher

---

* Some species theorists propose the elimination of the species rank (along with other ranks) from classification based on phylogeny. See, for example, Mishler and Wilkins (2018).

the proportion of good species among the entire stock of species in an area, the more
informative comparing the biodiversity among the habitats in the area by the number
of species will be. It may be true that the concept of good species is, at best, a 'rough
and ready' estimator of biodiversity. Still, it would help conservationists when no
precise and unified definition is available. In other words, Primack's attitude will
be justified when the conservation area contains a high proportion of good species.

Regarding the point made by David Hull (1997) that if species is an artifact, then
there will be no point in conserving endangered species, although we do not have a
precise and non-arbitrary definition of species to offer, our arguments so far do not
imply that the species category is simply an artifact. For example, one of the key features of a purely artificial category is that classifying things under it does not serve
further inductions and explanations; however, good species do serve such functions.
If an individual organism belongs to a good species *Xus bus*, and many of its members
have a property *F*, then that organism is expected to have *F*, and that expectation will
often be fulfilled. In addition, many good species have a relatively clear and non-arbitrary boundary, because they satisfy most or all of the major species criteria. This also
implies that good species are not mere mental artifacts. Of course, this is not to deny
that some of the taxa that are deemed to be species by taxonomists may be artifacts.
Nevertheless, the fact that we do not have a precise and unified definition of species
does not render conservation efforts of endangered species biologically meaningless.

2.5.2 IDENTIFICATION

As we have seen, it is often argued that the species problem is important because
misidentifying different species could have grave practical and conservational consequences. Authors such as Sigwart (2018) and Cracraft (2000) often cite actual
cases, such as that in which similar-looking snake species have different venoms, to
make this argument. A closer look at those cases, however, reveals that what really
matters there is that those forms should be distinguished at *some* rank of the classificatory hierarchy but not necessarily at the species level.

In the snake case, what is important is to recognise that these snakes belong to
different biological or taxonomic groups with different properties despite the fact
that their morphology is, by and large, similar. However, whether they are different
species is not particularly relevant to our purpose.* Even if these snakes were techni-
cally two subspecies of the same species, it would still be very important to differentiate them so that doctors can find the right antivenom for patients. The matter can
also be considered from another angle. Species classification, classifying biological
groups at the species level, is logically comprised of two steps: grouping organisms
and assigning them to the species rank. In the cases species theorists have cited to
argue for the practical importance of the species problem, what is really relevant is
the importance of differentiating organisms into biologically meaningful groups,
not assigning them to the species rank. This is what these theorists have overlooked.

---

* Barker (2019) makes the same point in his criticism of Philip Kitcher's species pluralism (Kitcher
1984).

It is worth noting that some of the taxonomists who emphasise the practical importance of the species problem do actually notice this. For example, Sigwart (2018) states that in some cases, what is important is to identify distinct lineages rather than their taxonomic ranks. Her example is sub-Antarctic rock hopper penguins. Rock hopper penguins, or the *Eudyptes chrysocome* species complex, have a convoluted history of classification. Historically, three species were recognised based on morphology in this complex: *Eudyptes chrysocome* (Southern Rockhopper Penguin), *Eudyptes filholi* (Eastern Rockhopper Penguin), and *Eudyptes moseleyi* (Northern Rockhopper Penguin). Later, the differences among them were seen as insufficient to count them distinct species, and these species were reclassified into a single species, *Eudyptes chrysocome*. Recently, however, genetic analysis, along with evidence from natural history and biogeography, has revealed that these three groups are different lineages with distinct properties, although there is disagreement over whether they should be treated as three distinct species. Thus, Sigwart (2018) concludes, 'The rank in this case is perhaps less important than the data that indicate it is a distinct lineage' (p. 21), suggesting that recognising a distinct lineage, rather than classifying it at the species rank, may be important for identification.

2.6 OBJECTIONS AND REPLIES

There may be concerns about the arguments offered so far. In this section, I will address some of them.* First, one may maintain that without a precise, non-arbitrary, and universally agreed-upon definition of species, there will be methodological problems on the foundation of conservation biology. For example, if one takes a look at the way biologists work with species, they actually use 'bad' as well as 'good' species. Since classification on such 'bad' species may vary depending on definitions of species, this makes it hard to understand what species really represents as a measure of biodiversity for conservation biologists. This is why some theorists, such as Jody Hey and his colleagues (Hey et al. 2003), believe that we should leave species counts behind us in conservation. Although, as observed in this chapter, biologists may not be too worried about these problems, they actually should be.

My reply to this concern is that I do not claim that the current practice in biology has no conceptual problems concerning species, but that these problems are often exaggerated, as little attention has been paid to the fact that factors discussed so far, such as good species and the general concept of species, alleviate various problems, including that of assessing biodiversity. Even if conservation biologists work on 'bad' species, there will not be many problems when their proportion out of the entire biota is relatively small. It is also worth noting that the solution proposed by Hey and others actually goes along with my argument. Recall that we formulated the argument for the importance of the species problem in terms of finding a precise, non-arbitrary, and universally agreed-upon definition of species (Section 2.4). If

---

* Some of the worries discussed here are originally voiced by Frank Zachos in his comments on the earlier version of the manuscript. I thank him for bringing my attention to them.

conservation biologists bypass species by using another unit of biodiversity, it makes it *less* important to find such a definition of species.*

The second concern is that many good species could be split by the diagnosability version of phylogenetic species concept (Nixon and Wheeler 1990).† In recent decades, many 'good' species have turned out to be made of local populations that can be recognised by DNA sequence and minute morphological traits. For example, Cracraft et al. (1998), following this definition, divided a species of tiger (*Panthera tigris*) into two species by a few diagnostic nucleotides in a single marker of the mtDNA and their phylogenetic relationship. According to them, tiger may be a good *taxon* but perhaps not a good *species*. As the resolution power of genetic analysis increases, more and more minute DNA differences can be detected between local populations. The combination of the technological progress and the preceding definition may imply that there will be significantly fewer good species once many taxonomists adopt this version of phylogenetic species concept.

Here is my reply to this concern. Our argument does not deny that there are disagreements to some degree on the right definition of species in the first place. However, they pose a serious threat to our project only when the disagreements are so profound that, for example, disagreements on the reference of species names constantly disrupts communication among biologists. Although that could happen in the future, it is not very clear whether this kind of thing is happening right now. In addition, some of the worries concerning the species problem will be alleviated once we have good taxa, if not good species. For example, identifying a good taxon of biological significance is often good enough for some practical purposes; if a snake with a particular kind of venom is identified, it helps us find the right antivenom even if it is not a species according to whichever definition one adopts.

2.7 CONCLUSIONS

In this chapter, we have shown that there are reasons to believe that the species problem is not as important as has commonly been thought. We have seen that biologists and philosophers have cited various reasons to believe that it is important to have a precise, non-arbitrary, and unified definition of species, which includes the concern for possible communication breakdown and the theoretical significance of the concept of species in biology.

Many of these concerns were exaggerated. We have reported several cases in which biologists have been able to successfully communicate their ideas without a unified

---

* One might want to add another reply. In a nutshell, a part of the concern here is that the biologists' nonchalant attitudes towards various conceptual problems of species are somehow wrong, and the species problem is important. A problem with this is that it may downplay the possibility that biologists may have reason to take such an attitude. Indeed, one can interpret what we found thus far as suggesting that various factors mentioned earlier, such as the shared reference of species names and the epistemic roles of the general concept of species, provide a reason to take such an attitude. In other words, if one wants to go against our argument, then they need to make a comparison between costs and benefits of maintaining the concept of species in conservation biology.

† Note that this version of phylogenetic species concept is different from the version cited earlier (de Queiroz and Donoghue 1988).

definition of species by sharing exemplars of species and keeping the reference of species names stable. It is even occasionally better to have a 'vague' understanding of species obtained through the 'umbrella' concept of species (the general concept of species) for effective communication rather than to seek a precise definition.

The theoretical significance of the concept of species in biology is also exaggerated. Even without a precise and unified definition of species, biologists can continue their research with the help of the components of the general concept of species. If we talk about a good species, a species taxon that is deemed to be a legitimate species by most species criteria or biologists, then we do not have to care about what the right definition of species is. The general concept also sets a research agenda for researchers by pointing to directions for future research. The use of the general concept could save us the time and mental effort needed to clarify what we really mean by 'species' when 'rough and ready' generalisations will do. Furthermore, we can bypass the discussion at the species level and go directly to population-level phylogenies if desirable.

Practical concerns regarding the lack of a precise definition of species can be alleviated. There are contexts in which the general concept of species is good enough to estimate biodiversity in a region or habitat. Good species satisfy most of the major species criteria and are, therefore, comparable to each other. Thus, if good species account for most of the biota in a region, then we can gauge the biodiversity there by counting the number of species. The concern that it would be meaningless to save endangered species if the species category turned out to be simply an artifact is also exaggerated. While one of the key features of a pure artifact category is that classifying things under the category does not serve further inductions and explanations, good species do facilitate inductions and explanations.

Biologists and philosophers have cited cases in which the correct identification of species could have practical importance, as in the case of similar-looking snakes. However, what is really important in these cases is to identify different biological groups, not necessarily different species. Even if these snakes were subspecies of the same species, it would still be important to differentiate them.

All we have argued here does not mean, however, that it is entirely meaningless to classify species and seek its definition. There are contexts in which a concise definition of species will help one's research. For example, if one studies the speciation of some groups, it might be helpful to have a 'working' definition of species at hand, because it may bring the main object of the research into sharper focus and help the researcher communicate their ideas more clearly. This is the position taken by Peter and Rosemary Grant in their book on speciation (Grant and Grant 2008). This book discusses the speciation process and the history of Darwin's finches in the Galapagos Islands. When it comes to a definition of species, the authors follow the biological species concept with minor modifications,[*] but they also note that this is merely a

---

[*] 'Species comprise one or more populations whose members are capable of interbreeding with little or no fitness loss. They are reproductively isolated from all other populations, either because there is no interbreeding, since members of each do not recognize each other as potential mates, or interbreeding is rare and usually results in relatively unfit (or no) offspring being produced' (Grant and Grant 2008, 111).

'working' definition for them (chapter 9). The reason is that what they are interested in is 'the process of speciation, its causes and consequences, [but] not … defining its product' (p. 110). We can see that having a working definition of species is clearly beneficial for their research, because it helps them focus on some aspects of the speciation process (such as hybridisation) while leaving out others.[*]

Our arguments do not deny that there are also occasions in which it is better for researchers with different backgrounds to share some definition of species rather than insisting on their own definitions. If different scholars use the word 'species' in radically different ways, miscommunication may well occur. If one researcher follows the biological species concept, and another uses a version of phylogenetic species concept (de Queiroz and Donoghue 1988), then they would disagree on whether or not a taxon is a species, which could have theoretical implications.

Our arguments also do not imply that it is meaningless to study the nature of species. For example, it is meaningful to take a closer look at good species and infer that some property F in them could be shared by other species, because it would foster our understanding of the species phenomenon. The observation made by Chan and Levin (2005) that we discussed earlier is worthwhile, because it can have theoretical implications for the study of species and speciation. We can say the same about a naming code as well. As we have seen in the debate between the Bush camp and the Coyne and Orr camp (Section 2.3.2), we cannot overstate the importance of keeping the references of species names stable to ensure effective communication among naturalists. Taxonomic nomenclatures clearly play a pivotal role in that regard.

In summary, our arguments do not require us to downplay the importance of seeking a definition in an appropriate context, studying the nature of species, and maintaining a naming code. What we argue for, however, is that having a precise and unified definition of species is not a logical consequence of accepting the importance of these activities.[†]

REFERENCES

Amitani, Y. 2013. "The Communication Puzzle of the Species Problem." *Annals of the Japan Association for Philosophy of Science* 21: 1–20.
Amitani, Y. 2015. "Prototypical Reasoning about Species and the Species Problem." *Biological Theory* 10: 289–300.

---

[*] One may think that this case presents a problem in regard to Guy Bush's point that choosing one's definition of species before commencing the study of speciation is always like placing the cart before the horse (Section 2.3.2). I largely agree with this, but there are two caveats to this reading. First, choosing a species definition could restrict our attention to unduly limited aspects of the speciation process in other cases. Second, the approach the Grants discussed in their book may not reflect how they conducted their research. Looking at the actual research process, the Grants may have chosen their definition *after* they suspected that reproductive isolation helps the populations evolutionarily diverging from each other.

[†] Acknowledgments: The author would like to thank John Wilkins and Frank Zachos for their valuable comments and suggestions to improve the manuscript. Igor Pavlinov helped us edit the manuscript. This work was supported by JSPS KAKENHI Grant Number 16K02137.

Amitani, Y. 2017. "The General Concept of Species." *Journal of Philosophical Ideas* 65s: 89–120.

Barker, M. J. 2019. "Eliminative Pluralism and Integrative Alternatives: The Case of SPECIES." *The British Journal for the Philosophy of Science* 70: 657–81.

Boyd, R. 1979. "Metaphor and Theory Change: What Is 'Metaphor' a Metaphor for." In *Metaphor and Thought*, edited by A. Ortony, 481–532. New York: Cambridge University Press.

Brigandt, I. 2003. "Species Pluralism Does Not Imply Species Eliminativism." *Philosophy of Science* 70: 1305–16.

Bush, G. L. 1969. "Sympatric Host Race Formation and Speciation in Frugivorous Flies of the Genus *Rhagoletis* (*Diptera, Tephritidae*)." *Evolution* 23: 237–51.

Bush, G. L. 1992. "Host Race Formation and Sympatric Speciation in *Rhagoletis* Fruit Flies (*Diptera: Tephritidae*)." *Psyche A Journal of Entomology* 99: 335–57.

Bush, G. L. 1993. "A Reaffirmation of Santa Rosalia, or Why Are There so Many Kinds of *Small* Animals?" In *Evolutionary Patterns and Processes*, edited by D. R. Lees and D. Edwards, 229–49. London: Academic Press.

Bush, G. L. 1994. "Sympatric Speciation in Animals: New Wine in Old Bottles." *Trends in Ecology and Evolution* 9: 285–88.

Bush, G. L. 1995. "Reply from G.L. Bush." *Trends in Ecology and Evolution* 10: 38.

Bush, G., and R. Butlin. 2004. "Sympatric Speciation in Insects." In *Adaptive Speciation*, edited by U. Dieckmann, M. Doebeli, and R. Leiden, 229–48. Cambridge: Cambridge University Press.

Bush, G., J. Feder, S. Berlocher, et al. 1989. "Sympatric Origins of *R. Pomonella*." *Nature* 339: 346–46.

Bush, G. L., and J. J. Smith. 1998. "The Genetics and Ecology of Sympatric Speciation: A Case Study." *Population Ecology* 40: 175–87.

Carson, H. L. 1989. "Sympatric Pest." *Nature* 338: 304.

Chan, K. M. A., and S. A. Levin. 2005. "LEAKY Prezygotic Isolation and Porous Genomes: RAPID Introgression of Maternally Inherited DNA." *Evolution* 59: 720–29.

Claridge, M. F., H. A. Dawah, and M. R. Wilson, eds. 1997. *Species – The Units of Biodiversity*. London: Chapman & Hall.

Coyne, J. A. 1994. "Ernst Mayr and the Origin of Species." *Evolution* 48: 19–30.

Coyne, J. A., and H. A. Orr. 2004. *Speciation*. Sunderland, MA: Sinauer Association Inc.

Cracraft, J. 2000. "Species Concepts in Theoretical and Applied Biology: A Systematic Debate with Consequences." In *Species Concepts and Phylogenetic Theory: A Debate*, ed. Q. D. Wheeler and R. Meier, 3–14. New York: Columbia University Press.

Cracraft, J., J. Feinstein, J. Vaughn, et al. 1998. "Sorting Out Tigers (*Panthera tigris*): Mitochondrial Sequences, Nuclear Inserts, Systematics, and Conservation Genetics." *Animal Conservation* 1: 139–50.

Eldredge, N., and J. Cracraft. 1980. *Phylogenetic Patterns and the Evolutionary Precess: Method and Theory in Comparative Biology*. New York: Columbia University Press.

Felsenstein, J. 2004. *Inferring Phylogenies*. Sunderland, MA: Sinauer Associates.

Grant, P., and R. Grant. 2008. *How and Why Species Multiply*. Princeton, NJ: Princeton University Press.

Grice, H. P. 1975. "Logic and Conversation." In *Syntax and Semantics, Volume 3*, edited by P. Cole and J. L. Morgan, 41–58. New York: Academic Press.

Hamada, N., and P. H. Adler. 1999. "Cytotaxonomy of Four Species in the *Simulium Perflavum* Species Group (Diptera: Simuliidae) from Brazilian Amazonia." *Systematic Entomology* 24: 273–88.

Hey, J., R. S. Waples, M. L. Arnold, et al. 2003. "Understanding and Confronting Species Uncertainty in Biology and Conservation." *Trends in Ecology and Evolution* 18: 597–603.

Holsinger, K. E. 1987. "Discussion: Pluralism and Species Concepts or When We Agree with One Another?" *Philosophy of Science* 54: 480–85.

Hull, D. L. 1997. "The Ideal Species Concept – and Why We Can't Get It." In *Species: The Units of Biodiversity*, edited by M. A. Claridge, H. A. Dewah, and M. R. Wilson, 357–80. London: Chapman & Hall.

Hull, D. L. 1999. "On the Plurality of Species: Questioning the Party Line." In *Species New Interdisciplinary Essays*, edited by R. A. Wilson. Cambridge, MA: The MIT Press.

Kaufman, S. B. 2018. "Nature, Nurture, and Human Autonomy with James Flynn." *The Psychology Podcast* [Audio Podcast]. Retrieved from https://scottbarrykaufman.com /podcast/nature-nurture-and-human-autonomy-with-james-flynn/.

Keller, E. F. 2000. *The Century of the Gene*. Cambridge, MA: Harvard University Press.

Kitcher, P. 1984. "Species." *Philosophy of Science* 51: 308–35.

Lewis, D. 1993. "Many, but Almost One." In *Ontology, Causality, and Mind: Essays on the Philosophy of D. M. Armstrong*, edited by J. Bacon, K. Cambell, and L. Reinhardt, 23–38. Cambridge: Cambridge University Press.

Mallet, J. 1996. "Reply from Mallet." *Trends in Ecology and Evolution* 11: 174–74.

Mayr, E. 1942. *Systematics and the Origin of Species from the Viewpoint of a Zoologist*. New York: Columbia University Press.

Mayr, E. 1957. "Species Concepts and Definitions." In *The Species Problem*, edited by E. Mayr, 1–22. Washington, DC: American Association for the Advancement of Science.

Mayr, E. 1963. *Animal Species and Evolution*. Cambridge, MA: Harvard University Press.

Mayr, E. 1969. *Principles of Systematic Zoology*. New York: McGraw-Hill.

Mayr, E. 1970. *Populations, Species and Evolution: An Abridgement of Aminal Species and Evolution*. Cambridge, MA: Harvard University Press.

McOuat, G. R. 1996. "Species, Rules and Meaning: The Politics of Language and the Ends of Definitions in 19th Century Natural History." *Studies in History and Philosophy of Science Part A* 27: 473–519.

Mishler, B. D., and J. S. Wilkins. 2018. "The Hunting of the SNaRC: A Snarky Solution to the Species Problem." *Philosophy, Theory, and Practice in Biology* 10: 1–18.

Neto, C. 2019. "What Is a Lineage?" *Philosophy of Science* 86: 1099–110.

Neto, C. 2020. "When Imprecision Is a Good Thing, or How Imprecise Concepts Facilitate Integration in Biology." *Biology & Philosophy* 35: 58.

Nixon, K. C., and Q. D. Wheeler. 1990. "An Amplification of the Phylogenetic Species Concept." *Cladistics* 6: 211–23.

Nosil, P. 2012. *Ecological Speciation*. Oxford: Oxford University Press.

Price, T. 2008. *Speciation in Birds*. Greenwood Village, CO: Roberts; Co.

Primack, R. 2012. *A Primer of Conservation Biology*. 5th ed. Sunderland, MA: Sinauer Association.

Pušić, B., P. Gregorić, and D. Franjević. 2017. "What Do Biologists Make of the Species Problem?" *Acta Biotheoretica* 65: 179–209.

de Queiroz, K. 1999. "The General Lineage Concept of Species and the Defining Properties of the Species Category." In *Species: New Interdisciplinary Essays*, edited by R. A. Wilson, 49–89. Cambridge, MA: MIT Press.

de Queiroz, K. 2005a. "A Unified Concept of Species and Its Consequences for the Future of Taxonomy." *Proceedings of the California Academy of Science* 56: 196–215.

de Queiroz, K. 2005b. "Ernst Mayr and the Modern Species Concept." In *Systematics and the Origin of Species: On Ernst Mayr's 100th Anniversary*, edited by J. Hey, W. M. Fitch, and F. J. Ayala, 243–63. Washington, DC: National Academic Press.

de Queiroz, K. 2007. "Species Concepts and Species Delimitation." *Systematic Biology* 56: 879–86.

de Queiroz, K., and M. J. Donoghue. 1988. "Phylogenetic Systematics and the Species Problem." *Cladistics* 4: 317–38.

Reydon, T. A. C. 2019. "Are Species Good Units for Biodiversity Studies and Conservation Efforts?" In *From Assessing to Conserving Biodiversity: Conceptual and Practical Challenges*, edited by E. Casetta, J. Marques da Silva, and D. Vecchi, 167–93. Cham: Springer International Publishing.

Rheinberger, H. J. 2000. "Gene Concepts." In *The Concept of the Gene in Development and Evolution: Historical and Epistemological Perspectives*, edited by P. J. Beurton, R. Falk, and H. J. Rheinberger, 219–39. Cambridge: Cambridge University Press.

Schluter, D. 2000. *The Ecology of Adaptive Radiation*. Oxford: Oxford University Press.

Seifert, B. 2020. "The Gene and Gene Expression (GAGE) Species Concept – An Universal Approach for All Eukaryotic Organisms." *Systematic Biology* 69: 1033–38.

Sigwart, J. D. 2018. *What Species Mean: A User's Guide to the Units of Biodiversity*. Boca Raton, FL: CRC Press.

Simpson, G. G. 1961. *Principles of Animal Taxonomy*. New York: Columbia University Press.

Stamos, D. N. 2003. *The Species Problem: Biological Species, Ontology, and the Metaphysics of Biology*. Lanham, MD: Lexinton Books.

Strickland, H. E. 1837. "Rules for Zoological Nomenclature." *Magazine of Natural History* 1: 173–76.

Strickland, H. E. 1843. "Report of a Committee Appointed 'to Consider of the Rules by Which the Nomenclature of Zoology May Be Established on a Uniform and Permanent Basis'." *Report of the 12th Meeting of the British Association for the Advancement of Science*, 105–21.

Van Valen, L. M. 1976. "Ecological Species, Multispecies, and Oaks." *Taxon* 25: 233–9.

Wilkins, J. 2018. "The Reality of Species: Real Phenomena Not Theoretical Objects." In *Routledge Handbook of Evolution and Philosophy*, edited by R. Joyce, 167–81. New York: Routledge.

Zachos, F. E. 2016. *Species Concepts in Biology*. Switzerland: Springer.

3 'Species' as a Technical Term

Multiple Meanings in Practice, One Idea in Theory

Thomas A.C. Reydon

CONTENTS

3.1 INTRODUCTION: MONISM VERSUS PLURALISM ABOUT SPECIES

Monism and pluralism are diverging perspectives on the nature of the species problem and the kind of solution that is being sought. The principal difference between these two perspectives consists in a fundamental assumption about the metaphysics of the species category. Monists assume that the proper names used in biology to refer to species, such as *Ornithorhynchus anatinus*, *Drosophila melanogaster*, *Dracaena cinnabari*, *Volvox globator*, or *Borrelia burgdorferi*, all refer to entities of one fairly homogeneous kind (the species category). That is, even though *Ornithorhynchus anatinus* as a species of mammal, *Dracaena cinnabari* as a species of plant, and *Volvox globator* as a species of algae are quite different kinds of groups, the assumption is that *as species*, these groups have sufficient commonalities to regard them as 'deep down' being the same kind of entities. The metaphysical assumption thus is that as a technical term in the biological sciences, 'species' refers to a category of entities in much the same way as technical terms such as 'Higgs boson' in particle physics and '^{40}Ar' in chemistry refer to categories of entities that share one or more traits that make them into the kind of entity they are.

This assumption is often bolstered by a realist view of species as entities existing in nature that can be discovered, identified, described, named, and studied. In

DOI: 10.1201/9780367855604-4

the same way as the various kinds of elementary particles and the various chemical elements, the assumption goes, species are parts of nature that exist independently of us. From a monistic perspective, the species problem then is primarily what kind of entities biological species *are*, what makes a group of organisms into a species (rather than a different kind of group), what makes individual organisms belong to one species rather than another, and what a definition of the scientific term 'species' would be that best covers these issues.

While monism about species seems a natural perspective to hold, its problems have long ago become clear, and several philosophers of biology (as well as a number of biologists) have suggested that a pluralistic perspective is more appropriate.[*] While pluralists do not necessarily deny that species are real entities in nature, their outlook on the species problem is fundamentally different from the monist's outlook. Pluralists pose similar questions as monists but without the expectation that for every question there will be one single answer that applies throughout all contexts of biological science. Dropping this expectation allows pluralists to approach the species problem in a number of different ways.

A moderate pluralistic approach is to think of the numerous definitions of the term 'species' (that is, the different so-called 'species concepts' available in the literature) as operational definitions of the same fundamental scientific concept. This approach acknowledges that '[t]here might [...] be theoretical and biological interests that can legitimately ground species groupings, and potentially do that in different ways. [...B]iodiversity should be conceived and divided up according to the interests guiding the classification' (Richards 2010, 116). On this version of pluralism, which goes by several names in the literature, different areas of biological research may use different ways of grouping organisms into species, depending on their research interests, and groupings used in different areas may be mutually incompatible and involve groups based on different grouping criteria. Indeed, different research contexts place diverging demands on the notion of species, and the demands of one research context often are not completely compatible with those of other research contexts (Kornet 1993; Hull 1997; Reydon 2004). This means that a definition of the term 'species' that meets the requirements placed on species groupings in one context of research may well fail to meet the requirements placed in other contexts, and therefore, a universally applicable definition is not feasible. Nevertheless, the thought is, all these different groupings ultimately are different ways of grouping organisms *into species*, such that all the extant different definitions of 'species' can be seen as operationalisations of one overarching definition, with each operationalisation well-tuned to the specific requirements of the research context in which it is used. Examples of this approach are Mayden's hierarchy of species concepts (Mayden 1997; 1999) and De Queiroz's General Lineage Concept (De Queiroz 1998; 1999; 2011; see also Richards 2010, 119–124).

---

[*] For more on the problems with monism about species, see (Dupré 1993; 1999; 2018; Hull 1997; 1999; Roselló-Mora 2003; Wilkins 2003; Reydon 2004; 2005; Richards 2010, 115–124 and 210ff.; Zachos 2016, 11–13; Reydon and Kunz 2019, 625–627). A defense of pluralism that strongly influenced philosophical thinking on the matter is provided by Dupré (1993; 1999).

Another pluralistic approach to the problem is more radical. It involves acknowledging that evolutionary processes work in different ways for different kinds of organisms – an evolving population of mammals is a very different entity from an evolving population of trees or an evolving population of algae – and accordingly, species are not necessarily the same kind of 'things' all across biodiversity. This version of pluralism (ontological pluralism (Richards 2010, 117–119)) thus denies that *Ornithorhynchus anatinus, Drosophila melanogaster, Dracaena cinnabari, Volvox globator,* and *Borrelia burgdorferi* all are ultimately the same kind of entities. And indeed, given the enormous differences between mammals, insects, angiosperms, green algae, and spirochete bacteria, why would we assume that these organisms all cluster into the same kind of groups? On this sort of radical pluralism, even though individual species can be seen as real entities in nature, there is no homogeneous category of entities called 'species'. In other words, while *Ornithorhynchus anatinus* is a real entity in nature that can be described and studied, as are *Dracaena cinnabari, Volvox globator,* and the referents of all other species names used in biology, there is nothing that compels us to think of them as entities of the same category (Ereshefsky 2010; 2011; 2014; Reydon and Kunz 2019). That is, the species category is not a real category, and there is no all-encompassing metaphysical account of what species are, or of what makes organisms members of their species. Radical pluralists about species accordingly deny that there would be an overarching definition of the term 'species' of the sort that Mayden and De Queiroz propose.

While at present, both versions of pluralism about species are widely endorsed by both philosophers of biology and practicing biologists, these perspectives leave one important question unanswered: if the term 'species' refers to a variety of kinds of entities, either in different contexts of research or in different parts of biodiversity (or possibly, both), what makes them all into *species*, and why do we treat them as somehow comparable kinds of entities? If there can be multiple answers to the questions of what kind of entities biological species *are*, what makes individual organisms belong to their species, and what an adequate definition of the scientific term 'species' would be, then how could all these answers pertain to the same category of entities? If in one research context, organisms are grouped in a different way than in another research context, why would both these ways of grouping count as groupings into species? And if *Dracaena cinnabari* and *Borrelia burgdorferi* are very different groupings with very different membership conditions, why would both count as species? To put the issue in other words, being a pluralist about something presupposes a certain degree of monism in the background.

Religious pluralism, for example, is the view that there is not just one legitimate religion but rather, a multiplicity of equally legitimate religions, that the diverging perspectives and beliefs offered by the various religions should be valued equally, and that we should exert humility when judging the – often incompatible – beliefs involved in the various religions (in particular when judging these from our own perspective). But, religious pluralism at the same time must involve a view about what makes a system of belief into a religion – that is, about the question of why all the various religions with respect to which one is a pluralist are conceived of *as religions* in the first place. In addition, religious pluralism cannot entail that just any belief system should be taken as a religion to be valued similarly to other religions – there must be some criterion by which the domain of religions (about which one is a pluralist) is delimited from the domain of other belief systems (that one values differently). In the same way, pluralism about species presupposes some criterion to delimit those groupings one counts as species (and about which one can then be a pluralist in the sense discussed here) from those that should not be counted as species. Thus, even pluralists about species cannot avoid the question of what species generally are, even if from a pluralistic perspective this is not as deep a metaphysical question as it is from a monistic perspective. For pluralists, the question is one of what may be called 'metaphysics light' – it is the question of why we treat diverse groupings as comparable kinds of entities without the expectation of a deep metaphysical answer that would explicate what all the various groupings called 'species' essentially have in common.

This chapter explores this 'light' metaphysical question by distinguishing between the *theoretical meaning* of 'species' and the *practical usages* of the term. I will provide an answer in the form of two theses, the *homonymy thesis* and the *species-as-status thesis*. According to the first thesis, the technical term 'species' is a homonymic term that in biological practice is associated with a multiplicity of meanings, while it is still possible to identify a theoretical idea that unifies these different meanings to a certain extent. This unifying theoretical idea is embodied in the species-as-status thesis: following a suggestion made by Theodosius Dobzhansky in the 1930s–1940s, I will suggest that the term 'species' does not denote a *kind* of entities but rather, a *status* that can be attributed to a group of organisms on theoretical grounds. I have introduced both theses in earlier work (Reydon 2004; 2005; 2019a) and will bring them together here into an answer to the species problem.

I will begin in Section 3.2 by highlighting how technical terms often are introduced into an area of science and subsequently undergo a diversification of meanings. I will use the case of the term 'gene' to illustrate this process of semantic diversification, because the history of this term can be traced particularly clearly from the early twentieth century to today. The case of 'species' as a technical term in biological science exhibits certain parallels with the case of 'gene', such that taking a closer look at the term 'gene' can provide us with an idea of what the – much more complicated – history of the term 'species' will have looked like. In Section 3.3, I then argue for the *homonymy thesis* on the basis of the considerations presented in Section 3.2. In Section 3.4, I elaborate the *species-as-status thesis* on the basis of Dobzhansky's views. Section 3.5 concludes by showing how the two theses together constitute a 'metaphysics light' answer to the species problem, drawing some implications for biological practice.

3.2   THE INTRODUCTION AND CHANGE OF TECHNICAL TERMS IN SCIENCE

In earlier work (Reydon 2004; 2005), I suggested that the species problem is mainly due to some intricacies of the process of conceptual change that commonly occurs as science progresses. Since the publication of Thomas Kuhn's *The Structure of Scientific Revolutions* (Kuhn 1962), the phenomenon of conceptual change has continuously been on the agenda of the philosophy of science. Kuhn famously argued that progress in science is not a matter of linear development in which knowledge is accumulated by simply adding new theories, explanations, concepts, terms, etc. to the already available body of knowledge. Rather, scientific progress often involves revolutionary phases in which old theories, explanations, concepts, etc. are replaced by new ones that sometimes compel us to see the world in a fundamentally different way. As Kuhn put it, scientists working in the new paradigm in some sense live in a different world from scientists working in the old paradigm, such that the two groups to some extent talk past each other (Kuhn 1962, 150; Oberheim and Hoyningen-Huene 2018). An important element of scientific progress thus is conceptual change, that is, semantic shifts in technical terms and the acquisition of novel meanings by terms that are already widely in use. Famous historical examples are the semantic shift of the term 'planet', which in the Copernican revolution came to include the earth and exclude the sun, and the diversification of the meaning of the term 'mass', which with the development of relativity theory came to mean the new, relativistic concept of mass but at the same time retained its old meaning of the Newtonian concept of mass (Thagard 1992; 2003; Oberheim and Hoyningen-Huene 2018). But note that while prominent examples of conceptual change are associated with scientific revolutions, conceptual change is a common aspect of non-revolutionary scientific progress too.

It should be noted that instances of conceptual change involve changes in the meaning of scientific terms as well as in their extension (i.e., the set of things in the world that a term refers to). The distinction between the sense or meaning (*Sinn*) of a term and its reference (*Bedeutung*) was introduced by mathematician and philosopher Gottlob Frege (1892) to clarify how identity statements can be non-trivially true. Frege separated the entity or entities in the world that a word is connected to (the word's reference) from the way in which it refers (the word's sense). The names 'Evening Star' and 'Morning Star', for example, have the same material reference (the planet Venus) but refer to that planet in different ways (conveying different information about the entity) and thus have different senses. This distinction is important in the present context, as it shows how technical terms can change in various ways: they can change their sense (their way of referring) but continue to refer to the same set of material entities, or they can change both their sense and their reference.* What can occur, then, is that at some point in the history of a science, one word comes to be used in a number of different senses and with a number of different material referents, where not necessarily *every* sense of the word is connected to a different material referent.

---

* Note that I am following the technical terminology of philosophy. Even though the German word '*Bedeutung*' usually translates as 'meaning', in the philosophical literature, Frege's terms are intended to resolve an ambiguity in the word 'meaning', leading to '*Sinn*' being translated as 'sense' and '*Bedeutung*' as 'reference'.

As this is a fairly abstract philosophical matter, let me illustrate it by considering a case that is considerably less complicated than the case of 'species'; namely, the case of the term 'gene'. The history of conceptual change that occurred with respect to the notion of the gene is considerably simpler than the species case, because the term 'gene' is considerably younger than the term 'species' and thus, has had less time to undergo semantic diversification. In addition, the term 'gene' has a clear point of origin in the history of biology – the year 1909 – and its conceptual history can be traced from there. Examining the case of 'gene' can help us see more clearly what happened in the development of the term 'species'.*

Even though there were precursor terms to the term (such as Darwin's term 'gemmule' or Hugo De Vries's 'pangene'), the history of the term 'gene' starts with the Danish botanist Wilhelm Johannsen. Johannsen formally introduced the term (together with the terms 'genotype' and 'phenotype') in his 1909 book, *Elemente der exakten Erblichkeitslehre*, and deliberately presented it as a theoretically neutral notion, 'a short and unprejudiced word for unit-factors in the – as to heredity – essential constitution of gametes and zygotes' (Johannsen 1923, 136; see also 1911, 132). As Johannsen put it, the term 'gene' should express 'the simple picture [...] that a property of the developing organism is or can be (co-) determined by "something" in the gametes. No hypothesis about the nature of this "something" should be formulated or supported' (Johannsen 1909, 124).† Johannsen emphasised that the newly introduced term should be hypothesis-free, i.e., that it should not be taken as having any implications regarding the nature of genes in general, leaving open the possibility that different genes have completely different natures. A gene, in Johannsen's view, could be *any* factor that co-determined an organism's phenotype and could be passed on from generation to generation.

Johannsen highlighted several reasons for using a theoretically neutral, hypothesis-free term. For one, there was a lack of clarity regarding the nature of hereditary factors, such that in Johannsen's view, any claims about their natures would be unwarranted speculation (Johannsen 1909, 317–323 and 376; 1911, 133). In addition, the technology at the time did not make it possible to know an individual organism's genotype, but it was possible to elucidate genotypic differences between organisms. Thus, the relation between organismal traits and 'something' in the gametes underlying them should be seen not as a direct relation between traits and the material factors that cause them, but as relations between trait *differences* and genotype *differences* (Johannsen 1911, 133 and 147–149; 1923, 135–137). Lastly, Johannsen held that the genomes of organisms *in fact* were not divisible into distinct units, such that organismal traits should be seen as determined by the entire genome rather than well-specifiable parts of the genome (Johannsen 1923, 137, 139). Johannsen held a thoroughly holistic view of the genotype, according to which genes could only be

---

* More detailed accounts can be found in Portin (1993; 2002); Rheinberger and Müller-Wille (2008); Griffiths and Stotz (2013). Roll-Hansen (2014) provides a historical account of Johannsen's work.

† My translation; the original reads: '[B]loß die einfache Vorstellung soll Ausdruck finden, daß durch "etwas" in den Gameten eine Eigenschaft des sich entwickelnden Organismus bedingt oder mitbestimmt wird oder werden kann. Keine Hypothese über das Wesen dieses "etwas" sollte dabei aufgestellt oder gestützt werden' (Johannsen 1909, 124).

identified as parts of the genotype of an organism in interaction with its environment (Roll-Hansen 2014, 2432).

Johannsen thus introduced the term 'gene' as a new technical term in biology intended as a purely instrumental accounting unit that could not be identified with a particular material or morphological structure in the organism (Mayr 1982, 737) and did not entail any theoretical commitments regarding the nature of those entities called 'genes'. For Johannsen, it was precisely this lack of theoretical commitment that made the notion useful in the context of the science of his time, as commitments to highly speculative views of what genes were might hinder rather than advance research (Mayr 1982, 736–737; Rheinberger and Müller-Wille 2008, 5). The concept of the gene, then, was introduced into biology explicitly and in a very intentional way, as Johannsen had specific ideas in mind regarding the concept's role and meaning.

Specific theoretical commitments regarding the nature of genes were added later as research progressed, as new lines of research were opened up and new investigative methods introduced. Most importantly, the clarification of the molecular structure of the hereditary material in the mid-twentieth century led to a conceptual change from the classical notion of the gene as referring to abstract genotypic differences between organisms that were responsible for phenotypic differences to a general notion of 'gene' as denoting continuous or discontinuous DNA-segments that specify the linear sequences of proteins and enzymes (i.e., molecules that coded for other molecules) – i.e., a transition from a *Classical Gene Concept* to a *Molecular Gene Concept* (Waters 1994). It soon became clear, though, that all the diverse functions that genes were thought to perform could not be mapped onto a category of structurally well-defined material entities, such that both meanings of the term 'gene' continued to be used in parallel. Moreover, the molecular conceptualisation of the gene diversified further: biologists repeatedly began to use the term to refer to different kinds of DNA-segments as new molecular biological techniques became available, such that which material entities were individuated as genes depended strongly on the investigative methods and techniques available as well as on the specific interests of researchers. Accordingly, Weber (2005, 227) argued that the term 'gene' possessed what he calls 'freely floating reference' – that is, instead of a fixed material reference, the term's reference came to depend on the specific context of research and changed with the progress of biological knowledge and the development of new methods. The latest stage of this history of conceptual change is that the term 'gene' is now also applied to functional aspects of genomes that are not strictly located somewhere on the organism's genome – genes on such a view are simply 'things you can do with your genome' (Griffiths and Stotz 2013, 223).

Among the philosophical lessons of the gene case are the following. First, conceptual change in science is often explicitly driven by, among other things, advances in scientific theorising and research practices, methodologies, and technologies. As Rheinberger and Müller-Wille (2008, 15) remarked, 'the history of twentieth-century genetics is characterised by a proliferation of methods for the individuation of genetic components, and accordingly, by a proliferation of gene definitions. [...] Major conceptual changes did not precede, but followed, experimental breakthroughs'.

Second, such advances do not necessarily lead to linear change in which the old meaning and/or reference of a technical term is simply replaced by a new and better one, which in turn can be replaced later by an even newer and better one, and so on. Conceptual change in science is much more complicated, because old concepts are often deeply entrenched in scientific theory and practice, such that they are retained and used alongside newly introduced concepts and meanings. Conceptual change thus often leads to a proliferation of related concepts, as in the case of the limited proliferation of gene concepts, yielding a situation in which one technical term ('gene') is used to refer to multiple concepts (i.e., is used with multiple senses and references) that often are thought of as competing concepts. This happened in the gene case, as the different concepts that have been introduced since the early twentieth century – genes for trait differences, genes as structurally well-defined DNA-sequences, genes as functionally defined parts of genomes – to some extent still play a role in biology.

Third, theoretical commitments can be introduced, abandoned along the way, and reinstated as part of a new concept at a later stage. Johannsen's whole-genotype view was a case in point of the latter: while for Johannsen, the genotype as a whole was responsible for the organism's phenotype (a view that he did not build into his notion of the gene, which was theory-free), the whole-genome view was initially replaced by a view of individual genes being responsible for individual traits, then by a view of multiple genes being responsible for individual traits and individual genes being connected to multiple traits, and then reintroduced around a century later into the notion of the gene as 'things you can do with your genome'. At the same time, the theoretical commitment of genes as well-delimited DNA-sequences, with its sequence determining the kind a particular gene token belongs to, lost traction as biological research progressed (Portin 1993; 2002; Griffiths and Stotz 2013). Yet, all the various theoretical views of what genes are to some extent are still found in current biology.

3.3   THE HOMONYMY THESIS

With the philosophical lessons from the developmental history of the term 'gene' in mind, some important aspects of the species problem can now be highlighted. Perhaps the most important parallel between the case of 'species' and the case of 'gene' is that in both cases, a term acquired new meanings as science progressed without the old meanings being completely abandoned, resulting in a single scientific term being used with a number of (sometimes more, sometimes less) distinct meanings in parallel (Reydon 2004). There are several important differences between the two cases, however, that make the case of 'species' considerably more complex than the case of 'gene'. While the conceptual history of the term 'gene' reaches back a little over a century, discussions on the term 'species' trace back several centuries and have yielded a vast body of literature. As several good overviews of the conceptual history of 'species' are available, including Richards (2010), Zachos (2016) and Wilkins (2018), and this history is much too extensive for me to be able to discuss it here in sufficient detail, I will only highlight some important aspects.

A notable difference from the case of 'gene' is that the term 'species' was never formally introduced into biological science but long predates the rise of biology as a science. The notion of species was cemented in natural history when Linnaeus introduced his five-level taxonomy, but the notion itself is much older – the Latin term has been present in scholarly writings at least since the Middle Ages, and the underlying idea has been around since Antiquity. The term 'species' as a classificatory and metaphysical term traces back (by way of the Greek term '*eidos*', which is usually translated as 'species') to classical Greek philosophy and occurs, for example, as a key term in Aristotle's works (Grene and Depew 2004, chapter 1). There, however, the term is used in a very different way than in contemporary biology and is connected to particular metaphysical commitments that most present-day species concepts no longer encompass. The notion of species thus did not have its roots in biological theory or biological research practices (as did the notion of the gene) but in ancient philosophical views that to a large extent cannot be upheld today – most importantly, Aristotle's commitment to forms instantiated in entities or the Platonic notion of essences existing in a realm of their own. Neither Plato nor Aristotle applied the notion of *eidos*/ species exclusively or even primarily to living organisms: for these philosophers, all natural objects could be grouped into species. This practice of applying the notion of species to both living and non-living entities continued far into the nineteenth century, as it was long common to speak of species of entities in general, including organisms, substances, minerals, etc.*

The notion of species, then, made its way into biology from a much more general context, carrying over metaphysical baggage from Plato's and Aristotle's worldviews, as well as later stages of thinking within as well as outside biology, into modern natural history. This baggage makes the species case considerably more complicated than the gene case. At least from the seventeenth century onward, naturalists began to be concerned with explicating the meaning of the notion of species as it is applied to the living world (i.e., with formulating what have come to be known as 'species concepts'), to some extent separating the notion from its metaphysical history (Mayr 1982, 254–275; Lherminer and Solignac 2000; Wilkins 2018).† As Wilkins puts it, '"species" develops a uniquely biological flavor around the seventeenth century' (Wilkins 2018, 3).

---

* See, for example, John Stuart Mill's *System of Logic* (1859, 81–89). Of the specific biological use of 'species', Mill remarks: 'By the naturalist, organised beings are never said to be of different species, if it is supposed that they could possibly have descended from the same stock. That, however, is a sense artificially given to the word, for the technical purposes of a particular science' (Mill 1859, 85). Mill here thus points out that the idea that members of a species must share common descent is an addition to the notion of species for the purposes of use in one particular science (accordingly implying that in other sciences, the term may have different meanings). Many natural history museums still have a mineral collection showing species of minerals next to the various collections of organisms, and today, it still is not uncommon to speak of species of minerals (e.g., Siddall 2002).

† Lherminer and Solignac provide a good article-length overview of 92 species concepts ranging from Aristotle and Plato via the Middle Ages and Early and Late Modernity to the present day (Lherminer and Solignac 2000). A book-length treasure trove for anyone interested in species concepts from Plato to the present is Wilkins' *Sourcebook* (2009).

Wilkins (2018, 3–4) argues that historically, the term 'species' can be seen as a marker for two closely related classificatory projects – the ancient tradition of searching for that what makes every entity, living as well as non-living, into the kind of entity it is (its essence; Wilkins calls this taxonomy in terms of universal categories) and the younger tradition of classifying organisms into groups in the context of modern natural history (biological taxonomy).

While it is unclear when authors first started to name and address the 'species problem', it is clear that well before Darwin's publications, prominent biologists such as Ray, Adanson, Linnaeus, Lamarck, Buffon, and Cuvier already offered explicatory definitions of the notion of species in attempts to clarify what, exactly, species are (Lherminer and Solignac 2000; Wilkins 2009; 2018). Darwin himself acknowledged the problem and tried to avoid it by considering the species level as conventional (Darwin 1859, 52; Ereshefsky 2010; 2011; Mishler 2010). In particular in the twentieth century, a veritable explosion of work on the notion of species is seen, authored by biologists as well as philosophers of biology (and sometimes, by biologists and philosophers in collaboration). Unfortunately, this literature has not yielded more clarity but rather, just more and more definitions of the term 'species' to choose from. The current situation with respect to the notion of species is confusing, to say the least: there is no generally accepted solution to the problem, numerous competing definitions of the term 'species' are available (none of which is generally accepted, and none of which can be applied throughout the whole of biodiversity and in all biological research contexts), and several book-length treatises on the problem have been published, none of which has been able to put the issue to rest once and for all.

In earlier work (Reydon 2004; 2005; 2008; Reydon and Kunz 2019), I have argued that a major cause of the persistence of the problem is this plethora of meanings of the same technical term – that is, the homonymic nature of the word 'species'. The *homonymy thesis*, as I call it, is the claim that

> 'Species' is a term that refers to multiple, distinct kinds of entities that feature in different roles in biological theorising and empirical research, and accordingly, the term 'species' is associated with multiple distinct concepts that cannot be subsumed under a single overarching concept.

That is, the term 'species' is similar in nature to a term like 'bank', which in different contexts can refer to the land alongside a river (the banks of the Allegheny River), a slope or mound (a bank of snow), an array (a bank of switches or batteries), or a financial institution (the Royal Bank of Scotland). It should be noted that the homonymy thesis is a metaphysical thesis and not merely a claim about the various senses and references of the word 'species'. That is, the homonymy thesis encompasses the claim that the word 'species' as a technical term in biology has multiple senses as well as multiple kinds of reference (i.e., multiplicity with respect to both '*Sinn*' and '*Bedeutung*'). I have suggested that 'species' is a homonymic and not a pluralistic term to clarify that the issue is not merely that multiple perspectives can be taken on the same biological entities (i.e., that all available definitions of 'species' talk about the same entities in nature but highlight different important aspects of these entities), but that the issue is considerably deeper than that.* Let me explain.

The principal aim of the homonymy thesis is to draw a contrast to pluralistic views of the species problem. As discussed earlier, pluralism requires a background unity of the things one is a pluralist about. In the case of species, pluralism with respect to the term 'species' would involve the claim that the term refers to different kinds of entities that notwithstanding their differences, ultimately all are members of the species category (i.e., ultimately, they all *are species*, albeit different kinds of species), and the term is associated with multiple distinct concepts that ultimately, all are species concepts. Mayden's suggestion that available species concepts constitute a hierarchy of operational concepts that all can be subsumed under one basic concept (Mayden 1997; 1999) and De Queiroz's General Lineage Concept (De Queiroz 1998; 1999; 2011) both are pluralistic positions in this sense – they recognise some degree of unity behind the plurality of species concepts. But, I think there are good reasons for doubting the conceptual unity these authors perceive there to be.

For one, there is *epistemological disunity* among the various meanings of the term 'species'. The notion of species features in biological science in at least three distinct epistemic roles: to denote units of classification, to denote units of generalisation, and to denote units of evolution (Reydon 2005). Units of classification are nothing more than fixed groups in a scientific reference system – basically, a set of named abstract boxes into which individual specimens are placed and that can serve as a stable information storage and retrieval system for scientific work (Mayr and Bock 2002). Note that such a storage and retrieval system does not presuppose that like is put with like into the boxes: all that is needed is a good catalogue that easily tells users in which box a specimen has been shelved, such that it can quickly be found and connected to other specimens. The boxes themselves do not need to contain similar items (in the same way as a library can simply store books in the order in which they are acquired, as long as there is a good catalogue available). Units of generalisation are groups of entities over which lawlike generalisations can be made. Kinds of elementary particles such as the Higgs boson or chemical isotopes such as ^{40}Ar are examples of such groups: all Higgs bosons share a set of properties, as do all ^{40}Ar atoms, such that it is possible to make inductive generalisations about the properties and the behavior of all members of the group on the basis of investigations of only a limited number of the group's members. In philosophy, such groups are called

---

* Note that the homonymy thesis should also not be interpreted as a *linguistic* claim about the etymology of the word 'species'. An informal response to the paper in which I first proposed the homonymy thesis (Reydon 2005) was that the correct claim should have been that the word 'species' is a *polyseme* (a word with multiple etymologically related meanings) and not a *homonym* (a word that happened to have acquired multiple, unrelated meanings). I am indebted to Eric Oberheim, who first made this point. This criticism was appropriate, I think, and probably, the claim should have been that 'species' is in part a polyseme and in part a homonym (many meanings are related, but it is not the case that all meanings of the term found in the literature are related in such a way that the term is a polyseme). But since I am not a linguist and the thesis is not meant as a claim about the precise etymology of the word 'species' or about the precise historical connections between the various contemporary meanings of the word (and also because the homonymy thesis has already appeared in print), for the sake of simplicity, I will retain the old (slightly incorrect) terminology.

'natural kinds', and the philosophy of biology encompasses a long-standing debate on the question of whether species can be conceived of as natural kinds in this sense (Reydon 2005, 138). I will return to this issue shortly, but here, it suffices to point out that there is no guarantee that the two epistemic roles will always coincide.

A common assumption is that scientific classifications tend to converge onto natural kinds, such that they can serve both as mere information storage and retrieval systems and as groups enabling inductive generalisations (see, for example, Bird and Tobin's review of the philosophical debate on the topic of natural kinds (Bird and Tobin 2018)). But in the case of species, there are at least three reasons to assume that units of classification are not necessarily coextensive with units of generalisation: (1) at any point in time, the members of a species will vary in numerous traits, such that grouping on the basis of one set of similarities typically conflicts with groupings based on other sets of similarities; (2) species evolve, such that later members of a species may be quite different from earlier members; and (3) in speciation events, the members of ancestral and descendant species share the same traits, as it takes some time for an ancestral and a descendant species to evolve away from each other phenotypically as well as genotypically. Thus, the stable, named clusters into which biodiversity is divided up will not generally allow inductive generalisations of the sort 'All and only organisms of species X exhibit the set of traits $\{t_1, t_2, t_3, \ldots\}$' – some generalisations will range only over a small part of the whole species, other generalisations will range far beyond the borders of the species, and hardly any (if any at all) generalisations will range over all and only the members of a species. The role of species as units of classification and as units of generalisation diverge, and different groups are required to perform the respective roles.*

Next to these two roles, the term 'species' adopted a third epistemic role with respect to the tracing of evolving entities (i.e., populations and metapopulations) in studies of evolutionary dynamics that came into focus with the development of the Modern Synthesis (Reydon 2005: 139). As units of classification and units of generalisation are required to be stable and unchanging, whereas units of evolution change in time, the latter epistemic role (tracing entities that undergo evolutionary change) cannot be performed by the same units that are supposed to be the building blocks of a stable information storage and retrieval system or stable groups over which inductive generalisations can be made. Thus, at least three epistemic roles are associated with the notion of species that cannot be simultaneously performed by the same kind of entities – in this sense, the term 'species' is associated with multiple referents.

This epistemological disunity is therefore connected to an *ontological disunity*. This disunity manifests itself in the metaphysical baggage that the term 'species' accumulated as biological science developed. While (as mentioned earlier) species used to be conceived of as kinds of entities, i.e., natural kinds in the same way as the chemical elements, in the mid-twentieth century, powerful arguments were made for

---

* I have discussed the problem of generalisation over species elsewhere in more detail (Reydon 2006; 2019b).

a view of species as concrete entities (individuals).* The term 'species' thus came to refer to two metaphysically distinct kinds of things: (natural) kinds and individuals (see Reydon 2005; Richards 2010, chapter 6; Zachos 2016, chapter 3). But, philosophers distinguish different kinds of individuals, and at least two metaphysical accounts of individuality are associated with the claim that species are individuals (Reydon 2008). An important metaphysical distinction to make, when asking whether species are individuals, is between *historical individuals* (i.e., lineages) and *process individuals* (i.e., populations or metapopulations that participate in evolutionary processes) (Lidén and Oxelman 1989; Reydon 2008; Zachos 2016, 8–11). In the technical language of metaphysics, the difference is that historical individuals are extended in space and time (they are four-dimensional entities) that *perdure*, while process individuals are extended in space only (they are three-dimensional entities) that exist at specific points in time and can be said to *endure* rather than perdure (Reydon 2008).

Practicing scientists may well think of such a metaphysical distinction as fanciful – a conceptual puzzle that keeps philosophers occupied but doesn't have any actual import for the sciences (or for everyday life, for that matter). Zachos (2016, 8–11 and 68–69), for example, holds the view (which is shared by several other practicing biologists) that lineages and populations are merely two different sides of the same coin. That is, thinking of a species as a (meta-)population amounts to taking a synchronic perspective, and thinking of a species as a lineage amounts to taking a diachronic perspective, but ultimately, these are just two perspectives on exactly the same entities. I think this amounts to a misunderstanding of the metaphysical subtleties involved in the issue, however. It is important to see that even though lineages and populations are connected, that doesn't mean they are the same kind. Every population that we find in nature has a lineage: its historical trace in time that includes all the ancestors of the current members of the population as well as all currently living members of the population. Conversely, however, not every lineage has a population – extinct lineages do not. A lineage and 'its'" population thus are not coextensive. Lineages have past and present organisms as their members (the lineage of which I am part includes my late grandparents, great-grandparents, and so on back in time), while populations only consist of currently living organisms (in March 2020, the human population was estimated to encompass about 7.8 billion living people; in 1927, it was estimated to encompass 2 billion living people). Moreover, lineages and populations encompass different part–whole relations. Organisms are connected to each other in lineages by unchanging part–whole relations (i.e., a lineage is a genealogical nexus of fixed parent–offspring relations), while organisms are connected to each other in populations by continuously changing relations of mating, cooperation, and competition (i.e., a population consists of organisms that interact with each other in various evolutionarily relevant ways and in which the interactions between organisms are changeable). Lineages are static parts of the pattern of the evolutionary history of life on Earth, while populations are dynamic groups of organisms that

---

* There is a wealth of philosophical literature on this issue. The classic papers arguing for a view of species as individuals are Ghiselin (1966; 1974) and Hull (1976; 1978).

participate in evolutionary processes (Lidén and Oxelman 1989). Metaphysically, then, lineages and populations are different kinds of individuals with different kinds of relations between the wholes and their constituent parts.

This distinction is important not only in the context of metaphysics, however, but also when we ask what kinds of entities our best scientific theories are talking about. While in systematic biology, which aims at reconstructing evolutionary history and building a stable classificatory system on the basis of the reconstructed pattern, lineages occupy center stage, in evolutionary biology, populations are in focus as the main entities in evolutionary processes. Moreover, while lineages and populations are connected, the populations that are denoted by species names do not necessarily map one-on-one onto the lineages that are denoted by species names. This is very clearly seen in the (sometimes very large) differences in species counts between species according to the Biological Species Concept (which defines species as populations) and versions of the Phylogenetic Species Concept (which define species as lineages). Studies have shown that using a version of the Phylogenetic Species Concept to individuate species in a particular area on average yields almost 150% of the number of species that are recognised using the Biological Species Concept (Agapow et al. 2004; Reydon and Kunz 2019) and sometimes even up to 200–250% the number of species (for studies on birds, see Peterson and Navarro-Sigüenza 1999; Barrowclough et al. 2016) – a phenomenon known as taxonomic inflation. This means that even though for every population a corresponding lineage exists, these are not necessarily the lineages that are ranked as species. Therefore, conceiving of a synchronic view of species and a diachronic view of species as different perspectives on the same entities is mistaken – from a synchronic (population) perspective, different entities are individuated as species than from a diachronic (lineage) perspective.

Taken together, the epistemological and ontological disunity discussed here supports my claim that the species problem is better understood as a case of homonymy (or more linguistically correct, perhaps, a mix of polysemy and homonymy) rather than pluralism. Pluralism amounts to acknowledging that 'species' as a technical term has multiple meanings while considering all those meanings as applying to the same category of entities. The claim that 'species' is a homonym takes the matter one step further by acknowledging that the disunity among the set of entities called 'species' and named with Linnaean binomials is much too large to think of all the different meanings of 'species' as applying to the same category of entities.[*]

---

[*] Zachos (2016) discusses my homonymy thesis, writing that the claim that 'species' is a homonym amounts to the observation that the 'term species is used with quite different meanings in different contexts which often causes unnecessary confusion' (Zachos 2016, 5). It should be clear, though, that this misses the principal difference between pluralism and the homonymy thesis. Note, too, that it has been suggested that the term 'species' is not special with respect to homonymy:

'Select,' 'adapt,' and 'fit' – so-called multivalent terms or homonyms – are ubiquitous elements of evolutionary language as a consequence of their recruitment from ordinary discourse [...]. The same word – 'adapt,' for example – typically refers to very different processes depending on its location within formal scientific vs. everyday language. To the frustration of many students, the multiple meanings of scientific terms are often subtle, implicit, and undefined.

(Nehm et al. 2010, 605)

This view can be clarified by comparing it to a somewhat similar view advanced in a recent paper by Novick and Doolittle (2021), to which my attention was drawn after the present paper had been submitted for publication (I am indebted to John Wilkins for bringing Novick and Doolittle's paper to my attention.). Novick and Doolittle, too, propose that 'species' is a polyseme. In addition, they propose that individual species concepts, such as the Biological Species Concept, are polysemes too (which I have not suggested). Novick and Doolittle reach their view by way of an argumentation that in a few aspects differs from mine. In particular, Novick and Doolittle argue that the polysemy of individual species concepts is due to the extension of available concepts to new domains of biodiversity and/or the convergence of historically independent uses when domains meet. (They consider how the application of the Biological Species Concept and the Ecological Species Concept to prokaryotes caused changes in usage and meaning of those concepts.) Novick and Doolittle develop their view in terms of 'patches', where a patch is a type of use of a concept and different uses for different domains of biodiversity constitute different patches. On Novick and Doolittle's view, individual species concepts are associated with several interconnected patches (as one concept is used and operationalized differently when applied to different domains of biodiversity), i.e., they are conceptual patchworks. In the same way, 'species' is a conceptual patchwork – a patchwork of patchworks. As they write, 'species' refers to 'a collection of interacting patchworks, generated by the application of various species subconcepts to new domains' (Novick and Doolittle 2021, 79). On their account, polysemy thus encompasses a diversity of uses, each for a different domain of biodiversity, and 'species' is doubly polysemous ('species' denotes a patchwork of species concepts, where each species concept is itself a patchwork too). In contrast, I have emphasized that theoretical and methodological changes in science can give existing concepts new meanings while the old meanings are retained. Here, polysemy is not due to the application of concepts in domains of biodiversity where they had not yet been applied, but rather to the advancement of knowledge about a domain and/or the advancement of investigative technologies. Moreover, my view allows for multiple uses for the same domain of biodiversity to group the same organisms differently for different context of research (Reydon and Kunz 2019). In addition, I emphasized the metaphysical baggage that the term 'species' carries from uses prior to the development of biology as a science. Novick and Doolittle's view and my view thus align, but only in part make the same claims and invoke the same kind of arguments.

Note that pluralism and homonymy/polysemy have different implications for the way we should approach the species problem and call for different lines of action. On the adoption of a pluralistic view, the concept of species (and the category to which it refers) is retained as part of the conceptual framework of biological science, and the main issue to be addressed is to provide an account of how its different subconcepts are connected under the one, overarching concept. In contrast, the diagnosis of homonymy/polysemy calls for abandoning the species concept as a unified scientific concept (and the category to which it is connected) and for a conceptual restructuring, i.e., a replacement of the old species concept by a set of new concepts and categories.

If the homonymy thesis is correct, and the set of entities called 'species' and named with Linnaean binomials is fundamentally disunified, what are we to make of 'species' as a technical term in the biological sciences? While some authors have called for abandoning the term altogether (e.g., Mishler 1999; 2010; Pleijel 1999; Hendry et al. 2000; Pleijel and Rouse 2000), in the next section, I will suggest a different approach that allows us to retain the technical term while at the same time abandoning the assumption that it refers to a category of entities that all are the same in some fundamental way.*

3.4   THE SPECIES-AS-STATUS THESIS

The 'metaphysics light' answer to the species problem that I want to suggest in this section starts from drawing the inevitable consequence from the homonymy thesis, namely, rejecting the species category as a real category of entities. As noted earlier in this chapter, this by itself is not a new viewpoint, and different authors have advanced different reasons for drawing this consequence (Mishler 1999; 2010; Pleijel 1999; Hendry et al. 2000; Pleijel and Rouse 2000; Ereshefsky 2010; 2011; 2014; Reydon and Kunz 2019). But, how can the term 'species' still be a meaningful technical term if it does not refer to a particular category of entities? In other words, does the question, What *are* species?, have an answer at all if the species category is rejected as not a real category of biological entities or units?

I want to suggest that the term 'species' can still be a meaningful technical term in biological science, and the aforementioned question does have an answer: 'species' refers to a theoretical idea, not to a category or kind of entities (see Novick and Doolittle 2021, for a similar suggestion). To elaborate this suggestion, let me begin by considering Darwin's writings.† In the *Origin of Species*, Darwin writes about taxonomic ranks that

> the *amount* of difference in the several branches or groups [...] may differ greatly, being due to the different degrees of modification which they have undergone; and this is expressed by the forms being ranked under different genera, families, sections, or orders. [...] [T]he degrees of modification which the different groups have undergone, have to be expressed by ranking them under different so-called genera, sub-families, families, sections, orders, and classes.
>
> **(Darwin 1859, 420 and 422; emphasis in original)**

---

* This is the reason why I did not discuss another distinction that might be thought to support the homonymy thesis. Mayden claims that

> [t]he term species has two different meanings that, when not clearly differentiated, will result in confusion and misunderstanding [...]. The term species is used to represent both a taxonomic category and those naturally occurring particulars that we discover, describe, and order into our classificatory system.
>
> (Mayden 1997, 387)

While Mayden is correct in saying that 'species' refers to the species category as well as species taxa (individual groups of organisms attributed species status), part of my argument is that there is no unified species category. This would leave 'species' as referring only to a manifold of entities – which on my view, still involves homonymy/polysemy of the term.

† I have also elaborated this suggestion in (Reydon 2019a).

On Darwin's view, evolution is descent with modification: evolution consists in the gradual divergence of organismal forms as generations follow each other (think of the famous figure in the *Origin of Species* in which evolution is represented as an ever-diverging bush of dotted lines). Against the background of this view, Darwin then suggests that *stages of divergence* can be marked by assigning them a taxonomic rank: smaller divergences can be given the status of varieties, larger divergences the status of species, still larger divergences the status of genera, and so on.

The term 'species', on such a view, has the meaning of a *status* that a group of organisms has acquired in achieving a particular level of modification in comparison to its ancestor group and to other groups in the same time-slice – a level of modification larger than that of a variety but smaller than that of a genus. Darwin rejected the reality of the species category (Ereshefsky 2010; 2011; 2014) and famously held that 'the term species [was] one arbitrarily given for the sake of convenience to a set of individuals closely resembling each other, and that it does not essentially differ from the term variety, which is given to less distinct and more fluctuating forms' (Darwin 1859: 52), but it is important to see that the view of 'species' as denoting a degree of divergence from an ancestral form as well as from closely related forms does not conflict with the non-reality of the species category or with a view of species status attribution being to some degree for the sake of convenience. Darwin described descent with modification as a slow, continuous, and gradual process, such that it is hardly possible to identify clear stages in the process. This means that we have to decide which degrees of modification should be counted as sufficiently large to attribute to a group the status of a variety, which is sufficiently large to attribute species status, and so on. Understanding the notion of species as referring to a level of evolution or a stage in the evolutionary process at which a particular degree of modification or divergence has been achieved thus is fully consistent with Darwin's view of 'species' as a term of convenience.

This view of what the technical term 'species' means can also be found in early work in the context of the Modern Synthesis, namely, in Theodosius Dobzhansky's writings on the topic, but seems to have lost traction since then. In his ground-breaking book that served as one of the foundational writings of the Modern Synthesis, *Genetics and the Origin of Species*, Dobzhansky observes that 'in the light of the evolution theories [...] such concepts as race, species, genus, family, etc., have come to be understood as *connoting nothing more than degrees of separation in the process of a gradual phylogenetic divergence*' (Dobzhansky 1937, 309; emphasis added), thus expressing a view very similar to Darwin's. In an earlier article that appeared in one of the first issues of what is now one of the most prominent journals in the philosophy of science, Dobzhansky expressed this view more specifically:

> Considered dynamically, the species represents that stage of evolutionary divergence, at which the once actually or potentially interbreeding array of forms becomes segregated into two or more separate arrays which are physiologically incapable of interbreeding. The fundamental importance of this stage is due to the fact that it is only the development of the isolating mechanisms that makes possible the coexistence in the same geographic area of different discrete groups of organisms. [...] [D]evelopment

of isolating mechanisms renders the differences between groups relatively fixed and irreversible, and permits them to dwell side by side without losing their differentiating characteristics. This, in turn, opens the possibility for the organisms dwelling together to become adapted to different places in the general economy of nature. The usage of the term 'species' can and should be made to reflect the attainment by a group of organisms of this evolutionary stage.

(Dobzhansky 1935, 354)

Here, several important components of Dobzhansky's view are visible. First, Dobzhansky takes a dynamic perspective on the notion of species, fitting the focus of evolutionary biology on studying the dynamics of evolutionary processes. Second, he sees the term 'species' as connected not merely to a stage of divergence in evolutionary processes but to a stage at which the divergence of forms has become fixated by mechanisms that prevent them from merging again. Third, he considers this stage to have theoretical importance with respect to one of the main *explananda* of evolutionary theory, namely, adaptation: achievement of the species stage by a group of organisms opens up the way for adaptation to its specific environment.

For Dobzhansky, then, the term 'species' does not refer to a category of entities but rather, to a status attribution. When an evolving population or metapopulation achieves a stage in the evolutionary process at which its divergence from other (meta-) populations in the 'array of forms' (i.e., its phenotypical and genotypical divergence) becomes fixated, it can be considered to have reached species status. As he explains, '[t]he stage of the evolutionary process at which this fixation takes place is fundamentally important, and the attainment of this stage by a group of organisms signifies the advent of species distinction' (Dobzhansky 1937, 312). As evolution is a gradual process, achievement of this stage is not a discrete transition but a gradual one: as groups of organisms gradually develop isolating mechanisms, they gradually become isolated from other groups. Dobzhansky emphasises that his definition 'lays emphasis on the dynamic nature of the species concept. Species is *a stage in a process, not a static unit*' (Dobzhansky 1937, 312; see also Dobzhansky 1940). Dobzhansky also holds this view for groups on their way to achieving the species stage. In the third edition of his book, for example, he writes that 'a race is not a static entity but a process' (Dobzhansky 1951, 177). The difference between subspecies/races and species on this view is gradual. An evolving population that has differentiated from other populations but for which this divergence has not yet become fixated by isolating mechanisms should be considered a subspecies, and as soon as the divergence is starting to become fixated, we can begin to think of it as a species. As he explains elsewhere, '[r]ace formation is essentially the development of genetic patterns which are adapted to a definite environment. Speciation is a process resulting in fixation of these patterns through the development of physiological isolating mechanisms' (Dobzhansky 1940, 316). Note that there is no discrete point in time at which the transition from subspecies to species takes place, and at the time of speciation, 'the classification of two forms as races or as species is arbitrary' (Dobzhansky 1940, 318).

Drawing inspiration from Darwin's and Dobzhansky's views, I want to suggest taking the term 'species' as denoting a particular stage in the evolutionary process that evolving groups of organisms – i.e., populations and metapopulations – can achieve. The species stage is characterised by comparative evolutionary indepen-dence from other evolving groups, that is, by phenotypic and genotypic divergence having become fixated in such a way that the evolving group can embark on its own trajectory of adaptation to the environment. Note that on such a view of what the term 'species' means, a species is not a kind of entity but rather, a property of an entity or a status that can be attributed to it under particular conditions. Put differ-ently, *the word 'species' is not a noun but an adjective.*[*]

Accordingly, I want to phrase the theoretical idea underlying the term 'species' in the form of what I call the *species-as status thesis*:

> Being a species is a status that can be attributed to entities that feature in evolution-ary processes on the basis of their having achieved a sufficient degree of evolutionary independence from other such entities.

Note that the species-as-status thesis is not intended as a detailed explication of what it means to be a species but as a more coarse-grained explicative definition.[†] While Dobzhansky only considered evolving populations in which isolating mechanisms had become established, in principle, one can attribute species status to all kinds of entities that feature in evolution, in particular to process entities as well as pattern entities. That is, populations, metapopulations, lineages, clades, and other sorts of organism groups can in principle be attributed the property of being a species if they can be thought of as being evolutionarily independent from other such entities.

For lineages, for example, evolutionary independence means something different than it does for populations: for lineages, it means being a separate branch in the tree or bush of evolutionary history, while for populations, it means having mechanisms that keep them reproductively separated from other populations. Also, for strains of bacteria, evolutionary independence will probably mean something different from what it means for sexually reproducing organisms, and for populations of trees, a still different meaning may be in play. Note, too, that evolutionary independence is not a strict either-or matter but a matter of degree: what is key in evolutionary processes is that parts of an evolutionary system become *sufficiently* independent from other parts of the system for novel traits to emerge and spread. The notion of evolution-ary independence has been a part of the notion of species at least since Simpson's Evolutionary Species Concept (see Simpson 1951; Mayden 1997, 395–396) but is not

---

[*] A very similar 'adjective, not noun' view has been developed with respect to natural kinds by Slater, who suggested that rather than exploring the metaphysical nature of natural kinds, we should think of 'natural kindness' as a kind of status that things can have on epistemological grounds (Slater 2013, 150; 2015, 378). On Slater's view, rather than thinking of 'natural kind' as referring to an ontological category of entities, we should think of 'being a natural kind as a sort of status that things or plurali-ties of things (from various ontological categories) can have' (Slater 2015, 407; 2013, 18). Accordingly, Slater understands 'natural kind' as an adjective (Slater 2013, 107 and 150). Novick and Doolittle (2021) recently also suggested that 'species concepts have a central role to play in studying processes of speciation' (2021, 78), but did not suggest that 'species' should in this context be seen as a status attribution.

[†] I first suggested this idea in Reydon 2019a, where I refer the reader to for more discussion.

part of all available definitions of the term 'species'. Also, the notion of evolutionary independence is imprecise, such that it remains unclear when exactly we can
think of a population, lineage, etc. as being evolutionarily independent from other
entities and how evolutionary independence can be measured (see Conix 2019 for a
recent discussion of this issue). Because of these issues, the species-as-status thesis
is merely a coarse-grained way of capturing the core theoretical idea underlying the
term 'species' – it cannot be a full-blown metaphysical thesis.

The attribution of species status to a group thus is principally an epistemological
attribution and in addition, at most a 'light metaphysical' one. To say that a group of
organisms is a species is to say that it is a group – whatever its precise metaphysical nature – that has achieved a stage of evolution that is of importance to us in the
light of evolutionary theory. That is, it is a stage highlighted *by us* because of its
explanatory importance. But, note that the theoretical idea embodied in the species
concept – evolutionary independence – does not have to be very specific to be useful.
It expresses an important aspect of the evolutionary process in a similar way as the
term 'friend', which is equally difficult to pinpoint exactly, expresses an important
aspect of processes of development of interpersonal relations.

3.5 CONCLUSION: WHAT DOES THIS MEAN FOR PHILOSOPHY AND FOR BIOLOGICAL PRACTICE?

The two theses discussed here, the homonymy thesis and the species-as-status thesis,
together answer the species problem in the following way. The homonymy thesis
shows us that the long-standing quest for an answer to the question of what species
are is misguided. It pursued the wrong question and should be replaced by two questions: What do we mean when we say that a group of organisms is a species? And,
is there a theoretically important idea underlying the various meanings we find in
practice? The homonymy thesis further shows that there is a manifold of meanings
associated with the term 'species' that diverge to such an extent that it becomes
implausible to think that there is one, fairly unified category of entities associated
with the term. The homonymy thesis enables us to understand the roots of the species problem as well as an important aspect of the current situation that should be
resolved. In other words, the homonymy thesis clarifies what the problem is that we
are faced with. Once the ontological assumption that there exists such a category has
been abandoned, the species-as-status thesis tells us how we can retain the technical term in a meaningful way without having to assume the existence of a species
category. Thus, the species-as-status thesis allows us to avoid the eliminativist view
of the radical pluralists, discussed earlier, that the term 'species' should be abolished
from the conceptual toolkit of biological science. The species-as-status thesis tells
us what the theoretical idea behind the term 'species' is, and in so doing, it tells
researchers what to look for when searching for species.

The solution to the species problem that is provided by the homonymy thesis and
the species-as-status thesis in combination is a 'light metaphysical' solution in the
sense that it does not specify an essence of the species category (or multiple essences
of multiple species categories) or pinpoints what species are deep down. But, it is

not devoid of any metaphysics either, as it say *something* about what it is to be a species. In the same way, the term 'friend' says *something* about the entities that can be attributed the status of friend (people, pets, and perhaps robots too?), but stops far short of specifying any essences for any ontological category of friends or pinpointing exactly what friends are deep down.

When it comes to the philosophy of biology, this solution provides new orientation for practice-oriented philosophy. The species problem, I suggest on the basis of my two theses, it is about clarifying why biologists single out some groups of organisms as species. This is a question that forces philosophers to pay close attention to biological practice instead of doing epistemology and metaphysics of biological science in the traditional way. This new way of seeing the problem thus is in line with the recent development of 'Philosophy of Science in Practice' as a movement (or approach, or way of thinking) in the philosophy of science. What we, as philosophers of science, should be examining under the name of 'The Species Problem' is *practices of attribution of species status* by biologists in the laboratory, in the field, in the museum, and in the (regrettably very few) institutes in which theoretical biology is practiced. Note that this is not a new suggestion. A few years ago, Kendig already argued that in the context of the 'practice turn' (Kendig 2016a, 3ff) in the philosophy of science, classificatory notions such as 'species' (but also other classificatory notions such as 'homologue' and 'natural kind') should be seen as denoting practices of grouping of kind-making (Kendig 2014; 2016a, b).* In other words, 'species' refers to certain practices that feature in the biological sciences, not to a category or several categories of entities. This way of thinking about classificatory terms, which was initiated by Kendig only fairly recently and still has to gain traction among philosophers of biology, precisely captures one of the main consequences of the two theses I presented here.

When it comes to biological practice, implications are less easy to draw, because the biological (and biomedical) sciences encompass a great many fields of research. Here, I want to end on a speculative note, though, and suggest that in all fields of research in which the technical term 'species' plays a role (the biological sciences, the biomedical sciences, but also such fields as anthropology and sociology when they examine ancestral species of *Homo sapiens*), it serves to guide research by specifying an important factor to look for. The factor of evolutionary independence is explanatorily important, as it features in explanations of how current traits are anchored in evolutionary history. But, the notion of evolutionary independence is imprecise and does not apply to a single biological phenomenon but to a 'heterogeneity of phenomena' (Conix 2019, 53). That is, researchers in different contexts of research mean different things when they talk about 'evolutionary independence'. But this heterogeneity notwithstanding, everyone knows what the basic idea is. The term 'species' thus can serve to guide research by conveying this basic idea. This might only be a small task to perform, but it is a crucial task in biological research nonetheless.

---

* See also Kendig's chapter in the present volume.

But – and here I return to the consequences for the philosophy of biology – practice-oriented philosophical research on the species problem can help to clarify what this task involves. The suggestion I want to end with, then, is that the further development of the term 'species' as a guide for biological practice must involve further philosophy-of-science-in-practice-style research on actual practices of species attribution. Besides presenting two theses as a joint solution to the 'old' species problem, I thus hope to have pointed to a 'new' species problem and in the direction in which further philosophical research on this latter problem should go.

REFERENCES

Agapow, P. M., Bininda-Emonds, O. R., Crandall, K. A., Gittleman, J. L., Mace, G. M., Marshall, J. C. and Purvis, A. 2004. The impact of species concept on biodiversity studies. *Quarterly Review of Biology* 79: 161–179.

Barrowclough, G. F., Cracraft, J., Klicka, J. and Zink, R. M. 2016. How many kinds of birds are there and why does it matter? *PLoS ONE* 11: e0166307.

Bird, A. and Tobin, E. 2018. Natural kinds. In Zalta, E. N. (Ed.), *The Stanford Encyclopedia of Philosophy (Spring 2018 Edition)*. https://plato.stanford.edu/archives/spr2018/entries/natural-kinds/.

Conix, S. 2019. Measuring evolutionary independence: A pragmatic approach to species classification. *Biology & Philosophy* 34: 53.

Darwin, C. R. 1859. *On the Origin of Species by Means of Natural Selection, or the Preservation of Favoured Races in the Struggle for Life*. London: John Murray.

De Queiroz, K. 1998. The general lineage concept of species, species criteria, and the process of speciation: A conceptual unification and terminological recommendations. In Howard, D. J. and Berlocher, S. H. (Eds.), *Endless Forms: Species and Speciation*. Oxford: Oxford University Press, pp. 57–75.

De Queiroz, K. 1999. The general lineage concept of species and the defining properties of the species category. In R. A. Wilson (Ed.), *Species: New Interdisciplinary Essays*. Cambridge, MA: MIT Press, pp. 49–89.

De Queiroz, K. 2011. Branches in the lines of descent: Charles Darwin and the evolution of the species concept. *Biological Journal of the Linnean Society* 103: 19–35.

Dobzhansky, T. 1935. A critique of the species concept in biology. *Philosophy of Science* 2: 344–355.

Dobzhansky, T. 1937. *Genetics and the Origin of Species*. New York: Columbia University Press.

Dobzhansky, T. 1940. Speciation as a stage in evolutionary divergence. *American Naturalist* 74: 312–321.

Dobzhansky, T. 1951. *Genetics and the Origin of Species (Third Edition, Revised)*. New York: Columbia University Press.

Dupré, J. 1993. *The Disorder of Things: Metaphysical Foundations of the Disunity of Science*. Cambridge, MA: Harvard University Press.

Dupré, J. 1999. On the impossibility of a monistic account of species. In R. A. Wilson (Ed.), *Species: New Interdisciplinary Essays*. Cambridge, MA: MIT Press, pp. 3–22.

Dupré, J. 2018. Processes, organisms, kinds, and the inevitability of pluralism. In Bueno, O., Chen, R.-L. and Fagan, M. B. (Eds.), *Individuation, Process, and Scientific Practices*. New York: Oxford University Press, pp. 21–38.

Ereshefsky, M. 2010. Darwin's solution to the species problem. *Synthese* 175: 405–425.

Ereshefsky, M. 2011. Mystery of mysteries: Darwin and the species problem. *Cladistics* 27: 67–79.

Ereshefsky, M. 2014. Consilience, historicity, and the species problem. In Thompson, R. P. and Walsh, D. (Eds.), *Evolutionary Biology: Conceptual, Ethical, and Religious Issues*. Cambridge: Cambridge University Press, pp. 65–86.

Frege, G. 1892. Über Sinn und Bedeutung. *Zeitschrift für Philosophie und philosophische Kritik* 100: 25–50.

Ghiselin, M. T. 1966. On psychologism in the logic of taxonomic controversies. *Systematic Zoology* 15: 207–215.

Ghiselin, M. T. 1974. A radical solution to the species problem. *Systematic Zoology* 23: 536–544.

Grene, M. and Depew, D. 2004. *The Philosophy of Biology: An Episodic History*. Cambridge: Cambridge University Press.

Griffiths, P. E. and Stotz, K. 2013. *Genetics and Philosophy: An Introduction*. Cambridge: Cambridge University Press.

Hendry, A. P., Vamosi, S. M., Latham, S. J., Heilbuth, J. C. and Day, T. 2000. Questioning species realities. *Conservation Genetics* 1: 67–76.

Hull, D. L. 1976. Are species really individuals? *Systematic Zoology* 25: 174–191.

Hull, D. L. 1978. A matter of individuality. *Philosophy of Science* 45: 335–360.

Hull, D. L. 1997. The ideal species concept – And why we can't get it. In Claridge, M. F., Dawah, H. A. and Wilson, M. R. (Eds.), *Species: The Units of Biodiversity*. London: Chapman & Hall, pp. 357–380.

Hull, D. L. 1999. On the plurality of species: Questioning the party line. In Wilson, R. A. (Ed.), *Species: New Interdisciplinary Essays*. Cambridge, MA: MIT Press, pp. 23–48.

Johannsen, W. 1909. *Elemente der exakten Erblichkeitslehre (Deutsche, wesentlich erweiterte Ausgabe in fünfundzwanzig Vorlesungen)*. Jena: Verlag von Gustav Fischer.

Johannsen, W. 1911. The genotype conception of heredity. *American Naturalist* 45: 129–159.

Johannsen, W. 1923. Some remarks about units in heredity. *Hereditas* 4: 133–141.

Kendig, C. E. 2014. Towards a multidimensional metaconception of species. *Ratio* 27, 155–172.

Kendig, C. E. 2016a. Introduction: Activities of kinding in scientific practice. In Kendig, C. E. (Ed.), *Natural Kinds and Classification in Scientific Practice*. Abingdon and New York: Routledge, pp. 1–13.

Kendig, C. E. 2016b. Homologizing as kinding. In Kendig, C. E. (Ed.), *Natural Kinds and Classification in Scientific Practice*. Abingdon and New York: Routledge, pp. 106–125.

Kornet, D. J. 1993. Permanent splits as speciation events: A formal reconstruction of the Internodal Species Concept. *Journal of Theoretical Biology* 164: 407–435.

Kuhn, T. S. 1962. *The Structure of Scientific Revolutions*. Chicago, IL: University of Chicago Press.

Lherminer, P. and Solignac, M. 2000. L'espèce: Définitions d'auteurs. *Comptes Rendus de l'Académie des Sciences (Paris), Sciences de la Vie* 323: 153–165.

Lidén, M. and Oxelman, B. 1989. Species – Pattern or process? *Taxon* 38: 228–232.

Mayden, R. L. 1997. A hierarchy of species concepts: The denouement in the saga of the species problem. In Claridge, M. F., Dawah, H. A. and Wilson, M. R. (Eds.), *Species: The Units of Biodiversity*. London: Chapman & Hall, pp. 381–424.

Mayden, R. L. 1999. Consilience and a hierarchy of species concepts: Advances toward closure on the species puzzle. *Journal of Nematology* 31: 95–116.

Mayr, E. 1982. *The Growth of Biological Thought: Diversity, Evolution, and Inheritance*. Cambridge, MA: Harvard University Press.

Mayr, E. and Bock, W. J. 2002. Classifications and other ordering systems. *Journal of Zoological Systematics and Evolutionary Research* 40: 169–194.

Mill, J. S. 1859. *A System of Logic, Ratiocinative and Inductive*. New York: Harper & Brothers.

Mishler, B. D. 1999. Getting rid of species? In R. A. Wilson (Ed.), *Species: New Interdisciplinary Essays*. Cambridge, MA: MIT Press, pp. 306–315.

Mishler, B. D. 2010. Species are not uniquely real biological entities. In Ayala, F. J. and Arp, R. (Eds.), *Contemporary Debates in Philosophy of Biology*. Chichester: Wiley-Blackwell, pp. 110–122.

Nehm, R. H., Rector, M. A. and Ha, M. 2010. "Force-talk" in evolutionary explanation: Metaphors and misconceptions. *Evolution: Education and Outreach* 3: 605–613.

Novick, A. and Doolittle, W. F. 2021. 'Species' without species. *Studies in History and Philosophy of Science* 87: 72–80.

Oberheim, E. and Hoyningen-Huene, P. 2018. The incommensurability of scientific theories. In Zalta, E. N. (Ed.), *The Stanford Encyclopedia of Philosophy (Fall 2018 Edition)*. http://plato.stanford.edu/archives/fall2018/entries/incommensurability/.

Peterson, A. T. and Navarro-Sigüenza, A. G. 1999. Alternate species concepts as bases for determining priority conservation areas. *Conservation Biology* 13: 427–431.

Pleijel, F. 1999. Phylogenetic taxonomy, a farewell to species, and a revision of *Heteropodarke* (*Hesionidae, Polychaeta, Annelida*). *Systematic Biology* 48: 755–789.

Pleijel, F. and Rouse, G. W. 2000. Least-inclusive taxonomic unit: A new taxonomic concept for biology. *Proceedings of the Royal Society of London B* 267: 627–630.

Portin, P. 1993. The concept of the gene: Short history and present status. *Quarterly Review of Biology* 68: 173–223.

Portin, P. 2002. Historical development of the concept of the gene. *Journal of Medicine and Philosophy* 27: 257–286.

Reydon, T. A. C. 2004. Why does the species problem still persist? *BioEssays* 26: 300–305.

Reydon, T. A. C. 2005. On the nature of the species problem and the four meanings of 'species'. *Studies in History and Philosophy of Biological and Biomedical Sciences* 36: 135–158.

Reydon, T. A. C. 2006. Generalizations and kinds in natural science: The case of species. *Studies in History and Philosophy of Biological and Biomedical Sciences* 37: 230–255.

Reydon, T. A. C. 2008. Species in three and four dimensions. *Synthese* 164: 161–184.

Reydon, T. A. C. 2019a. Are species good units for biodiversity studies and conservation efforts? In Casetta, E., Marquez da Silva, J. and Vecchi, D. (Eds.), *From Assessing to Conserving Biodiversity: Conceptual and Practical Challenges*. Cham: Springer, pp. 167–193.

Reydon, T. A. C. 2019b. Taxa hold little information about organisms: Some inferential problems in biological systematics. *History and Philosophy of the Life Sciences* 41: 40.

Reydon, T. A. C. and Kunz, W. 2019. Species as natural entities, instrumental units and ranked taxa: New perspectives on the grouping and ranking problems. *Biological Journal of the Linnean Society* 126: 623–636.

Rheinberger, H.-J. and Müller-Wille, S. 2008. Gene concepts. In Sarkar, S. and Plutynski, A. (Eds). *A Companion to the Philosophy of Biology*. Malden, MA: Blackwell, pp. 3–21.

Richards, R. A. 2010. *The Species Problem: A Philosophical Analysis*. Cambridge: Cambridge University Press.

Roll-Hansen, N. 2014. The holist tradition in twentieth century genetics. Wilhelm Johannsen's genotype concept. *Journal of Physiology* 592: 2431–2438.

Roselló-Mora, R. 2003. The species problem, can we achieve a universal concept? *Systematic and Applied Microbiology* 26: 323–326.

Siddall, R. 2002. The composition of the Earth: Rocks and minerals. In De Vivo, B., Grasemann, B. and Stüwe, K. (Eds.), *Encyclopedia of Life Support Systems, Geology, Vol. 1*. Oxford: EOLSS Publishers.

Simpson, G. G. 1951. The species concept. *Evolution* 5: 285–298.

Slater, M. H. 2013. *Are Species Real? An Essay on the Metaphysics of Species.* Basingstoke: Palgrave Macmillan.

Slater, M. H. 2015. Natural kindness. *British Journal for the Philosophy of Science* 66: 375–411.

Thagard, P. 1992. *Conceptual Revolutions.* Princeton, NJ: Princeton University Press.

Thagard, P. 2003. Conceptual change. In: Nadel, L. (Ed.). *Encyclopedia of Cognitive Science (Volume 1).* London: Macmillan, pp. 666–670.

Waters, C. K. 1994. Genes made molecular. *Philosophy of Science* 61: 163–185.

Weber, M. 2005. *Philosophy of Experimental Biology.* Cambridge: Cambridge University Press.

Wilkins, J. S. 2003. How to be a chaste species pluralist-realist: The origins of species modes and the synapomorphic species concept. *Biology & Philosophy* 18: 621–638.

Wilkins, J. S. 2009. *Defining Species: A Sourcebook from Antiquity to Today.* New York: Peter Lang.

Wilkins, J. S. 2018. *Species: The Evolution of the Idea (Second Edition).* Boca Raton, FL: CRC Press.

Zachos, F. E. 2016. *Species Concepts in Biology: Historical Development, Theoretical Foundations and Practical Relevance.* Cham: Springer.

4 What *Should* Species Be?
Taxonomic Inflation and the Ethics of Splitting and Lumping

Jay Odenbaugh

CONTENTS

4.1 BIOLOGY'S CURRENCY AND THE SPECIES DEBATE

Species are of the chief currency of biology. When we wonder how diverse life is on our planet, we learn that there might be as many as 5–100 million species (Erwin, 1982; May, 1990; Stork, 1993; May, 2011; Mora et al., 2011). When we ask what we are losing, we are told that the species extinction rate is 100–1,000 times the background extinction rate found in the fossil record (May et al., 1995). When conservation biologists prioritise places conservation, they do so on the basis of species richness and evenness. When we think about conservation in the United States, one of its most important tools is the Endangered Species Act (ESA) (Rohlf, 1989; Czech et al., 2001; Burgess, 2003). When we extol our successes, we talk of the recovery of species such as the brown pelican, the Steller's sea lion, gray wolf, bald eagle, and so on. Our species concept is extraordinarily important for how we think about living things and their diversity.

DOI: 10.1201/9780367855604-5

Biologists are deeply divided over the species category (Ereshefsky, 1992b; Wheeler and Meier, 2000; Coyne and Orr, 2004; Wilkins, 2009). There are a variety of legitimate species concepts with no clear frontrunner. Consider Ernst Mayr's famous biological species concept (BSC). He writes, 'Species are groups of interbreeding natural populations that are reproductively isolated from other such groups' (Mayr, 1963, 89). There are lots of worries about the BSC, but for our purposes, we need only note a few of them (Ehrlich and Raven, 1969; Sokal and Crovello, 1970; Van Valen, 1976; Wiley, 1978). Asexual organisms do not interbreed. Thus, on the BSC, there are no species of asexual organisms. Many species exhibit some introgression and thus, are not obviously distinct species. Last, the BSC is extremely difficult to apply to the fossil record, since sex organs do not normally fossilise, and thus, reproductive behavior is difficult to corroborate.*

In light of the problems with the BSC (and for other reasons), biologists have put other species concepts on the taxonomic table. Here are the most prominent.†

Ecological Species Concept Species are lineages of organisms that occupy the same ecological niche.

(**Van Valen, 1976**)

Evolutionary Species Concept Species are a single lineage of ancestor–descendant populations which maintains its identity from other such lineages and which has its own evolutionary tendencies and historical fate.

(**Wiley, 1978**)

Phylogenetic Species Concept Species are the smallest diagnosable population of organisms that share a common ancestor.

(**Cracraft, 1983**)

Species pluralism is the claim that there is no single correct species concept that classifies organisms in exactly the same way; rather, there are several correct species concepts (Ereshefsky, 1992a). That is, for groups of organisms, different species concepts will correctly place them in distinct species. *Species monism* is the claim that there is a single correct species concept. Some allege that species pluralism is temporary, because we will eventually find the single best concept (Hull, 1999). Others argue that it is a permanent state of affairs in taxonomy and systematics. From a practical point of view, we are all species pluralists for now.‡

---

* This is not to say that defenders of the BSC do not have responses to these criticisms. They do. My point here is to note that many biologists are not convinced by those responses.

† Though I discuss three different species concepts, these concepts are placeholders for families of more fine-grained concepts.

‡ One might interpret this practical pluralism as meaning that every biologist uses more than one species concept in their research. I mean something more modest; different groups of biologists use different species concepts. There may even be biologists who use none.

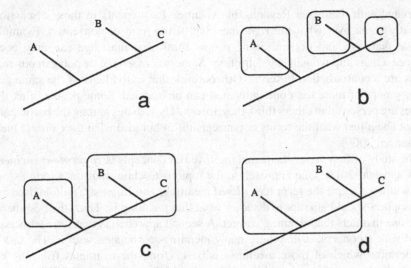

FIGURE 4.1   Species pluralism (Ereshefsky, 1992a, 675).

To see how this pluralism works, consider insects that live on the side of a mountain with three populations, *A*, *B*, and *C* (Ereshefsky, 1992a). Suppose that each population forms a single monophyletic group; *B* and *C* share an ecological niche with *A* having its own; *A* and *B* can interbreed but are isolated from *C*. Thus, the BSC classifies our species (*AB*)*C*, the ESC classifies our species *A*(*BC*), and the Phylogenetic Species Concept (PSC) classifies our species *A*,*B*,*C*. We can depict the situation as follows.

Figure 4.1a provides a phylogenetic tree of our taxa *A*, *B*, and *C*. Figure 4.1b provides the classification of species according to the PSC. Figure 4.1c provides the classification of species according to the ESC. Figure 4.1d provides the classification of species according to the BSC. One might wonder if this is just a conceptual possibility or an empirical reality, and it is an empirical reality (Ereshefsky, 1992a). And, as we shall see, it is a worrisome reality too.

Regardless of whether species monism or species pluralism is correct, there are many species concepts that are used by biologists. As I said, we find ourselves with a 'practical pluralism', and it is not disappearing anytime soon. In this chapter, I explore what reasons we have for choosing the species concept or concepts that are best for the biological sciences. Our question is this: are *ethical* reasons relevant to the selection of species concepts?

4.2   CONCEPTUAL ENGINEERING

Sometimes those in a conversation disagree over the meaning of terms or concepts.* In fact, the disagreement may reveal that those terms have many different meanings

---

* Here, I am distinguishing between words and concepts that are denoted by words. For example, the English term 'species' and the French term '*espèce*' denote the same concept of species.

associated with them (see Reydon, this volume). Participants in those discussions typically argue over what the *right* meaning of the term or concept is. Examples include marriage and person. Some people think that marriage can only occur between a man and woman 'by definition'. Same-sex marriage or polygamous marriages are a contradiction in terms. Others think that individuals of the same gender or groups of more than one individual can be married. Some people think that fetuses are persons, but others think they are not. The debates among the participants are not about just what the terms or concepts mean but also what they *should* mean (Haslanger, 2000).

The study of how we evaluate and improve our concepts is *conceptual engineering* (Cappelen, 2018). One approach to the topic is to claim that for a term associated with a concept, the term has a fixed meaning, and linguists, philologists, and philosophers should attempt to discover what that meaning is. Thus, the best theory is the one that gets that meaning correct. A second approach is that for a term associated with a concept, it may have many meanings associated with it. The task is to determine which of those meanings is best. Using the examples from the last paragraph, the questions should be, 'What should "marriage" mean? What should "person" mean?' Conceptual engineers advocate for the second view. Their argument is roughly this. If a word has several possible meanings, we should pick the best meaning of those available. Many important terms have different meanings. Thus, we should pick the best meaning of those available. This selection of the best meaning just is conceptual engineering.[*]

Conceptual engineering might sound odd to some. There are several questions that come to mind: How can meanings be defective? If we change the meaning of terms, will we be able to consistently communicate? Are we able to change the meanings of words? Let's consider reasons why we would do so (Burgess and Plunkett, 2013). One way in which a term might be defective is that its uses are *intrinsically bad*. The most obvious case here would be that of hate speech. If we use a term that derogates another, we have disrespected them. On some moral views, this is morally wrong by itself. Since this is not relevant to the purposes of this chapter, I will leave this candidate behind. A different case would be that it is *instrumentally bad*.

If we use a term with a certain meaning, our utterances will have overall negative effects. These effects might be mostly ethical. But they need not be. For example, certain terms are excessively vague, which is problematic for scientific progress. Vagueness could be one reason to conceptually engineer our terms, but we might also think ethical consequences apply here too.

One might worry that if we change the meanings of terms, won't we simply misunderstand one another? After all, if we think we mean the same thing when we do not, we will simply be equivocating. Consider the topic of this chapter; that of species. It is likely that the plurality of species concept caused confusion at various points in the

---

[*] As a scientific example of conceptual engineering, Rudolf Carnap (1950) argued that our concept of probability was problematic. In order to explicate it, one had to make it more precise and useful for various scientific tasks. This would require that it be an improvement and thus, different from our original concept.

history of biology. However, as more reflection and care was used, biologists became aware of the different concepts at work.* They became more explicit about how they were using the term 'species' and thus, avoided some of the possible confusions. Additionally, one might argue that these communicative disruptions are productive. Miscommunication can show something is not working, and participants are given reason to do the conceptual work needed. This has occurred in the biological sciences, though it is interesting that no single species concept has been deemed best.

One crucial assumption of conceptual engineering is that we can purposively change the meanings of our terms. For if we cannot, then the recognition of conceptual confusion will not help us get clearer in our thinking. In this chapter, I will assume that conceptual engineering is possible. Moreover, I assume that it actually occurs in the sciences. To see how this is possible, consider a simple theory of kind terms like 'species' (Putnam, 1975; Sterelny, 1983). Philosophers of language and philosophers of science distinguish between *reference fixing* and *reference borrowing*. We can imagine earlier speakers recognising that organisms are not randomly associated; they form groups of similar organisms. They were probably mystified as to why such groupings occurred, but they recognised them. An early speaker might have said (in their language of course), 'By "species"', I mean any organisms that form such a group'. Subsequent speakers then deferred to such a speaker and ultimately borrowed their usage from this ur-speaker (or ur-speakers) by intending to do so. Thus, the reference of the term 'species' was fixed by these activities, and subsequent speakers meant what they did by a causal chain of intentions reaching back to that initial group. Now, as new speakers appear, we can continuously reground (or refix) our terms.† We can thus purposively change the meaning of our terms. This most clearly happens in conceptual debates in the sciences. I take debates over species to be a case in point.

My chapter in a sense is an examination of ethical conceptual engineering in the biological sciences. That is, when we have a term like 'species' being used in various ways, we might choose to revise it for a variety of reasons. Are we justified in doing so for ethical reasons? I now want to consider just such a case; that of taxonomic inflation.

4.3 TAXONOMIC INFLATION

With the rise of molecular data and methods, the PSC has been used in biology with greater frequency. We can characterise the current PSC as follows: a species

---

* Marc Ereshefsky (1992a) has even argued that we should drop the term 'species' in favor of specific terms like biospecies, ecospecies, and phylospecies to clearly indicate communicative intentions.
† One of the innovations of Sterelny (1983) is the recognition that we are continually refixing and regrounding natural kind terms. Their meaning is not permanently fixed by the ur-speaker(s). This particularly helps to address problems of reference change.

is a smallest group of organisms that share at least one uniquely derived character.* Agapow et al. (2004) examined 89 studies in which groups were classified with
both the BSC (or morphospecies concepts) and the PSC. There are three possibilities
that can result from reanalysis. First, a BSC is broken into several PSCs; they are
nested in a BSC. Second, populations in distinct BSCs are rearranged into distinct
PSCs; they are non-nested. Third, distinct BSCs are placed in a single PSC; they are
reverse-nested. Their analysis found that by and large, distinct PSCs are nested in .
BSCs. Specifically, they found the studies that used the BSC had between 1,245 and
1,282 species; when reclassified with the PSC, these studies had between 1,912 and
2,112 species. This was a 48.7% increase. In fungi, there was a 300% increase; in
lichens, there was a 259% increase; in reptiles, there was a 137% increase; in mammals, there was a 87% increase; in arthropods, there was a 77% increase; in birds,
there was a 88% increase. Put simply, the use of the PSC multiplied the number of
species beyond that of the BSC. The question, though, is this: does the PSC multiply
the number of species beyond necessity? Taxonomic inflation occurs when existing
populations or subspecies are raised to the species level. But why worry about this –
after all, might there simply be more species than we thought?

4.4   TAXONOMIC INFLATION AND CONSERVATION

Taxonomic inflation is considered problematic because it increases the number of
endangered species. And if we increase the number of endangered species, we make
conservation that much harder (Agapow et al., 2004; Isaac et al., 2004; Zachos,
2013). First, the number of species increases, as we saw in the previous section.
Second, the abundance of each species becomes smaller. This means that they are
more vulnerable to demographic stochasticity, genetic stochasticity, inbreeding, and
thus, extinction. Third, the amount of money for inventory and recovery that must
be spent increases as well. For example, under the ESA, if it costs on average $2.76
million to completely recover an endangered species, then it would cost $4.6 billion
to save all the currently endangered species. With the PSC, this would increase to
$7.6 billion, which is the entire annual budget of the United States Fish and Wildlife
Service (Agapow et al., 2004, 169).

We can summarise the argument as follows.† If taxonomy is inflated, then the
number of endangered species will be increased. However, if the number of endangered species is increased, then this makes conservation goals harder to meet. We
should not make conservation goals harder to meet. Therefore, we should not inflate

---

* A *homology* is a trait shared between species and their common ancestor, whereas a *homoplasy* is a
  trait shared between species but not by a common ancestor. A *shared derived homology* is one that
  is found in the ancestor of a species and all their descendants, whereas a *shared ancestral homology*
  is one had by the ancestor of a species and only some of their descendants. Shared derived traits, or
  synapomorphies, are special, because only they indicate monophyletic groups. It is worth noting that
  on some versions of the PSC, the character may be plesiomorphic, provided that it is diagnosable and
  is confined to this group. ·
† No biologist or philosopher has explicitly offered this argument. However, some have come close, and
  it is certainly an argument that is on the minds of biologists.

taxonomy. We are engaged in an example of conceptual engineering. Let's work through the argument premise by premise, labeling each section accordingly for ease of reading.

4.4.1 PREMISE 1: TAXONOMIC INFLATION INCREASES THE NUMBER OF ENDANGERED SPECIES

The first premise is supported by the above-mentioned empirical evidence. One might argue that inflating taxonomy does not increase the number of endangered species, because the PSC is not a legitimate species concept. For example, the PSC has difficulty with bacteria and horizontal gene transfer (though see Staley, 2006). But this is consistent with the PSC correctly classifying *some* groups of organisms. One might also argue that there are no species to begin with. There is more that could be said here about the reality of species, but we will leave these worries to the side (Coyne and Orr, 2004; Mishler, 1999; Slater, 2013). Rejecting the species category as real would have even more drastic implications than anything I consider in this chapter.

4.4.2 PREMISE 2: INCREASING THE NUMBER OF ENDANGERED SPECIES IS DETRIMENTAL TO CONSERVATION

The second premise has been challenged by some biologists and philosophers. Here are two objections to it. First, it is not detrimental because species should not be the unit of conservation. Second, it is not detrimental because not all species are of value. Let's take each objection in turn.

Some suggest that instead of preserving species, we should preserve phylogenetic or functional diversity (Agapow et al., 2004; Isaac et al., 2004). Maybe so, but how do we conserve diversity and higher taxa and not conserve species? Consider one example of important functional diversity, pollinators (Nabhan and Buchmann, 1997). Most flowers require pollinators for reproduction (of 240,000 plant species, 200,000 require an animal pollinator). This includes 70% of the crop species that feed the world. Over 100,000 species of bats, bees, beetles, birds, butterflies, and flies provide these services. Approximately one-third of our food is derived from plants pollinated by wild pollinators (Nabhan and Buchmann, 1997, 135). There is, then, no easy way to preserve pollination without preserving the species that accomplish it. Thus, even if species are not the only unit of conservation, they are one of them and probably the most important legally. Since species is a unit of conservation, then increasing the number of endangered species is detrimental to conservation.

One might argue that not every species, much less endangered species, has final or instrumental value (Sandler, 2012). Many ethicists talk of intrinsic value, but this suggests that the value of something is determined by its intrinsic properties. Thus, some ethicists prefer to use the term 'final value' to avoid this suggestion. Something has final value if it is valuable independent of the value of other things. Here is how Ronald Sandler articulates objective final value.

Something possesses *objective final value* (hereafter, just *objective value*) if its value is
independent of any actual preferences, attitudes, judgments, emotions, or other evaluative states regarding it … If objective value exists, there are properties or sets of properties that, when they are instantiated in any entity, experience, act, or state of affairs,
have (or confer) value. Moreover, valuers ought to recognize and respond to value.

(Sandler, 2012, 18)

In environmental ethics, the most prominent way to understand whether species
have objective final value involves the following considerations. If something exhibits goal-directed behavior, then it has interests. If something has interests (i.e., can
be benefited or harmed), we should protect it. Species exhibit goal-directed behavior. So, species have interests. Therefore, we should conserve them. However, there
are serious problems with this argument. Population biologists have long recognised
that birds often have fewer viable offspring than they can. Wouldn't evolution by
natural selection select for the greatest number of offspring? One hypothesis is that
birds forgo having more offspring for the good of the group to avoid overshooting
the carrying capacity of the environment (Wynne-Edwards, 1962). Thus, individual
sacrifice was in the interest of the species. Clutch size evolved by group, or species, selection. Since 1947, the great tit (*Parus major*) has been studied in Wytham
Woods around Oxford, United Kingdom, initially by David Lack (1954). Most of
the breeding pairs have eight to nine offspring. However, if more eggs are added,
they can incubate them with success. Still, as the number of hatchlings in the brood
increases, their average weight decreases. This is due to the hatchlings receiving
less and lower-quality food (e.g., caterpillars). Heavier chicks have a greater probability of survival and reproduction. In experiments, it has been demonstrated that
the optimal clutch size is approximately eight to nine eggs. Lack and others argued
that individual selection explained clutch size; group selection is not required. Most
evolutionary biologists think that group selection can occur under certain restrictive
circumstances and that it has occurred in the history of life occasionally. However,
it is a general consensus that it occurs rarely (though see Sober and Wilson, 1999). If
this is correct, then species rarely exhibit teleological behavior of their own. At best,
any teleological behavior exhibited is a by-product of that of its constituent organisms. As evolutionary biologist George Williams (2008) pointed out, there is a big
difference between a *fleet herd* of deer and a herd of *fleet deer*. Thus, we should be
skeptical that species have objective final value.

A second way to understand final value is as subjective final value. Here is Ronald
Sandler again.

Something has *subjective final value* (hereafter, just *subjective value*) if its value is
dependent upon valuers having some evaluative stance regarding it. Subjective value
is created by valuers through their evaluative attitudes, judgments, and preferences. It
does not exist prior to or independent from them. There are a wide variety of things that
are valued noninstrumentally – for example, personal mementos, cultural and religious
artifacts, ceremonies and rituals, accomplishments, performances, and historical sites.

(Sandler, 2012, 19)

Put simply, something has subjective final value if it is valued for its own sake. It is surely true that there are species that we value for their own sake. However, it is also true that not every species is so valued. For example, only a fraction of the species that exist are known, and thus, it is highly unlikely that anyone values those species. Likewise, there are species that are known but that we in fact disvalue – think of parasites and pests. So, it is highly unlikely that all species have this sort of inherent value.

However, critics also contend that not every species has instrumental value. Instrumental value is the value something has as a means to some other valuable end. Consider the most famous illustration of this point – *Percina tanasi* or the snail darter (Plater, 2013). This is a species in the perch family which was discovered in East Tennessee in 1973 and was listed as endangered in 1975 under the U.S. ESA of 1973. Its listing led to the Supreme Court halting the completion of the Tellico Dam. This three-inch fish with camouflage dorsal patterns that feeds on insects in creeks and rivers doesn't seem to have much value aesthetically or commercially. But we should be careful not to overgeneralise. Species often have important instrumental values to human well-being (Daily et al., 1997; Tallis et al., 2011). They provide food, fuel, fiber, and medicine. The flowering plant rosy periwinkle (*Catharanthus roseus*) has been used for treating diseases including diabetes, malaria, and Hodgkin's lymphoma. The annual world fish catch is about 100 million metric tons, valued between $50 and $100 billion, and the commercial harvest of just freshwater fish in 1990 was 14 million tons, valued at $8.2 billion. We use about 7,000 plant species for food, but about 70,000 plant species are known to be edible. Of the top 150 prescription drugs used in the United States, 118 are based on natural sources. Pharmaceuticals in the developed world are valued at $40 billion per year.

As one more concrete example, let's consider the instrumental values of salmonid species. Salmon has enormous value for fishermen, processors, distributors, restaurants, suppliers, boat-builders, tour operators, fishing guides, and charter boat operators.[*] As of 1988, there were an estimated 62,750 salmon-dependent jobs in the Pacific Northwest, which generated about $1.25 billion to the regional economy. In the 1990s, the actual economic value of Columbia-based salmon fisheries dropped as low as $2 million. Salmon encourages recreation and tourism to the Pacific Northwest in the United States and Alaska. Additionally, they are important to sport fishing and angling. Salmon serve as a regional symbol and are found represented in art and souvenirs. They also serve as a flagship species for other species in the region. Salmon are incredibly important to Native American life and ceremonial rituals. Additionally, young salmon are a rich source of food for fish and birds, given their lipid content. Adults provide carbon, phosphorus, and nitrogen from the ocean to nutrient-poor lakes and streams. Their carcasses provide food for invertebrates like algae, fungi, and bacteria, and for vertebrates like bears, foxes, wolves, ravens, and eagles. It is unlikely that every species has instrumental value, but many do, and thus, increasing the number of endangered species would be detrimental.

[*] The information on salmon and their life history along with the ecosystem services they provide is taken from Gende et al. (2002); Quinn (2011); Trout (2001); Woody et al. (2003).

To summarise my arguments, I agree that not every species has final value, nor does every species have instrumental value. However, from the fact that not all species are of value, it does not follow that no PSC is of value. So, these claims, even if correct, do not show that taxonomic inflation is a problem.

4.4.3   PREMISE 3: WE SHOULD NOT MAKE CONSERVATION GOALS HARDER TO MEET

The third premise, 'We shouldn't do what is detrimental to conservation', seems uncontroversial, but in fact, it raises two important worries. First, there is a more basic legal point. In the majority opinion made by Chief Justice Burger on June 15, 1978, he wrote,

> One would be hard pressed to find a statutory provision whose terms were any plainer than those in 7 of the Endangered Species Act. Its very words affirmatively command all federal agencies 'to insure that actions authorized, funded, or carried out by them do not jeopardize the continued existence' of an endangered species or 'result in the destruction or modification of habitat of such species ...' 16 U.S.C. 1536 (1976 ed.). This language admits of no exceptions.

Justice Burger's point is that the ESA doesn't only protect those endangered species with final or instrumental value. It protects them, period. The ESA would be revised in 1978 so that a group of seven senior officials could exempt a federal agency if (a) the federal project is of regional or national significance, (b) there is no 'reasonable and prudent alternative', and (c) the project as proposed 'clearly outweighs the alternatives'.

Second, it is true that classifying more species makes the job of conservation biologists, policymakers, and environmentalists harder. But if many species are of great value, and in fact, there are more of them than we thought, then our job *should* be that much harder. As Agapow et al. (2004, 170) note, 'Rejecting the PSC solely because of (apparently) unpleasant biodiversity implications smacks of expedience'. Consider an analogy. At various times in history, legal rights were extended to marginalised groups. This increased the moral demands on those in their community and country. We could have made our moral lives easier by not extending those rights. This, however, would be the wrong response. Expediency often takes a second seat to morality (Hale, 2016).

4.5   TAXONOMIC INFLATION AND TRIAGE

Thus, in light of the arguments I have surveyed, there are three options. First, we deny that what the PSC classifies are genuine species. In light of species pluralism, this seems problematic, but maybe it is correct. Second, we recognise that conservation becomes more difficult, but if species should be protected, they should be. Prima facie, we should increase the amount of money the federal government spends on protecting endangered and threatened species. Third, we question whether every species should be protected and have the difficult conversations concerning which

species matter and why (Rolston, 1985; Russow, 1981; Sandler, 2012). In this political climate, this third option, though troubling, is what honest conservation requires.
I would argue that we forcefully pursue the second option, but I also recognise that
the third option is important, because triage is currently unavoidable.

Ecologist Leah Gerber (2016) has argued that we should employ decision theory
to determine how we should allocate federal money to endangered species. Currently,
less than 25% of the $1.21 billion per year needed for implementing recovery plans
for 1,125 species is actually allocated to recovery (Gerber, 2016, 3565). She modeled
what would happen if we moved funds from 'overfunded' to 'underfunded' species.
Reallocation of surplus funding from these 50 overfunded recovery efforts would erase
deficits in funding for up to 182 underfunded species (Gerber, 2016, 3566). As one
of Gerber's example, she considers the northern spotted owl (*Strix occidentalis cau-
rina*). She calculated that between 1989 and 2011, $4.5 million was spent to recover the
northern spotted owl, but it has been declining by approximately 4% per year. There
are two assumptions Gerber makes that are worth challenging. First, she assumes
that the federal budget for the ESA is constant. Second, the success of listing a species is solely a function of recovering that very species. The first claim is problematic
largely because the Trump Administration was looking to cut billions of dollars from
the Department of Interior,* and of course, the new Biden Administration might very
well increase it. Additionally, given the value of species, we should be advocating
for increasing federal spending on the ESA listings. Additionally, one can argue that
even when a listed species is not recovered, this listing was successful insofar as other
species were protected in its habitat. By protecting the northern spotted owl and the
old-growth temperate rainforest it lives in, we also protect many other species, like soil
arthropods, spiders, insects, mites, millipedes, lichen, fungi, mosses, small mammals,
and bats. Gerber defines success too narrowly, I think. Still, given the problem of triage, taxonomic inflation makes it even harder.†

4.6 CONCLUSION

In this chapter, I have considered species pluralism and the taxonomic inflation that
results. We also considered conceptual engineering and how many philosophers
think we should be improving our concepts when they are found problematic. Some
might argue that because this inflation has serious and negative implications for conservation, we should not inflate the number of species. It would be morally wrong,
one might argue. That is, we should dismiss the PSC for ethical reasons. I evaluated

---

* See www.reuters.com/article/us-endangered-species-triage-idUSKBN19A1DK
† One important recent approach to triaging species for their conservation value is the EDGE of
Existence of program run by the Zoological Society of London. The EDGE program is based on giving priority to those species that are 'Evolutionarily Distinct' and 'Globally Endangered' (Isaac et
al., 2007). It couples information regarding a species' evolutionary distinctness with its International
Union for Conservation of Nature (IUCN) threat category. The EDGE approach is especially interesting for philosophers, because it appears to assume that each species has some value, and species with a
higher EDGE score should be prioritised for conservation over those with lower EDGE scores. As we
have seen, not every species may be valuable. Additionally, there are interesting questions as to why
phylogenetic diversity matters when it does.

a variety of arguments for this conclusion and find them all to be flawed. However, taxonomic inflation will make triage even worse. Thus, we should be looking carefully not only at how to increase federal spending on endangered species but also how to protect those most valuable taxa when we must selectively choose.

REFERENCES

Agapow, P.-M., O. R. Bininda-Emonds, K. A. Crandall, J. L. Gittleman, G. M. Mace, J. C. Marshall, and A. Purvis. 2004. The impact of species concept on biodiversity studies. *The Quarterly Review of Biology 79*(2), 161–179.

Burgess, A., and D. Plunkett. 2013. Conceptual ethics I. *Philosophy Compass 8*(12), 1091–1101.

Burgess, B. B. 2003. *Fate of the Wild: The Endangered Species Act and the Future of Biodiversity.* University of Georgia Press.

Cappelen, H. 2018. *Fixing Language: An Essay on Conceptual Engineering.* Oxford University Press.

Carnap, R. 1950. *The Logical Foundations of Probability.* University of Chicago Press.

Coyne, J. A., and H. A. Orr. 2004. *Speciation.* Sinauer Associates, Inc.

Cracraft, J. 1983. Species concepts and speciation analysis. *Current Ornithology 1*(4), 159–187.

Czech, B. and P. R. Krausman. 2001. *The Endangered Species Act: History, Conservation Biology, and Public Policy.* JHU Press.

Daily, G. C., ed. 1997. *Nature's Services,*. Washington DC, Island Press.

Ehrlich, P., and P. Raven. 1969. Differentiation of populations. *Science (New York, NY) 165*(3899), 1228–1232.

Ereshefsky, M. 1992a. Eliminative pluralism. *Philosophy of Science 59*(4), 671–690.

Ereshefsky, M. 1992b. *The Units of Evolution: Essays on the Nature of Species.* MIT Press.

Erwin, T. L. 1982. Tropical forests: Their richness in coleoptera and other arthropod species. *The Coleopterists Bulletin 36*(1), 74–75.

Gende, S. M., R. T. Edwards, M. F. Willson, and M. S. Wipfli. 2002. Pacific salmon in aquatic and terrestrial ecosystems pacific salmon subsidize freshwater and terrestrial ecosystems through several pathways, which generates unique management and conservation issues but also provides valuable research opportunities. *BioScience 52*(10), 917–928.

Gerber, L. R. 2016. Conservation triage or injurious neglect in endangered species recovery. *Proceedings of the National Academy of Sciences 113*(13), 3563–3566.

Hale, B. 2016. *The Wild and the Wicked: On Nature and Human Nature.* MIT Press.

Haslanger, S. 2000. Gender and race: (What) are they? (what) do we want them to be? *Noûs 34*(1), 31–55.

Hull, D. 1999. On the plurality of species: Questioning the party line. In *Species: New Interdisciplinary Essays,* edited by R. A. Wilson. pp. 307–315. MIT Press.

Isaac, N. J., J. Mallet, and G. M. Mace. 2004. Taxonomic inflation: Its influence on macroecology and conservation. *Trends in Ecology & Evolution 19*(9), 464–469.

Isaac, N. J., S. T. Turvey, B. Collen, C. Waterman, and J. E. Baillie. 2007. Mammals on the edge: Conservation priorities based on threat and phylogeny. *PloS One 2*(3), e296.

Lack, D. 1954. *The Natural Regulation of Animal Numbers.* Oxford University Press.

May, R. M., John H. Lawton, and Nigel E. Stork. 1995. Assessing extinction rates. In R. May and J. Lawton (Eds.), *Extinction Rates, Chapter 1,* pp. 1–24. Oxford University Press.

May, R. M. 1990. How many species? *Philosophical Transactions of the Royal Society B 330*(1257), 293–304.

May, R. M. 2011. Why worry about how many species and their loss? *PLoS Biology* 9(8), e1001130.

Mayr, E. 1963. *Animal Species and Evolution*. Belknap Press.

Mishler, B. D. 1999. Getting rid of species. In *Species: New Interdisciplinary Essays*, edited by R. A. Wilson, pp. 23–48. MIT Press.

Mora, C., D. P. Tittensor, S. Adl, A. G. Simpson, and B. Worm. 2011. How many species are there on earth and in the ocean? *PLoS Biology* 9(8), e1001127.

Nabhan, G. P., and S. L. Buchmann. 1997. Services provided by pollinators. In G. C. Daily et al. (Eds.), *Nature's Services*, pp. 133–150. Island Press.

Plater, Z. J. 2013. *The Snail Darter and the Dam: How Pork-Barrel Politics Endangered a Little Fish and Killed a River*. Yale University Press.

Putnam, H. 1975. The meaning of 'meaning'. *Minnesota Studies in the Philosophy of Science* 7, 131–193.

Quinn, T. P. 2011. *The Behavior and Ecology of Pacific Salmon and Trout*. UBC Press.

Rohlf, D. J. 1989. *The Endangered Species Act: A Guide to Its Protections and Implementation*. Stanford Environmental Law Society.

Rolston, H. 1985. Duties to endangered species. *BioScience* 35(11), 718–726.

Russow, L.-M. 1981. Why do species matter? *Environmental Ethics* 3(2), 101–112.

Sandler, R. L. 2012. *The Ethics of Species: An Introduction*. Cambridge University Press.

Slater, M. 2013. *Are Species Real?: An Essay on the Metaphysics of Species*. Springer.

Sober, E., and D. S. Wilson. 1999. *Unto Others: The Evolution and Psychology of Unselfish Behavior*. Harvard University Press.

Sokal, R. R., and T. J. Crovello. 1970. The biological species concept: A critical evaluation. *The American Naturalist* 104(936), 127–153.

Staley, J. T. 2006. The bacterial species dilemma and the genomic–phylogenetic species concept. *Philosophical Transactions of the Royal Society B: Biological Sciences* 361(1475), 1899–1909.

Sterelny, K. 1983. Natural kind terms. *Pacific philosophical quarterly* 64(2), 110–125.

Stork, N. E. 1993. How many species are there? *Biodiversity and Conservation* 2, 215–232.

Tallis, H., T. H. Ricketts, G. C. Daily, and S. Polasky. 2011. *Natural Capital: Theory and Practice of Mapping Ecosystem Services*. Oxford University Press.

Trout, O. 2001. *Oregon Salmon: Essays on the State of the Fish at the Turn of the Millennium*. Oregon Trout.

Van Valen, L. 1976. Ecological species, multispecies, and oaks. *Taxon*, 25(2/3), 233–239.

Wheeler, Q. D., and R. Meier. 2000. *Species Concepts and Phylogenetic Theory: A Debate*. Columbia University Press.

Wiley, E. O. 1978. The evolutionary species concept reconsidered. *Systematic Biology* 27(1), 17–26.

Wilkins, J. S. 2009. *Species: A History of the Idea*. University of California Press.

Williams, G. C. 2008. *Adaptation and Natural Selection: A Critique of Some Current Evolutionary Thought*. Princeton University Press.

Woody, E., E. C. Wolf, and S. Zuckerman. 2003. *Salmon Nation: People, Fish, and Our Common Home*. Oregon State University Press.

Wynne-Edwards, V. C. 1962. *Animal Dispersion in Relation to Social Behaviour*. Oliver & Boyd.

Zachos, F. E. 2013. Taxonomy: Species splitting puts conservation at risk. *Nature* 494(7435), 35–35.

5 The Good Species

John S. Wilkins

CONTENTS

5.1 INTRODUCTION

> Nor shall I here discuss the various definitions which have been given of the term species. No one definition has satisfied all naturalists; yet every naturalist knows vaguely what he means when he speaks of a species.

> **(Darwin 1859, p. 37)**

In recent years, there have been attacks on the notion of there being an objective rank (or category, as Mayr called it) that all species, and only species, occupy in the living world. It is sometimes called a 'level of organization' or a 'natural kind'.* As we know from extensive work by historians and scientists themselves, while the ranks of species and genus were formally instituted in taxonomy by Linnaeus in the 10th edition of the *Systema Naturae*, they were in use in various similar ways since at least the publication of Caspar Bauhin's Πιναξ *[PINAX] theatri botanici* in 1651 (Cain 1994). Moreover, although *species* was simply defined by Ray in 1686 as 'signs or indications of their specific distinction ... distinct propagation from seed', it was not defined further, nor were there universally accepted criteria for identifying and diagnosing species, Linnaeus notwithstanding.†

---

* Biologists sometimes refer to species as an objective level of organisation in the biological world. Philosophers often refer to species as 'natural kinds', following Mill's *System of Logic* (1843). A natural kind in Mill's sense is something that has shared unique sets of properties. A taxonomic rank like species, or sometimes genus, is a natural kind if and only if it is specified by a shared set of properties, usually called an 'essence' (Wilkins 2013). There are still advocates for this, or related, views (Devitt 2018, LaPorte 2018).
† Ray 1686; trans. Lazenby 1995. Ray does expressly reject the number of flowers and the size of plants as specific *discriminata*.

DOI: 10.1201/9780367855604-6

The question of what a species is has been answered in various ways, but one way is to deny that species are anything singular and state instead that a pluralism of species 'concepts'* is required, leading to issues regarding the choice of metrics, the uses of these conceptions in conservation biology, horticulture, and agriculture, and so forth. Brent Mishler, for instance, has long championed the idea that there is no species concept and that if we were being true to evolution and the consequent phylogenetic approach to taxa, we should replace it with a 'smallest clade' idea.† Others have argued that this would cause massive taxonomic inflation. In this chapter, I am going to suggest a (practice-based) solution or rather, a description of how biologists come by the notions they employ. Famously, C. Tate Regan quipped that a species is whatever a suitably qualified biologist chooses to call a species, which has been called the 'cynical species concept'.‡ I will argue that it is not cynical at all. I shall rely upon several claims:

1. That a taxonomist is a 'prepared mind' when describing and delimiting a species;
2. That taxonomists have a prototypical notion of a 'good species' for the group§ of organisms they study;
3. That this 'good species' notion is subject to revision by empirical experience as well as professional conventions; and
4. That a good species is defined by the scale of species criteria in that group.

Scale is something most often applied to ecological systems in biology as well as to microbial and microscopic processes and entities. It is not usually applied to species. Instead, taxonomists, and biologists in general, usually speak of the 'level' of species, or of species themselves being instances of a 'level of organisation'. This smacks of the Great Chain of Being, wherein taxa are assigned to a ladder of increasing virtue (complexity, intelligence, and so on), but rank has fallen into great disrepair in the last few generations as concepts like *family*, *phylum*, and *genus* are seen to be arbitrary and non-natural. The one level that holds out is *species*. The stubbornness of this calls for an explanation, and it may in fact teach us about how biology classifies the world it encounters.

---

* I stand firm: there is only one species *concept*, but many species *conceptions* (or definitions, or specifications, etc.). Biology has many specific applications of the term *species* (see Wilkins 2011), but the term and concept are what is being applied, defined or modified. This is analogous to Mayr's category–taxon distinction. Each conception of species, such as the General Lineage Concept or the Mate Recognition Concept, is an application of the concept *species*.

† In which I concur (Mishler and Wilkins 2018).

‡ The details are to be found in my 2018, p. 261. Philip Kitcher coined the term 'cynical species concept' (Kitcher 1984).

§ In this chapter, when I refer to a group, I mean the close relatives of the specimens being observed, such as the group Psittaciformes (parrots) when dealing with a specific possible species (such as a newly observed lorikeet). Groups may be named in the Linnaean schema, or they may be a clade, or both.

5.2   PROTOTYPES AND 'GOOD SPECIES'

> I swallowed [my master's] incidental teachings and gradually got to know the 'good' and 'bad' species of my homeland through a kind of tradition, but I frankly admit that even then I could not be quite clear how to accept a given case without tradition and recognise whether one was dealing with a 'good' or a 'bad' species.
>
> **(Kerner 1866, p. 5)**

Despite a long history of naturalists and later biologists calling this or that species a 'good species', the importance of this conception was not picked up in the philosophical literature, as far as I can tell, until Yuichi Amitani proposed his prototypical notion of good species (2015). What did naturalists mean by it? As Amitani presents it, naturalists take some unproblematic species as a prototype of what a species was meant to be [implicitly: in that group], and they apply it until it breaks down in ambiguity or impracticality. For instance, in Notebook B, Darwin, in 1837 or 1838, wrote:

> A. B. C. D. (A) crossing with (B), and (B) being crossed with (C) prevents offspring of A becoming a **good species** well adapted to locality, but it is instead a stunted & diseased form of plant, adapted to A. B. C. D. Destroy plants B. C. D. & A will soon form **good species**!*

In the *Origin* itself, he said:

> When the views advanced by me in this volume, and by Mr. Wallace, or when analogous views on the origin of species are generally admitted, we can dimly foresee that there will be a considerable revolution in natural history. Systematists will be able to pursue their labours as at present; but they will not be incessantly haunted by the shadowy doubt whether this or that form be a **true species**. This, I feel sure and I speak after experience, will be no slight relief. The endless disputes whether or not some fifty species of British brambles are **good species** will cease. Systematists will have only to decide (not that this will be easy) whether any form be sufficiently constant and distinct from other forms, to be capable of definition; and if definable, whether the differences be sufficiently important to deserve a specific name.
>
> **(1859, p. 425f, emphases added)**

The phrase 'good species' in English seems to have been adopted widely in botany in the first decade of the nineteenth century,† and it is no coincidence that Darwin uses

---

* Page 211, emphasis added. From Darwin Online, at http://darwin-online.org.uk/content/frameset ?itemID=CUL-DAR121.-&pageseq=38&viewtype=side, accessed December 22, 2020.
† The term *'gute Arten'* in German seems to have come via the English phrase rather than vice versa. See the Müller-Wille and Hall (2016) translation of Mendel's *Versuche* (1866), commentary page 39. This has now been published (Mendel 2020). Austrian botanist Anton Kerner von Marilaun, whom they cite, published a series of 10 articles under the title *'Gute und schlechte Arten'*, later published as a booklet (Kerner 1866). The term was used in the 1820s in German botany. I could not find the term in French before the English, except simply to denote a species that is good to graft or eat or otherwise use in horticulture. Many thanks to Staffan Müller-Wille for his assistance with tracking this down.

it in a botanical context. It seems to have been required as systematists adopted the Linnaean 'natural system'.* That such a concept was needed so soon after adopting a ranked system (by 1840) indicates the fact that naturalists were aware of the variation within species and genera quite early on, and the consequent vagueness of many species. As Amitani notes:

> the concept of *good species*, a prototype of *species*, mediates the non-definitional, or perhaps implicit mode of understanding of species. A good species is a taxon judged to be a species according to more than one species criterion—such as reproductive isolation and phylogenetic properties—or a taxon judged generally to be a species by competent biologists, whether the phrase merely expresses one's epistemic confidence in the taxonomic judgment or has even an ontological implication that the species category is divided into good and not-so-good species.

Amitani notes that for some, a good species can be whatever all systematists would accept is a species, irrespective of the definitions used. There are undoubtedly some of these, but it would not be the rule, I think. Many systematists have historically treated good species as a morphological cluster (e.g., Rensch 1959, pp. 24–28, introducing his notion of *Rassenkreise*) or a monophyletic group (see Mishler and Wilkins 2018), and these are not universally accepted conceptions even within a subdiscipline (e.g., lichenology; see Kendig, this volume) or within a single group. However, Amitani's point about a good species being one that is judged by more than one species *criterion* is a good point to make. Species do not have a singular criterion for being assigned *specieshood* (as Amitani – and Pavlinov, this volume – calls it), despite the mistaken assumption that single key diagnoses were required by the Linnaean scheme.† Instead, they always have been, and increasingly are, delimited by a large number of criteria, including (but not restricted to) molecular, structural, adaptive, developmental, and behavioral characters.‡

Any use of single key attributes (for example, DNA barcoding) is either a methodological necessity made an epistemic virtue due to a lack of available data, or it must follow on from the identification of prototypical good species in that group (Whewell's 'type species' for a genus). For without a type species or good species prototype, how can we know what the single key is identifying? If one finds that by a singular metric, one has *n* OTUs,§ can one say that one has *n* species? First, we'd need to know that the singular metric identifies good species before we could be

---

* William Kirby, in his *Monographia Apum Angliae* of 1802, several times refers to 'good genera' as well. Whewell later described a genus as being based around a 'type' species, which is an exemplary species within the genus (Whewell 1840, vol 2, pp. 517–519); see Wilkins (2013), p. 225.

† The Linnaean scheme had single key 'essential characters' but only to differentiate the type specimens from other related species. They were identification keys, not delimitation characters.

‡ Thomas Reydon notes in review that this is connected to the relatively recent shift towards 'integrative taxonomy', which makes use of more than a few types of evidence (Padial et al. 2010, Will et al. 2005, Will and Rubinoff 2004). However, Fujita and Leaché (2010, p. 2) note: 'Many good species will not exhibit fixed morphological differences or fixed DNA differences across multiple loci. An integrative taxonomy should not force systematists to search for diagnostic attributes that may have no relation to the data or process used for species delimitation.'

§ Operational taxonomic units; see Sneath and Sokal (1973).

confident in using it for as yet unknown species. Once we acquire that confidence, then it may be useful to use that criterion, such as a COI sequence in mitochondria or an *rbcL* gene from chloroplasts, or an sRNA sequence in fungi and single-celled organisms. This then implies that we know ahead of time what a good species is for each group and subgroup, and so on.[*]

So, the question at issue is how we know what the prototypical form of species-hood is in any given group, in order to test whether a particular identifier technique or data type reliably identifies species. And in turn, this raises the question of how biologists construct their prototypes of specieshood.

5.3   THE MAKING OF A GOOD SPECIES

> The limits of the majority of species are so undefinable that few naturalists are agreed upon them; to a great extent they are matters of opinion, even amongst those persons who believe that species are original and immutable creations; and as our knowledge of the forms and allies of each increases, so do these differences of opinion; the progress of systematic science being, in short, obviously unfavourable to the view that most species are limitable by descriptions or characters, unless large allowances are made for variation.

> **(Hooker 1859, p. iii)**

Biologists in the field encounter not species but organisms. Sometimes, organisms are difficult enough to delimit on their own, as in clonal stands, biofilms, obligate endosymbionts, superorganisms like eusocial hives, and so on.[†] Most of the time, however, a biologist knows more or less what an organism of the group they study is. The term 'organism', which supplanted the older 'organised beings' in the early nine-teenth century, connotes a system that maintains itself in its environment (Cheung 2006) and is the benchmark unit in most fields (not so much in clonal or colonial species). Most organisms have something like a developmental system, what used to be called a 'life cycle', and the development can be often observed more or less empirically without much in the way of ancillary theoretical assumptions, as the existence of agriculture and horticulture demonstrates. So, how do biologists get from organisms to species?

---

[*] An excellent example of this occurs frequently when barcoding insects – for example, ichneumon or braconid wasps (Sharkey et al. 2021) or pygmy grasshoppers (Lehmann et al. 2017). The numbers of possible species for the former groups are massive, and so, in order to avoid what they call the 'taxonomic impediment', they do barcoding for 'accelerated taxonomy'. However, in order to calibrate the sensitivity of COI differences used for species delimitation and subsequent description, they had to test it out first on known 'valid species'. Lehman et al. specifically note that alpha taxonomy is still required, and Krell (2004) call the resulting taxa 'parataxonomic units' (PUs), not yet species. They are effectively OTUs or *phenons* (Sokal and Rohlf 1962). My thanks to Igor Pavlinov for pointing these references out to me and suggesting that this may be a new case of good species criteria. I think that barcoding does not *supplant* good species, and in the absence of knowing what counts as ichneumon or braconid species, barcoding would be largely uninformative and probably uninterpretable.

[†] See, for instance, Queller and Strassmann (2016).

The first and most obvious way is that biologists get their ideas from their training and education. That is, *species* are cultural transmits. But, this merely pushes the problem a bit further back, leading to a chain of transmission that is indefinite. Moreover, while it is now agreed that science is a social process, that is not, nor could it be, the only thing it is. Social construction of taxa is not the whole story. So, let us consider a simplified sequence of species construction, beginning in the premodern period. All human cultures identify types of living organisms. Some are quite exact and commensurate with modern scientific types, while others are vague and generalised. I have previously suggested that precision in taxonomising the living world depends a lot upon how economically important that knowledge is for the taxonomisers (Wilkins 2018a, p. 352f). For systematists, that importance is obvious (their living consists in being precise), while for hunter-forager societies, success in classification is literally their meal ticket. People inveterately wrong in their classifications have a tendency not to flourish. So initially, taxonomy begins with folk taxonomies, and these are not infallible, merely good enough. For instance, the cassowary was classified as a human for totemic reasons among the Karam people in Papua New Guinea.* But, they would certainly have known a lot about their prey species and food plants.

In my historical review of the species concept, I noted that the idea of a concept for living things evolved initially out of folk botany, in the form of herbals, and folk zoology, in the form of bestiaries, but that is not the whole story. There are horticultural reasons for finding natural units among potentially commercially significant plants such as ginger, pepper, and so on,† and there are, of course, breeding reasons for commercially significant animals such as horses, cattle, and sheep. Some noted that sheep and goats, horses and donkeys, and so on could interbreed with varying degrees of success and fertility of the progeny, but in the main, breeders knew to interbreed within species. As late as the nineteenth century, Darwin himself was still relying upon breeders of pigeons, horses, and other livestock for information on species.

As naturalism developed into a scientific endeavor, folk taxonomies gave way to expert taxonomies. As Joachim Jung (1587–1657) said, botanists prefer finding novel stocks to making dry formal systems (paraphrasing his *De Plantis Doxoscopiae* 1.1.5),‡ and he was one of the first to make a close study of the plants themselves rather than only using prior works to assemble his classifications.§ He was, of course,

---

* Bulmer (1967). See also Berlin (1973), Atran (1998), and Media and Atran (1999) for further context, and Kendig (2020) for a neat case study among the Samí and other boreal cultures.
† 'Spice' in English, and cognate terms in French (*espice*), Portuguese, and Spanish (*la especia*), Italian (*spezia*), etc., are in fact derived from the Latin *species* (kinds) beginning in the fourteenth century. See the etymology for the entry 'Spice' in the *Oxford English Dictionary*, which has also an obsolete usage (sense 3) in which 'spice' still meant 'kind' until the seventeenth century.
‡ Jung (1747), p. 69. Igor Pavlinov introduced me to this passage, and classicist Tim Parkin helped me interpret Jung's Latin.
§ Jung was preceded by the work of Conrad Gesner (1516–1565), who unfortunately died before finishing his *Historia Plantae*, and so his work remained unpublished until the 1750s. Gesner's lovely watercolors can be seen in the library at the University of Erlangen–Nuremberg. I am indebted to Thony Christie for showing me these *objets d'art*.

followed by Ray, and then Linnaeus, and all the taxonomists who had thereafter taken up the mantle of botanist. Each of them studied large numbers of specimens (the taxonomic equivalent of an organism), described them in the context of the group they were part of, and named them, leading to Linnaeus' genus–species epithet nomenclature and the broader system of the *Systema Naturae*. By the nineteenth century, botanists began to specialise either by flora (such as Hooker's *Florae*) or by group (order, class, or family).* The professionalisation of botany and zoology that began in the late nineteenth century brought with it the formation of subdisciplines like the study of diatoms, flowering plants, fungi, onychophora, bacteria, and so on. Each subdiscipline had its own ways of identifying, describing, and diagnosing species, dependent upon the properties of the group members themselves.

5.4   THE CRITERIA FOR IDENTIFYING GOOD SPECIES

So, the development of a notion for each field of study that diagnosed species was in part dependent upon folk taxonomies, then a tightening of methodologies, specialisation, and novel techniques. For instance, the taxonomy of fungi (mycology) depended a lot on the morphology of the fruiting bodies and the ecological niche in which they were found until molecular techniques came along. Fields like lichenology had to deal with two or more commensal organisms from distinct domains of the tree of life, and hence, the notion of a lichen 'species' is fraught when using criteria that were developed to deal with organisms that reproduce sexually and have a single zygote (Kendig, this volume).

Each specialty in biology – zoology, botany, mycology, bacteriology, and virology – and indeed, each evolutionary group within those specialties has its own shared set of criteria for identifying a good species. These do not always apply, however, and when they do not, then specialists refer to 'cryptic species', 'species complexes', 'micro species', 'ring species', 'quasispecies', and when all else fails, 'subspecies', 'varieties', and so on. But, there would be an exemplary form of species in that group. Of course, many groups do not match the exemplar even in a minority of cases. In many animals, parthenogenic species can form within a sexual clade, secondarily losing sexuality, or some aspects of it, via cross-species hybridisation and/or polyploidy. One case – the whiptail lizard *Cnemidophora* (now *Aspidoscelis*) *lemniscatus* – 'may actually represent a complex of independently formed hybrids, further complicating the situation in this "species"' (Gregory and Mable 2005). Other examples include snakes, other lizards, stick insects, wasps, cockroaches, salamanders, snails, fishes, nematodes, and rotifers, to mention a few. No naturally occurring parthenogenic bird or mammal species appears to be known, although parthenogenic eggs have been laid by turkeys in farms.

So, in the case of the whiptail lizard, for instance, how do herpetologists identify an asexual group as a species, when lizards are usually identified as 'biological species' *sensu* Mayr? It obviously cannot be reproductive isolation, since every individual in the asexual 'species' ('agamospecies' or 'paraspecies') is reproductively

---

* Kleinman (2018).

isolated, and nobody would take them all to be individually distinct species. Also, they are typically clones of each other, so they form a 'clonal species'.* The answer must lie in the assay criteria used to group them. In the main, these differentiating or identifying criteria will be morphological, ontogenetic, and molecular. Each forms a slightly different grouping (since even clones individuate during development and behavior), and so, something like a clustering notion must be used. The thresholds used will vary according to what is 'usual' or 'reasonable' in the case of the other species in the clade. The competency in Amitani's quote does indeed lie with the systematists. For example, in single-celled organisms like lobose amoebae (Smirnov 2009), until the development of molecular methods, only morphology was available, which led to the loss of identification criteria, as initial taxonomy was highly individualised. Smirnov notes:

> Naked lobose amoebae are among the most difficult protists to differentiate. Because they are believed to be agamous (clonal) organisms, the biological species concept, which involves defining species based on their reproductive isolation, is not applicable. The general consensus is that for such taxa the morphospecies concept is the only one practically available. However, analysis of the morphological differences between amoeboid protists is rather difficult, and conclusions are often unreliable, especially for closely related species. This is partly because the shape of an amoeba is dynamic; in stained preparations after fixation and dehydration specimens are often no longer representative. So, there is no way to preserve a type specimen of an amoeba – a holotype, so important in traditional biological systematics. Many amoebae species are culturable, and therefore type strains can be deposited in culture collections. However, this practice became widely used only after the 1960s and there are still many examples where strains deposited with the culture collections were lost. So, until the advent of microphotography, the only tools to document amoebae species were line drawings and text descriptions, both of which tended to be rather author-specific. For example, despite careful descriptions provided by E. Penard in his fundamental monograph published in 1902, and a large number of stained preparations left by him, many of his 'species' are now unrecognizable.

Smirnov 2009, p500

This case represents all the issues of identifying species over time. The assay techniques – microphotography, molecular analysis, and electron micrography – permitted finer (more precise and accurate) and more repeatable taxonomic discrimination. One might almost say that the species are made by way of these techniques, if not for the obvious fact that identifying something is epiphenomenal to the real things so identified. In recent years, for example, the use of 'barcoding' techniques has teetered perilously on the edge of making the species whatever it is that has a unique COI gene (or whatever else is used in plants, etc.) rather than the species being putatively identified by the unique marker. In short, the sign is not the thing signified, as

---

* Like the 'genets' of other species such as the North American Quaking Aspen (*Populus tremuloides*). See Mitton and Grant (1996), p. 27 and Kendig this volume.

any logician knows.* Identifying a species based on a single criterion relies on a prior correlation between the criterion and known species.

That said, it is important to note that there are several scales involved in identifying species from early modern times to now. Morphology, of microscopic organisms in particular, involves being able to observe and record the forms of things at a *mesoscale*: that is, within the region of size where light waves are able to identify structure (even under a microscope). There is also a *macroscale*: the broader environment and distribution of the organisms. There is also, of course, the *microscale*: the size at which structures are no longer visible using ordinary light and must be assayed using electron micrography (EM) or DNA–DNA hybridisation and so on.† This scale is obviously one of size. The criteria used to identify species have become increasingly smaller.

But, scale applies to more than size. It also applies to functionality. Reproductive isolation is obviously one such. It occurs at the organism level and both above and below it. Reproductive isolation requires functional behavior at the genetic (very small), cytological (small), anatomical (medium), and organism (the benchmark level for that species) levels. Species delimitation also involves ecological, social, and geographical and meteorological scale criteria (*X s* can live in arctic or tropical conditions, for example). Molecular techniques rely upon the persistence, spread, and functional role (or lack, in the case of pseudogenes) of the assayed sequences and structures. The morphology of molecules is only significant for diagnosing species if it happens to be aligned with species identity. Subspecific haplotype groupings, for example, are not significant for species identification, although they are for population and subspecific structural identification. So, for a molecule (a 'speciation gene' as well as a 'barcode') to act as a *discriminatum* for a particular species, it must have some 'inutility metric' that underpins the inference (as characters used in taxonomy are very often nonfunctional or nonadaptive, to avoid convergences).

Finally, scale applies also to time: the duration of a species at either geological scales (*macro*), population scales (*meso*),‡ or short-term lineages (*micro*) is relevant. As Darwin noted, varieties are more temporary; species are more permanent. Here again, reproductive isolation is a good case. Reproductive isolation requires microscale proximity and activity (I cannot reproduce with Anne of Cleves, for example, for very different reasons than Henry VIII). Whether or not a species is reproductively distinct over geological time is merely speculative, but there are cases of persistent reproductive reach between species isolated for millions of years, such as the snapping shrimp and sea urchins isolated by the isthmus of Panama for around

---

* Something not always recognized by taxonomists, who sometimes tend to be wedded to the classifications of tradition and treat novel taxonomic techniques and results as *doing* something to their organisms (not generally, but I have seen it). Occasionally, it is overlooked by philosophers too.

† EM is used for very small organisms that are below the resolution of light microscopy. DNA–DNA hybridisation was used, for example, by Sibley and Ahlquist to measure genetic similarity, which at the time was not otherwise observable (1983).

‡ Igor Pavlinov has drawn my attention to *mesospecies* in Dillon (1966), which he defines as 'The second stage [of a species' life cycle], or mesospecies, has stable range boundaries, a relatively great abundance, and extensive subspeciation'. I do not mean the same thing here.

three million years.* Moreover, many species will hybridise with other species, allowing the introgression of genes, as displayed in the Neanderthal and Denisovan genes in non-African migratory populations of humans (Stringer 2012).

Each of these scalars effectively defines an abstract phase space, which will vary in its particular components for each species grouping. In keeping with the points made earlier, what these spaces encompass depends on the learned experience not only of the singular taxonomist but of all those who work on the groups being taxonomised. In short, the subdiscipline will always have an idea of 'good species' for that group, which may be revised to be more or less inclusive as new data and re-analyses come in ('lumping and splitting') or simply because a group has few specialists, and the personal preferences of one or a few systematists skewed the approach taken.

This is what one might call an 'operative concept' in that the driving forces are the use of operative data (unlike, say, the 'biological' species concept, which would require large-scale breeding experiments to employ it, as Ehrlich noted in 1961), only it is not a concept but a practice.†

5.5   PHENOMENA AND EXPLANATIONS

Theoretical concepts of kinds and other generalisations in science are seen by many to be metaphysical universals that cover the domain of phenomena needing explanations. In short, if something is a species, then it has a certain number of ideal properties as specified by the theory. This is sometimes known as the *Ramseyfication of theories* after Frank Ramsey's formalisation method for theories (Melia and Saatsi 2006, Wilkins 2018b). In short, the idea is that a theory is reduced to logical sentences with variables that are quantifiable. The variables then denote the kinds or classes of objects in the theory (Quine sloganised this as 'to be is to be the value of a bound variable', 1948). I am arguing that instead, some general terms in science form out of practical considerations and specialist experience, with a strong admixture of sociopolitical interests (see my forthcoming chapter). As it happens, the initial motivation for there even *being* a natural concept of species was intimately connected with a more literal interpretation of the Noah's Ark story in the Bible (*Genesis* [*Bereshit*] chapters 6 to 9; see my 2018a, 56–62). In order to fit all living kinds that breathed within the specifications of the Ark (*Gen.* 6:15–16), it was necessary to give 'kind' (*min*, מין) a more precise definition (many later-named species were therefore considered local variants of the created kind). When John Ray gave his famous 'definition', he was also doing so to meet the needs of exploration and colonisation, and so on.

---

* Genetic incompatibility does rise as separation persists, but there is still a shared developmental system inherited from the original populations. See Knowlton et al. (1993), Lessios and Cunningham (1990), Ma et al. (1978), Marko 2002, Ma et al. (1978), Schultz et al. (2000).
† This is what Kendig has called *kinding*, the process of making kinds via the practices of science (Kendig 2016). Brigandt (2003, p. 1309) called species 'investigative kinds': 'An investigative kind is a group of things that are presumed to belong together due to some underlying mechanism or a structural property'.

But if species are not quantified variables in a Ramsey Sentence of biological theory, does this mean they are not something that can be treated as scientific? Surely, twenty-first-century biologists are unaffected by fifteenth-century theologians? Of course, they (in the main) are – biologists name species because their forebears named species, using operational criteria their forebears developed, not because of Athanasius Kircher's writings on the Ark. But, the *need* for a rank or theoretical and metaphysical conception is not based entirely upon scientific tradition or experience. Instead, it really is a holdover of older religious and philosophical goals and intentions (see Wilkins forthcoming), which has been bequeathed to us via Mill and Venn as 'natural kinds'.

As I have argued previously,* species are best thought of not as concepts that explain (*explanantia*) the phenomena, but as phenomena that stand in need of explanation themselves (*explicanda*).† The question then arises – what makes a phenomenon? This takes us back, apparently, to the problem of the criteria of specieshood.‡ In order to know whether the differentiating criteria used are sufficient to identify good species, we need to know what good species are, and in order to know what good species are, we need reliable differentiating criteria. However, in order to know a phenomenon, does it also follow that we need to know what makes something a phenomenon in the first instance? Consider Newton's list of phenomena at the end of the second edition (1713) of the *Principia*:

Phenomenon 1: The circumjovial planets, by radii drawn to the centre of Jupiter, describe areas proportional to the times, and their periodic times – the fixed stars being at rest – are as the 3/2 powers of their distances from that centre.

Phenomenon 2: The circumsaturnian planets, by radii drawn to the centre of Saturn, describe areas proportional to the times, and their periodic times – the fixed stars being at rest – are as the 3/2 powers of their distances from that centre.

Phenomenon 3: The orbits of the five primary planets – Mercury, Venus, Mars, Jupiter, and Saturn – encircle the sun.

---

* Wilkins (2018a chapter 14, 2018b).
† Ingo Brigandt (2003) was first to argue that species are phenomenal objects in need of explanation, but Kim Sterelny (1999) had earlier mentioned what he called 'phenomenological species', which he defined as 'recognizable, reidentifiable clusters of organisms' [(p. 119). Sterelny does not think these are sufficient for the acceptance of groups as species, as he instead argues for an evolutionary conception, a lineage undergoing changes in ecological conditions. He says that

it has to be shown, not assumed, that phenomenological species constitute a biological kind. Minimalists about species do not expect theoretical biology to vindicate the species category: they think it is merely a phenomenological kind. We can clump organisms into species, but the biology explaining the clumping pattern is so diverse that *species* does not name a biological kind.

I agree with this but do not feel the need for a universal conception of *species* as Sterelny does.
‡ This is a form of *diallelus*, in which the problem of choosing a criterion for identifying something itself needs a criterion, ad infinitum; see Chisholm (1982).

Phenomenon 4: The periodic times of the five primary planets and of either the sun about the earth or the earth about the sun – the fixed stars being at rest – are as the 3/2 powers of their mean distances from the sun.

Phenomenon 5: The primary planets, by radii drawn to the earth, describe areas in no way proportional to the times but, by radii drawn to the sun, traverse areas proportional to the times.

Phenomenon 6: The moon, by a radius drawn to the centre of the earth, describes areas proportional to the times.[*]

These are not the phenomena of Bogen and Woodward (1988, see Bogen 2011), who specify that phenomena are 'patterns in data'. There *are* patterns implicit in Newton's list (the 3/2 powers identified by Kepler, for example), but there are many patterns out there to be had, and so, the selection rests with the state of the debate and the questions of how to explain some patterns and not others (for instance, with Newton, Ptolemaic patterns).[†] A phenomenon is a pattern that is *of interest* or in the view of the scientific field of experts, needs explanation. Species are exactly that. This is an epistemic notion of *species*, but not one divorced from the biological world.[‡]

The acquisition of these criteria for matters of interest is a matter of professional and epistemic development: science students are given conceptual tools along with the empirical tools and protocols during their initial education, becoming more detailed as they progress from laity to expert. This means that the majority of specialists in the natural sciences will acquire scattershot instruction, mostly via mimesis, and thus, will have at best partial understanding of the major conceptions utilised in their discipline outside the fields or topics in which they specialise. In short, they cannot say all they know (for much of it is mimetic), and they cannot know all of their field. The construction of phenomena will therefore be a process of proposal and criticism or rejection by peers and instructors, as Kerner in the epigram earlier described.

The role that good species play, as opposed to bad species, lies in the adequacy of the phenomenon to require explanation. Prototypical species will not need much explanation beyond their individual properties; but bad species will require much more investigation and explanation than good species, until either they have been reconciled with the prototypical notions, or they refine the criteria used by the field (or a new prototype is chosen).

This approach to species has a number of virtues, in my opinion: for a start, it makes sense of species pluralism – each species is the result of historical contingency in the mechanisms that isolate it from other species (as understood by those working in that field or group); there may be underlying genes or developmental systems for speciation in that group, but there is no reason to think there will be universally shared such mechanisms across all groups, even when restricted to, say,

---

[*] I am indebted to Kirsten Walsh for bringing this passage to my attention.
[†] There is also observational data listed in the full appendix.
[‡] See Barker, this volume, for a critique of this approach.

animals or plants. Each species gets an explanation based on whether or not it meets shared criteria or processes for that group (Wilkins 2003).

Moreover, species as phenomena explains why so many biologists are committed to the reality of the groups they study – they really do exist as phenomena. Whether or not they are more than conventional recognition of patterns depends on the group itself. It also resolves (or dissolves) the natural kinds issue. Species are natural groups, which may or may not share properties or mechanisms that cause them to group in the way they do. Some will – they have some sort of 'essence'. Some will not – but they may share a homeostatic property mechanism. Some will be multiply realised (as in respeciation events, Turner 2002). And so forth.

I am not aware of any reason to think, however, that species are a level of biological organisation or a fixed rank in biology. *A* species may be real, but *species* is not (Mishler and Wilkins 2018). In the end, what counts as 'real' or 'biologically meaningful' in biology depends upon the choice of criteria for naturalness, and the scale of the organisms and their *differentia*. I plump for monophyly, myself, but that is not surprising (Vanderlaan et al. 2013).[*] Monophyly is regarded as the 'real' structure of living things based on actual patterns in the data by most biologists trained since the 1980s. However, species can be, but often are not, monophyletic, and restricting species to monophyletic lineages means having some prior notion of where to stop subdividing. So, a diagnosable group of organisms is a real group in the sense that the data is not merely subjectively interpretable, but a species is not, in the cladistic sense, 'natural' unless it is also monophyletic.

A comment about 'bad species', or 'not-good species': these are basically species that do not fit the consensus criteria for that group. As there are an indefinitely large number of possible ways to be not-good, I'm unsure what to say about them (how many ways are there to not fit a set of criteria?). Dillon's stages are supposed to be ontic, while 'good species' are at least epistemic representations of specieshood in a group. A heterogeneous meta-population might be difficult to diagnose as a species using the usual criteria. That doesn't make it a bad species. A bad species is one that has been misdiagnosed. This places the entire weight of the concept on the criteria being used, and since that varies by the experience of the discipline, as well as the preferences of different specialists, the term is clearly based on disparate and often incommensurable epistemic criteria.[†]

From this, we can make the general comment, following Cracraft (1989), about concepts like *species*: they have an *ontic* aspect and an *epistemic* aspect, and generally, one aspect dominates their use. For species, we have previously considered, during the twentieth century, that the category had a theoretical ontic aspect predominating. I would argue that by now, we have to admit that *species* is primarily an epistemic notion.

---

[*] Monophyly is regarded as a single cut of a branch of an evolutionary tree. However, there is a debate over whether it includes the common ancestral taxon or not, with those who adopt the latter definition using *holophyly* for the former. This usage is not the consensus view, however. See the discussion in the citation.

[†] I do not mean Kuhnian incommensurability here, but merely that different uses of *species* use different metrics of identification, and that the metrics may not be interconvertible.

5.6   THE EVOLUTION OF THE CONCEPT

Species is what I call an 'operative concept', which is to say a concept formed from experience within the theoretical framework and methodological practices of a scientific discipline, but which is not formulated in terms of either the theories that predominate or which underpin that discipline (Weber's 'ideal-type concepts'*), or the assay techniques used to identify them (including such techniques as the statistical clustering of data points). It is not a simple inductive empirical class, if such things exist, but neither is it a theory-derived natural kind (Figure 5.1).[†]

Amitani's prototype notion of 'good species' leaves out the reasons why certain prototypical species are chosen to be, as it were, the exemplary type of species of the group. I have tried to fill this out in in terms of education, experience, and cultural context. Experienced observers of species have at their disposal a range of tools: practical training by prior experts; knowledge of related species (and thus, expectations about what will be seen, so that novel features are highlighted); whatever

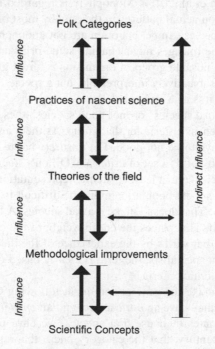

FIGURE 5.1   A simplified diagram of the development of operative concepts out of folk usage.

---

* Drysdale (1996) gives a good treatment of this.
† An extremely useful historical review of the distinction between lawlike ('nomothetic') and particularistic or token-based ('idiographic') concepts in science can be found in Rieppel (2006). As far as I can tell, at least within philosophy of science, operative concepts have not otherwise been given a proper treatment. A nice general treatment of tradition and social construction of categories is given by Stewart (2014).

theoretical prior information there may be about the group in question; and cognitive ancillary tools, such as what counts as a good explanation in that field.

Operative concepts* as I conceive of them are a kind of work-in-progress category that sciences and disciplines must construct as they develop the overall theory and protocols of their field. As such, they often begin in ordinary language. This is what happened to *species* – it began life in early modern Europe as a vernacular term for 'kind' in the academic language of the day, Latin, with biblical and logical connotations. Even when Ray first defined botanical and zoological kinds, it was a special case of the broader notion. Such operative concepts are very often revised (e.g., fishes), or abandoned (reptiles) as more precise and natural notions develop (clades).

For historical and professional reasons, *species* has not undergone this revision or abandonment, due to its being fixed in protocols of naming and classification, and more recently, ecological and conservational practice (Figure 5.2). But it should be, at least as a category. We name kinds in our everyday lives, but nobody thinks that kinds of car such as *cabriolet* need to have an underlying theoretical definition or usage. In biology, we can name and describe, and maybe even define, species as

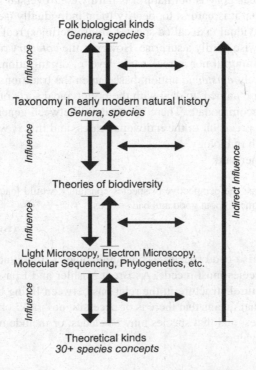

FIGURE 5.2   A simplified sequence of the formation of species concepts.

---

* Not to be confused with *operationalisation* as proposed by Percy Bridgman (1927), in which the meaning of a scientific term is defined solely by the operations of measurement and instrumental analysis of the science.

kinds without needing to think they have anything more than an operative – empirical and conventional – nature.

Operative concepts are picked up by individuals from those who teach them (see the epigram earlier by Anton Kerner) through instruction, imitation, and the affordances of the discipline, such as instruments, techniques, and locales. They can be acquired either consciously or unconsciously. When an individual reaches a certain social standing, their (usually retrospective) definitions of the operative concept become both explicit and constraining on how the term is used, leading often to conflicts with equally authoritative individuals in the race for funding and resources. But, they are not theoretical terms, although the objects in the world that they pick out are susceptible to theoretical accounts. Basically, they are social, professional, and practical terms of art, which developed from ordinary language and psychology.

5.7   CONCLUSION

That *species* as a category is not natural is hardly controversial these days, even if many still think that it is natural to identify them individually (see also Dupré, this volume). That individual taxa called species are (sometimes) real objects (groups of organisms) is likewise hardly a surprise. However, the *category* of species is neither a biologically meaningful nor a 'real' rank, level of organisation, or even a type of lineage. It is a purely *epistemic* notion, based upon the traditions of the disciplines in which species are named, to deal with the prototypical sorts of actual kinds each specialty has to accommodate. There is a tradeoff between generality in the group being studied and precision in the individual cases, and this is what each specialty has to contend with in biology.

In sum, as Kerner said:

> Abandoning at last the perspective of species constancy, would [cause us to] give up the childish argument about good and bad species.

(Kerner 1866, p. 8)

Once we characterise good species as epistemic objects, the attachment we have to the category of species must recede. As Brent Mishler and I have suggested, if we are looking for natural structure in the relations between living beings, monophyly is sufficient, and that means that there is of necessity no 'rank' or 'level' that makes something a species, and that species may be clades, or include many subclades, or be, in fact, polyphyletic.

5.8   ACKNOWLEDGEMENTS

I am most grateful to Igor Pavlinov, Frank Zachos, and Thomas Reydon for their critical responses to the drafts of this paper, which have made it much better than it was. Yuichi Amitani wrote the initial inspiration for this chapter and set up the historical and methodological issue (2015). I also acknowledge help from Staffan Müller-Wille. Of course, none of my many colleagues are responsible for anything silly I say here.

REFERENCES

Amitani, Y., 2015. Prototypical reasoning about species and the species problem. *Biological Theory*, 10 (4), 289–300.

Atran, Scott. 1998. "Folk Biology and the Anthropology of Science: Cognitive Universals and the Cultural Particulars." *Behavioral and Brain Sciences*, 21 (4), 547–609.

Berlin, Brent. 1973. "Folk Systematics in Relation to Biological Classification and Nomenclature." *Annual Review of Ecology and Systematics*, 4 (1), 259–71. https://doi.org/doi:10.1146/annurev.es.04.110173.001355.

Bogen, J., 2011. 'Saving the phenomena' and saving the phenomena. *Synthese*, 182 (1), 7–22.

Bogen, J., and Woodward, J., 1988. Saving the phenomena. *The Philosophical Review*, 67 (3), 303–352.

Bridgman, P.W., 1927. *The logic of modern physics*. New York: The Macmillan Company.

Brigandt, I., 2003. Species pluralism does not imply species eliminativism. *Philosophy of Science*, 70 (5), 1305–1316.

Bulmer, R., 1967. Why is the cassowary not a bird? A problem among the Karam of the New Guinea highlands. *Journal of the Royal Anthropological Institute*, 2 (1), 5–25.

Cain, A.J., 1994. Rank and sequence in Caspar Bauhin's *Pinax*. *Botanical Journal of the Linnean Society*, 114 (4), 311–356.

Cheung, T., 2006. From the organism of a body to the body of an organism: Occurrence and meaning of the word from the seventeenth to the nineteenth centuries. *The British Journal for the History of Science*, 39 (03), 319–339.

Chisholm, R.M., 1982. The problem of the Criterion. *In*: *The foundations of knowing*. Minneapolis: University of Minnesota Press, 61–75.

Cracraft, J., 1989. Speciation and its ontology: The empirical consequences of altering species concepts for understanding patterns and processes of differentiation. *In*: D. Otte and J. Endler, eds. *Speciation and its consequences*. Sunderland, MA: Sinauer, 28–59.

Darwin, C.R., 1859. *On the origin of species by means of natural selection, or the preservation of favoured races in the struggle for life*. London: John Murray.

Devitt, M., 2018. Individual essentialism in biology. *Biology & Philosophy*, 33(5), 39.

Dillon, L.S., 1966. The life cycle of the species: An extension of current concepts. *Systematic Zoology*, 15 (2), 112–126.

Drysdale, J., 1996. How are social-scientific concepts formed? A reconstruction of Max Weber's theory of concept formation. *Sociological Theory*, 14 (1), 71–88.

Ehrlich, P.R., 1961. Has the biological species concept outlived its usefulness? *Systematic Zoology*, 10 (4), 167–176.

Fujita, Matthew K., and Adam, D. Leaché. 2011. "A Coalescent Perspective on Delimiting and Naming Species: A Reply to Bauer et Al." *Proceedings of the Royal Society B: Biological Sciences*, 278 (1705), 493–95. https://doi.org/10.1098/rspb.2010.1864.

Gregory, T.R., and Mable, B.K., 2005. Polyploidy in animals. *In*: T.R. Gregory, ed. *The evolution of the genome*. Burlington, VT: Academic Press, 427–517.

Hooker, J.D., 1859. *On the flora of Australia, its origin, affinities, and distribution; being an introductory essay to the Flora of Tasmania. Reprinted from pt 3 of The botany of the Antarctic expedition, Flora Tasmania*, vol. 1. London: Lovell Reeve.

Jung, J., 1747. *Opuscula botanico-physica*. Cobvrgi: Svmtibvs et typis Georgii Ottonis Typogr. Dvcal. Priv.

Kendig, C., 2016. Activities of kinding in scientific practice. *In*: C. Kendig, ed. *Natural kinds and classification in scientific practice*. Abingdon and New York: Routledge, 1–14.

Kendig, C., 2020. Ontology and values anchor indigenous and grey nomenclatures: A case study in lichen naming practices among the Samí, Sherpa, Scots, and Okanagan. *Studies in History and Philosophy of Science Part C: Studies in History and Philosophy of Biological and Biomedical Sciences*, 84, 101340.

Kerner, A., 1866. *Gute und schlechte Arten*. Innsbruck: Wagner.

Kitcher, P., 1984. Species. *Philosophy of Science*, 51, 308–333.

Kleinman, K., 2018. Genera, evolution, and botanists in 1940: Edgar Anderson's 'Survey of Modern Opinion'. *Studies in History and Philosophy of Science Part C: Studies in History and Philosophy of Biological and Biomedical Sciences*, 67, 1–7.

Knowlton, N., Weigt, L.A., Solorzano, L.A., Mills, D.K., and Bermingham, E., 1993. Divergence in proteins, mitochondrial DNA, and reproductive compatibility across the isthmus of Panama. *Science*, 260 (5114), 1629–1632.

Krell, F.-T., 2004. Parataxonomy vs. taxonomy in biodiversity studies – pitfalls and applicability of 'morphospecies' sorting. *Biodiversity & Conservation*, 13 (4), 795–812.

LaPorte, J., 2018. Modern essentialism for species and its animadversions. *In*: R. Joyce, ed. *Routledge handbook of evolution and philosophy*. Abingdon, UK: Routledge, 182–193.

Lazenby, E.M., 1995. *The Historia Plantarum Generalis by John Ray: Book I – a translation and commentary*. Thesis for Doctor of Philosophy. Newcastle, UK: University of Newcastle upon Tyne.

Lehmann, A., Devriese, H., Tumbrinck, J., Skejo, J., Lehmann, G.U.C., and Hochkirch, A., 2017. The importance of validated alpha taxonomy for phylogenetic and DNA barcoding studies: A comment on species identification of pygmy grasshoppers (Orthoptera, Tetrigidae). *ZooKeys*, 679, 139–144.

Lessios, H.A., and Cunningham, C.W., 1990. Gametic incompatibility between species of the sea urchin Echinometra on the two sides of the isthmus of Panama. *Evolution*, 44 (4), 933–941.

Ma, N.S., Rossan, R.N., Kelley, S.T., Harper, J.S., Bedard, M.T., and Jones, T.C., 1978. Banding patterns of the chromosomes of two new karyotypes of the owl monkey, Aotus, captured in Panama. *Journal of Medical Primatology*, 7 (3), 146–55.

Marko, P.B., 2002. Fossil calibration of molecular clocks and the divergence times of geminate species pairs separated by the isthmus of Panama. *Molecular Biology and Evolution*, 19 (11), 2005–2021.

Medin, Douglas L., and Scott, Atran. 1999. *Folkbiology*. Cambridge MA: MIT Press.

Melia, J., and Saatsi, J., 2006. Ramseyfication and theoretical content. *The British Journal for the Philosophy of Science*, 57 (3), 561–585.

Mendel, G.J., 1866. Versuche über Plflanzenhybriden. *Verhandlungen des naturforschenden Vereines in Brünn*, Bd. IV für das Jahr, 1865 Abhandlungen (3–47).

Mendel, G.J., 2020. *Experiments on plant hybrids (Versuche über Pflanzen-Hybriden). New Translation with Commentary*. Brno, Czech Republic: Masaryk University Press.

Mill, J.S., 1843. *A system of logic, ratiocinative and inductive: Being a connected view of the principles of evidence, and methods of scientific investigation*. London: John W. Parker.

Mishler, B.D., and Wilkins, J.S., 2018. The hunting of the SNaRC: A snarky solution to the species problem. *Philosophy, Theory, and Practice in Biology*, 10(1). doi:10.3998/ptpbio.16039257.0010.001

Mitton, J.B., and Grant, M.C., 1996. Genetic variation and the natural history of quaking aspen: The ways in which aspen reproduces underlie its great geographic range, high levels of genetic variability, and persistence. *BioScience*, 46 (1), 25–31.

Müller-Wille, S., and Hall, K., trans., 2016. Experiments on plant hybrids (1866). [online]. *Experiments on Plant Hybrids (1866). Translation and commentary by Staffan Müller-Wille and Kersten Hall. British Society for the History of Science Translation Series*. Available from: http://www.bshs.org.uk/bshs-translations/mendel [Accessed 28 December 2020].

Padial, J.M., Miralles, A., De la Riva, I., and Vences, M., 2010. The integrative future of taxonomy. *Frontiers in Zoology*, 7(1), 16.

Queller, D.C., and Strassmann, J.E., 2016. Problems of multi-species organisms: Endosymbionts to holobionts. *Biology & Philosophy*, 31 (6), 855–873.

Quine, W.V.O., 1948. On what there is. *Review of Metaphysics*, 2(5), 21–38.

Ray, J., 1686. *Historia Plantarum Species hactenus editas aliasque insuper multas noviter inventas & descriptas complectens: In qua agitur primò De Plantis in genere, Earúmque Partibus, Accidentibus & Differentiis; Deinde Genera omnia tum summa tum subalterna ad Species usque infimas, Notis suis certis & Characteristicis Definita, Methodo Naturæ vestigiis insistente disponuntur; Species singulæ accurate describuntur, obscura illustrantur, omissa supplentur, superflua resecantur, Synonyma necessaria adjiciunctur; Vires denique & Usus recepti compendiò traduntur*. London: Clark.

Rensch, Bernhard. 1959. *Evolution above the Species Level*. New York: Columbia University Press. Originally published in 1954 by Ferdinand Enke Verlag, Stuttgart, as Neure Probleme der Abstammungslehre.

Rieppel, O., 2006. On concept formation in systematics. *Cladistics*, 22 (5), 474–492.

Schultz, P.W., Zelezny, L., and Dalrymple, N.J., 2000. A multinational perspective on the relation between Judeo-Christian religious beliefs and attitudes of environmental concern. *Environment and Behavior*, 32 (4), 576–591.

Sharkey, M.J., Janzen, D.H., Hallwachs, W., Chapman, E.G., Smith, M.A., Dapkey, T., Brown, A., Ratnasingham, S., Naik, S., Manjunath, R., Perez, K., Milton, M., Hebert, P., Shaw, S.R., Kittel, R.N., Solis, M.A., Metz, M.A., Goldstein, P.Z., Brown, J.W., Quicke, D.L.J., Achterberg, C. van, Brown, B.V., and Burns, J.M., 2021. Minimalist revision and description of 403 new species in 11 subfamilies of Costa Rican braconid parasitoid wasps, including host records for 219 species. *ZooKeys*, 1013, 1–665.

Sibley, C.G., and Ahlquist, J.E., 1983. Phylogeny and classification of birds based on the data of DNA-DNA hybridization. *In:* R.F. Johnston, ed. *Current ornithology*. New York: Springer US, 245–292.

Smirnov, A., 2009. Amoebas, Lobose. *In:* M. Schaechter, ed. *Encyclopedia of microbiology*. 3rd ed. Oxford: Academic Press, 558–577.

Sneath, P.H.A., and Sokal, R.R., 1973. *Numerical taxonomy: The principles and practice of numerical classification*. San Francisco, CA: W. H. Freeman.

Sokal, R.R., and Rohlf, F.J., 1962. The comparison of dendrograms by objective methods. *TAXON*, 11 (2), 33–40.

Sterelny, K., 1999. Species as evolutionary mosaics. *In:* R.A. Wilson, ed. *Species, new interdisciplinary essays*. Cambridge, MA: Bradford/MIT Press, 119–138.

Stewart, J., 2014. An enquiry concerning the nature of conceptual categories: A case-study on the social dimension of human cognition. *Frontiers in Psychology*, 5. https://www.frontiersin.org/articles/10.3389/fpsyg.2014.00654/full

Stringer, C., 2012. What makes a modern human. *Nature*, 485 (7396), 33–35.

Turner, G.F., 2002. Parallel speciation, despeciation and respeciation: Implications for species definition. *Fish and Fisheries*, 3 (3), 225–229.

Vanderlaan, T.A., Ebach, M.C., Williams, D.M., and Wilkins, J.S., 2013. Defining and redefining monophyly: Haeckel, Hennig, Ashlock, Nelson and the proliferation of definitions. *Australian Systematic Botany*, 26 (5), 347–355.

Whewell, W., 1840. *The philosophy of the inductive sciences: Founded upon their history.* London: John W. Parker.

Wilkins, J.S., 2003. How to be a chaste species pluralist-realist: The origins of species modes and the Synapomorphic Species Concept. *Biology and Philosophy*, 18, 621–638.

Wilkins, J.S., 2011. Philosophically speaking, how many species concepts are there? *Zootaxa*, 2765, 58–60.

Wilkins, J.S., 2013. Biological essentialism and the tidal change of natural kinds. *Science & Education*, 22(2), 221–240.

Wilkins, J.S., 2018a. *Species: The evolution of the idea.* 2nd ed. Boca Raton, FL: CRC Press.

Wilkins, J.S., 2018b. The reality of species: Real phenomena not theoretical objects. *In*: R. Joyce, ed. *Routledge handbook of evolution and philosophy.* Abingdon, UK: Routledge, 167–181.

Wilkins, J.S., forthcoming. God, species, and politics. *In*: B. Swartz and B. Mishler, eds. *Speciesism.* Berkeley, CA: University of California Press.

Will, K.W., Mishler, B.D., and Wheeler, Q.D., 2005. The perils of DNA barcoding and the need for integrative taxonomy. *Systematic Biology*, 54(5), 844–851.

Will, K.W., and Rubinoff, D., 2004. Myth of the molecule: DNA barcodes for species cannot replace morphology for identification and classification. *Cladistics*, 20(1), 47–55.

Section 2

Practice and Methods

6 Species in the Time of Big Data

The Multi-Species Coalescent, the General Lineage Concept, and Species Delimitation

Aleta Quinn

CONTENTS

6.1 INTRODUCTION

This chapter will not offer any particular solution to the seemingly never-ending 'species problem' – indeed, I suspect that the nature of the problem has been misunderstood along similar lines as in cases analyzed by Lennox (2001) and that the solution falls along the lines argued by Haber (2019). Instead, I will analyze what precisely the claims made in species delimitation studies amount to when they are based on modern computational phylogenomic methods.

I begin (Section 6.2) by summarising some of the recent historical context of these species delimitation studies, briefly discussing the implementation of the multi-species coalescent (MSC) model in phylogenetics. Section 6.3 analyzes a recent challenge to MSC-based species delimitation based on simulation study (Sukumaran and Knowles, 2017). Section 6.4 illustrates the arguments via recent field studies and connects this recent debate about MSC-based species delimitation to claims about the sufficiency of the general lineage concept (GLC) (de Queiroz, 1998). One aim of this

DOI: 10.1201/9780367855604-8

chapter is to enrich discussion of cases beyond a traditional list of known, 'abnormal' types of cases by demonstrating that we should expect genuinely ambiguous cases to be fairly common (but not at all ubiquitous). Modern phylogenomic methods have enabled us to detect many genuinely ambiguous cases without affecting the reality of many un-ambiguous 'good species'. I conclude (Section 6.5) by reflecting on the possibility of yet another revolution in species delimitation that, like the advent of the genomic era, promises a rich, new source of data at a previously unimaginable scale.

6.2   BIG DATA IN SYSTEMATICS

The rapid growth of phylogenomics from the 1990s through the 2010s was supported by a huge increase in funding (and interest), largely spurred by potential medical and epidemiological applications (Hillis, 1994; Apetrei et al., 2004; Papa, 2012; Scaduto et al., 2010). This growth encompassed multiple axes.

New methods of sequencing have enabled researchers to study and use genome-scale data. First-generation Sanger sequencing (which relies on polymerase chain reaction [PCR] amplification) remains valuable as a component within an overarching phylogenomic workflow and is invaluable to institutions with limited funding (Jennings, 2016).* Next-generation sequencing (NGS, or next gen)† enables researchers to rapidly examine millions of base pairs. New methods of alignment, and methods of checking alignment and other data quality issues, enable researchers to work with thousands of data-points.

Many of the new methods require substantial computational power, and so, the growth of phylogenomics also depended on rapid advances in computing theory and technology. Phylogenetic analyses also harnessed increased computing power, for example to apply and test models of sequence evolution. These models in turn are implemented within methods for (1) delimiting species and (2) inferring phylogenies. The latter includes improved heuristic methods for sampling tree-space using Markov Chain Monte Carlo (MCMC) methods, rendering a philosophical problem (the Bad Lot Argument – van Fraassen, 1980) an empirical challenge (Quinn, 2016).‡ Species delimitation has also benefitted from implementation of Bayesian approaches. With respect to both species delimitation and the phylogeny problem, Bayesian methods are frequently used in conjunction with methods based on a likelihood framework; thus, phylogenetic inference is not easily collapsed into a dichotomy between Bayesian and likelihoodist epistemologies but rather, provides fruitful ground for philosophical exploration of these ideas (Sober, 2002).

---

* Jennings argues compellingly that first-generation sequencing should not be viewed as obsolete. In addition to current uses, Jennings argues that it is likely that the ideas built into Sanger sequencing will spur ideas for answering novel problems.
† The term 'next-generation' is a bit inelegant, because the 'next' suggests that it cannot be succeeded. Divisions of second, third, and fourth generations have been proposed, each of which falls under the umbrella term NGS (Jennings, 2016).
‡ The claim that phylogenetic inference is a form of abductive inference has been defended by Sober (1988; 1991) and by Fitzhugh (2006).

To summarise: in the 1990s–2010s, systematists were flooded with powerful new tools in the form of genome-scale data and increased supercomputer power. Researchers (increasingly, computer scientists and mathematicians) have devised a plethora of methods to use the Big Data.

Ideally, uses of this data will fit into established best practices for species delimitation, following an overarching logic of research methodology. A now well-known injunction is to use multiple sources of evidence when delimiting species (Carstens et al., 2013), treating claims about species and the naming of species within a framework of integrative taxonomy (Dayrat, 2005; Will et al., 2005). Researchers indeed are well advised to recognise the advantages, and sometimes critical importance, of using multiple sources of evidence in science in general (Whewell, 1858; Hesse, 1968; Fisch, 1985; McMullin, 1992; Snyder, 2006). The challenge is in the interpretation of data as evidence: evidence of what? Evidence that genetic diversity is unequally distributed among a sample of organisms across a geographical region may be evidence that two or more populations are evolving independently. But, it may not. Genetic diversity may be differentially distributed within a (meta)population for any of a number of reasons and at least arguably, for no reason at all. Sometimes, the events on the ground just happen to turn out in such a way that patterns emerge without some unique, identifiable process producing the pattern in question.

A first point is to be cognisant that methods that can be used for species delimitation should not be treated as methods of species delimitation. The methods do not, in themselves, produce evidence for hypotheses about species. They produce data that in conjunction with a variety of assumptions, may constitute evidence for some species hypotheses and against others. Those assumptions should be borne in mind. This can be done in a critical mode, in requiring further evidence for or in criticising hypotheses about species. More productively, awareness of the assumptions can and should guide the overall methodology of any species delimitation study. The point is reflected in the title of Carstens et al.'s (2013) contribution, which is not 'How to delimit species' but 'How to fail at species delimitation': sometimes, the methods fail to produce compelling evidence for any particular species hypothesis, and in that case, researchers have successfully failed to delimit species. Proposing and naming taxa are scientific actions that should be accompanied by evidence for the relevant hypotheses about species (O'Hara, 1992; 1993; de Queiroz, 2011).

A second point is that distinct lines of evidence may speak to distinct aspects of the empirical situation (O'Hara, 1988). Evidence about differential distribution of genetic diversity may speak to particular species hypotheses. Evidence about the current rate of gene flow between two (sub)populations may or may not test the same species hypotheses. When seeking to combine distinct analytic methods, researchers must be aware of the particular questions that each method addresses and the extent to which the relevant hypotheses overlap.

Even when abundant empirical data is available and is analyzed in keeping with rigorous logic, the deeper challenge still looms: evidence for some species hypothesis is evidence of what, precisely? The answer would appear to be evidence that there is some number n of existing species in a region with some level of confidence.

This straightforward answer is problematic for reasons that, I argue in the next section, can no longer be ignored.

6.3   MSC-BASED SPECIES DELIMITATION AND ITS DISCONTENTS

Some emerging methods implement the MSC model, which was devised in the theoretical context of population genetics. In a widely cited essay in *Evolution*, Scott Edwards (2009) argued that implementation of the MSC has spurred a paradigm shift in phylogenetics. In broad terms, phylogeneticists can now reconstruct evolutionary relationships by accounting for population effects that had formerly been treated as unresolvable noise, more or less detrimental to particular methods of phylogenetic analysis. In particular, MSC-based methods use the empirical phenomenon of discordance (mismatch between gene trees and species trees) as a source of evidence. Use of MSC-based inference spread rapidly in studies of relationships between species (Carstens and Knowles, 2007; McCormack et al., 2012; Song et al., 2012; Xi et al., 2014) and for studies delimiting populations into species (Yang and Rannala, 2010; Zhang et al., 2011; 2013; Flouri et al., 2018).

Edwards (2016) argued that the paradigm shift constitutes a theoretical unification of two formerly distinct domains: population geneticists' study of variation within and across populations, and phylogeneticists' study of relationships across species (Hennig, 1966). As Avise et al. (1987) argued, such a unification would require not only use of population genetics concepts by systematists but use of phylogenetic concepts and methodology in population genetics – or, to put the point another way, theoretical unification broader and deeper than incorporation of a single new form of data or method of analysis into existing approaches. It is unsurprising, then, that MSC-based methods at the bridge of phylogenetics and population genetics run head-on into the species problem.

At present, MSC-based methods applied to species delimitation rely on several assumptions known to be problematic: the Jukes–Cantor model of molecular evolution and the assumption of no gene flow between lineages Yang (2015). In phylogenetic inference, Jukes-Cantor is now employed mainly to provide a base hypothesis that is typically rejected in favor of a more sophisticated model of molecular evolution; indeed, a recent study (Abadi et al., 2019) proposed to standardise the GTR+Γ+I model as a default.[*] Some recently developed methods are claimed to be robust to gene flow between hypothesised lineages (e.g., Chifman and Kubatko, 2014). Perhaps, in the near future, the MSC will be implemented for species delimitation in a way compatible with GTR+Γ+I and incorporating gene flow.

MSC-based methods for providing evidence about species delimitation are known to be faulty in known ways; of course, this is an unsurprising situation that recurs widely in model-based research (Box, 1976). A deeper conceptual problem looms, however, and has been made particularly salient by a recent empirical demonstration

---

[*] Abadi et al. (2019) argue that model-testing may not be a necessary step at all. However, the simulation analysis that suggested this result contained a problematic circularity that likely stacked the deck in favor of GTR+Γ+I.

that one commonly used MSC-based model (BPP; Yang and Rannala (2010)) is prone to overestimating species (Sukumaran and Knowles, 2017). Sukumaran and Knowles argue that MSC-based methods detect population structure rather than species.

Critics have pointed out that Sukumaran and Knowles' empirical study could only produce that result. Indeed, that is precisely the point: the method's assumptions guarantee overestimation of species unless a highly permissive concept of 'species' is adopted. Critics have also questioned the sense of 'species' employed by Sukumaran and Knowles. It might be thought, for example, that what Sukumaran and Knowles (2017) meant by 'structure' others would interpret as 'species'.

Sukumaran and Knowles referred repeatedly to 'true species', distinguishing these from sets of organisms delimited on the basis of genetic structure with populations. What, for Sukumaran and Knowles, are 'true species'? Sukumaran and Knowles explicitly declined to adopt any particular species concept (though their methods and results are compatible with many candidate species concepts, including the GLC). It might have been better to eschew the phrase 'true species' to avoid confusion. What the true species amount to in Sukumaran and Knowles' study are those lineages that are simulated to be true on the basis of a chosen model of speciation (implemented via ProtractedSpeciationProcess, Etienne and Rosindell, 2012; Etienne et al., 2014; Lambert et al., 2015). Each model posits that sub-populations become isolated from other sub-populations over time, that some of these sub-populations rejoin the general population (or become extinct), that some sub-populations go on to become full-fledged species, and that full-fledged species do not merge with other full-fledged species. These assumptions are operationalised by five parameters that are applied to full-fledged species and to 'incipient species'. An incipient species evolves independently from the population from which it diverged, in the sense that the incipient species is assigned its own parameter values, and its behavior is simulated independently from the behavior of its parent population. An incipient species has assigned probabilities of giving rise to a full-fledged species, merging with the parent population, or going extinct. The parent population may be either a 'true species' or an incipient species. A full-fledged species may go extinct, produce new incipient species, and/or merge with a child incipient species.

In each model, a 'true species' means a species that has no possibility of merging with a population from which it has descended (or any other population, except incipient species that arise from the population after its emergence as a 'true species'). One could interpret 'evolves independently' in the sense that each model treats an entity as having its own unique trajectory that will lead to a unique 'fate' (O'Hara, 1993; 1997). This fate may be extinction or persistence, and in the case of incipient species, may be reunion with the parent population. Alternatively, one could interpret 'fate' such that only full-fledged species have unique fates, i.e., reunion with a parent population does not count as a fate but rather, is the loss of the possibility of having a unique fate.

Each model is simplified to exclude any degree of gene flow between modeled entities. The conceptual situation may be muddied by cases involving different degrees of gene flow at different times (see Section 6.4). Even without this complication, however, a conceptual challenge arises if one wants to identify a set number n of species at a specified time.

Interpretation 1: Incipient species just are species. Each incipient species is an independently evolving lineage with its own unique fate, in the sense that it might become extinct, reunite with its parent lineage, or continue evolving independently indefinitely. The number n of species at any specified time is equal to the sum of lineages identified as 'true species' and 'incipient species' in the ProtractedSpeciationProcess model.

Interpretation 2: Populations identified as 'incipient species' belong to the 'true species' from which they have become isolated. The number n of species at any specified time is equal to the sum of lineages identified as 'true species' in the model. 'Incipient species' have no special ontological status; claims about 'incipient species' may be predictions about the future of identifiable populations. There is no fact of the matter as to whether 'incipient species' have unique fates; probably, some will come to have unique fates, while others do not.

Interpretation 3: There is no fact of the matter as to whether populations identified as 'incipient species' are distinct species or belong to the lineages from which they have split. There is no fact of the matter as to what is the number n of species at any specified time (except, perhaps, in the case that there are presently no 'incipient species', i.e., no populations are significantly isolated). This is not merely an epistemic fact about what we can know but an ontological fact (Sober, 1984, pp. 339–341). One might designate a number p that expresses an estimate of how many species there will be at some future time (recognising that it is unlikely that there will be a fact of the matter as to how many species there are at that future time, unless there happen to be no incipient species at the future time).[*]

Note that the interpretation of species can affect what some data says. The very same set of data might be evidence that (1) a newly isolated population is a recently divergent independent species or that (2) the newly isolated population is not presently a unique species. The dataset might also (3) be silent with respect to whether the population in question is a unique species at the present time.

Sukumaran and Knowles' (2017) study relied on model simulations. In the next section, I illustrate the modeled situation via empirical studies with field data and address the suggestion that Sukumaran and Knowles' pockets of 'structure' just are species.

---

[*] Interpretation 3 leads to a puzzle. Let the present time be t_1. An observer at t_1 correctly assesses that there is no fact of the matter as to whether each incipient species at t_1 is a species or not, and that there is no fact of the matter as to the number n of species at t_1. Suppose that in the future, say time t_{100}, all incipient species that were present at t_1 have either gone extinct, reunited with their parent lineages, or become true species. Is there now at t_{100} a fact of the matter as to how many species were present at t_1? Can there come to be a fact of the matter as to what happened in the past, and how can this come to be (potentially) long after the past occurred? Would an observer at t_{100} say that the observer at t_1 was correct and came to be wrong, or just correct, or just wrong?

6.4   THE MSC AND GLC IN APPLICATION

Researchers sometimes claim to implement the GLC of species, referencing de Queiroz (2007) or Wiley (1978) (or sometimes Simpson, 1951 or Simpson, 1961). De Queiroz has argued convincingly that being an independently evolving lineage is a necessary condition for being a species (de Queiroz, 1998; 1999; 2005a; 2007). However, he has not provided support for the claim that it is a sufficient condition (c.f. de Queiroz, 2005c; Pigliucci, 2003; Freudenstein et al., 2017). Robert O'Hara has pointed out that 'population' (and relatedly, 'metapopulation' and 'subpopulation') is not fully defined in terms of necessary and sufficient properties (de Queiroz, pers. comm.), and so, it may not be possible to fully specify sufficient conditions in a definition of 'species'. The population concept serves to specify the level of lineage in consideration (Haber, 2012). With respect to species, the problem is in specifying what counts as 'independently evolving'. The solutions require adopting some further substantive conceptual baggage (such as some degree of reproductive isolation), possibly allowing that species can split and subsequently merge (de Queiroz, 2005c).

De Queiroz successfully urged that species as lineages be adopted as a conceptual framework. Further empirical information must be applied in order to delineate species (Rieppel, 2009; 2010). That further empirical information is theory-laden in ways not fully specified by the GLC itself (why reproductive isolation [Cuvier, 1817; Whewell, 1840; Mayr, 1942; de Queiroz, 2005b; Wilkins, 2009], or mate recognition [Paterson, 1985], or some other thing?). The GLC does rule out many types of information as irrelevant and unifies many ideas about species that have been labelled distinct 'concepts'. Many of the proposed ideas worth accommodating have survived scientific scrutiny over decades, and the GLC provides not so much a replacement as a shared theoretical framework. The framework highlights the need for explicit justification of proposed criteria, for example of the idea that species be defined on the basis of plurality of concordance of gene trees (Velasco, 2019).

For years, the GLC appeared to be sufficient for species delimitation in practice, despite philosophical concerns (e.g., Brigandt, 2003; Ereshefsky, 2001, 2010; Wilkins, 2003). The situation has changed as a result of Big Data and the proliferation of computational methods for examining this data. This should not be surprising, given that biological theory predicts the occurrence of truly ambiguous situations (Darwin, 1859). I will now analyze one ambiguous case to show how the three distinct interpretations of 'species' presented in the preceding section can be implemented such that the same empirical situation yields distinct taxonomic results.

Chan et al. (2017) presented three possible interpretations of populations within the western lineage of Malaysian Torrent Frogs (*Amolops larutensis*). The empirical situation is hypothesised as follows (p. 5446):

> populations from the western lineage were highly structured and showed varying levels of historical and contemporary migration consistent with a complex history involving gene flow between recently diverging lineages.

If being an independently evolving lineage is sufficient to constitute a species, any recently diverging lineages are distinct species. (They may merge into a single species in the future, or they may not). Of two populations in particular, Chan et al. note that (p. 5446)

> gene flow between Larutensis on the northwestern mountain range and the geographi-
> cally proximate W1 and W2 populations on the central range were the most reduced
> (Nm = 0.3). We interpret this as an indication of incipient speciation triggered by
> recent and rapid human development and the disruption of habitat corridors along the
> Bintang-Kledang range, a small mountain range situated between the northwestern
> and central mountain ranges (Jamaluddin, Pau, & Siti-Azizah, 2011). Given sufficient
> time, Larutensis's lower long-term migration rates could lead this population to qualify
> as a species separate from the other western populations.

The newly isolated population is here identified as an incipient species but not currently a separate species. The authors continue (p. 5446):

> However at this point, the data do not support this split and we therefore consider these
> populations as belonging to a single species.

Given that the data do not support the split, the authors might have adopted an agnos-
tic position with respect to whether the seemingly isolated population constitutes a
species.

Three interpretations of the data are possible, corresponding to the three interpretations described earlier.

> **Interpretation 1:** The newly isolated population is a recently divergent inde-
> pendent species.
> **Interpretation 2:** The newly isolated population may be labelled an 'incipient
> species', but it is not presently a species.
> **Interpretation 3:** The data are silent with respect to whether the population
> in question is a species.

The third interpretation is compatible with the view that there is no fact of the matter
as to how many species of *Amolops* there are. Under the first and second interpretations, there is a fact of the matter. What further data or information would distinguish between the three interpretations? It may be that no further data would settle
this claim. What, then, is the difference among the three interpretations? All three
possibilities appear to be consistent with the GLC. It is unclear what should guide
our choice of interpretation.

Let us consider some other possible empirical situations. It should be clear
that none of the following possibilities is far-fetched. Moreover, techniques have
advanced to such a point that it is possible to reliably identify some instances of
each case in practice. I do not mean that we can reliably identify all cases of each
type. I mean that for each case, it is reasonable to expect that some instance(s) can be
detected in practice. The possibility of cases like these has long been recognised (and
sometimes debated). What is new is the reality that some of our inquiries conclude

with confidence that the case obtains in reality. In each case, at issue is what counts as 'independent' evolution.

Case 1: Two independently evolving lineages contribute to a sink hybrid population from which no individuals migrate out. The sink population has some unique alleles, which do not occur in either parent population. The majority of evolution in the sink population results from the inputs of each source population, but the presence of the unique alleles indicates that the sink population is undergoing some evolution of its own.

Case 2: Two independently evolving lineages with a sink hybrid population from which no individuals migrate out. The sink population lacks unique alleles. However, allele frequencies in the sink population are unique compared with either source population and change over time as a result of inputs from each source population and also as a result of mating and adaptation within the hybrid population. If evolution is defined 'as any change in the frequency of alleles within a gene pool from one generation to the next' (Curtis and Barnes, 1989, p 974; see also Endler, 1992; Hartl, 1988; Hull, 1992; Wilkins, 2001; Wilson, 1992a), then the hybrid population is evolving, and its evolution is to some degree independent of the evolution of the source populations.

Gottscho et al. (2017) concluded that the fringe-toed lizard population referred to as *Uma rufopunctata* is a hybrid sink between *U. cowlesi* and *U. notata*. The sink population may represent an instance of case 1 or case 2. As Gottscho et al. noted (p. 113), 'Further study to determine the biological nature of this hybrid zone is warranted'.

Case 3: There is currently a great deal of gene flow across two populations, but historically, they were largely isolated. Evidence indicates the occurrence of past hybridisation, but this hybridisation had little effect on either population until recently.

Case 4: Two populations exchange gene flow freely for a few generations and then, are reproductively isolated for many generations. This cycle repeats, during which time the populations evolve to look differently, have different mate choices, and so on for various species delimitation criteria. The populations subsequently lose these criteria and then regain some of the criteria, perhaps in different iterations. This cycle repeats more or less stably.

Red Wolves (*Canis rufus* [Kelly et al., 2008] or *Canis lupus rufus* [Wilson and Reeder, 2005]) hybridise with Coyotes (*Canis latrans*) in large parts of their present range, possibly as a result of anthropogenic habitat change (Paradiso and Nowak, 1972; Wilson et al., 2000), and may correspond to case 3. However, Red Wolves may have a more complex history of hybridization, involving both Coyotes and Gray Wolves (*Canis lupus*) (von Holdt et al., 2011), and may instead correspond to case 4, perhaps with the added component of a third population. Polar Bears (*Ursus maritimus*) and Brown Bears (*Ursus arctos*) are good candidates for case 3 in the event that climate change leads to increased contact and hybridisation between the two currently recognised species (Pongracz et al., 2017).

It should be noted that there may be versions of each case that involve more than two candidate independently evolving lineages, (perhaps) as in the Red Wolf case. The very same species delimitation criteria can be established and then lost, and the criteria can be acquired in this or that order and then lost in this or that order.

The list of cases easily could be extended. Some readers might propose simply recognising distinct species in all of the above cases,[*] and so, it is worth pointing out cases in which scientists have resisted recognising separate species. Raccoons occur on several islands in the Caribbean, and the 2003 International Union for Conservation of Nature (IUCN) Red List identified five island raccoon species as Endangered (Zeveloff, 2003). Helgen and Wilson (2003) and Helgen et al. (2008) provided compelling evidence that these raccoons were all introduced within the last few centuries, with historically documented independent introductions occurring in 1650–79, in the early eighteenth century, and in 1932–33. Other introduction events are also possible. Raccoon populations on Bahamian islands – previously recognised as Bahamas Raccoons (*Procyon maynardi*) – are now considered invasive; conservation strategies have switched from trying to preserve these populations to trying to extirpate them. Raccoons on Guadalupe were demoted from *Procyon minor* to a subspecies, *Procyon lotor minor*, yet retain their Endangered status. It is unsurprising that conservation decisions would reflect non-epistemic values, but on what grounds should the taxonomic decisions be made? Many researchers would balk at recognising a distinct species on the basis of under a century's worth of independent evolution. If not 100 years, then perhaps 350 years is enough; but perhaps, the judgment should consider the fact that the population resulted from human intervention (or perhaps not). There are, of course, other cases further along the continuum of time, such as the populations of *Lampropeltis californiae* on Gran Canaria that originated from pet kingsnakes in the early 2000s. As of 2014, several populations were genetically distinct with respect to California populations and with respect to each other (Monzón-Argüello et al., 2015).[†] The populations are also morphologically distinctive, having descended from albino and other morphs popular among snake-breeders. Cases that do not involve human intervention include incidents of salamanders colonising previously unoccupied mountains, caves, or drainage systems.[‡] Such colonisations may occur at any point in time.

The complexities of biological reality force the questions: what timescale and what degree of independence constitute 'independent evolution'? At present, there is no clear answer, and we may never settle on an answer. This does not mean that the species category is not 'real' (Wilkins, 2003); nor does it mean that there are no 'real' and well-defined instances of species. It does mean that scientific practice should recognise the existence of genuinely ambiguous cases in addition to clear-cut

---

[*] Several readers of an earlier version of this manuscript suggested this resolution.

[†] Monzón-Argüello et al. identified genetic differences between populations as evidence of distinct founder events. There are now records at additional localities, which may represent additional distinct populations (Brian Hinds, pers. comm).

[‡] The effects of anthropogenic climate change may count as human intervention.

cases. The type method and the absence of explicit definitions of species enable the active nomenclatural Codes (and the PhyloCode) to function notwithstanding ambiguous cases.

6.5 CONCLUSIONS

Systematists and taxonomists face challenges of many kinds, and I want to be clear that I am not critiquing the value of studies that necessarily use very limited data to advance claims about cryptic biodiversity in regions at (sometimes extreme) risk of habitat destruction. The decision to recognise species may have immediate, important practical effects beyond the scientific community. Naming species on the basis of limited evidence may contribute to particular conservation goals but runs a longer-term risk of eroding credibility of the taxonomic community (and by extension, the legal and political framework of conservation, which depends on the concept of species [Agar, 2018]). That genetic structure has been discovered is an important signal that further research is needed, and the nature of the structure provides information about where and how to conduct that research. The discovery of genetic structure alone, however, does not support a claim to have delimited species. It is not the evidence that would be necessary to confirm or disconfirm such a claim.

There have been hints, here and there, that more traditional forms of biological practice may help to resolve some of the challenges discussed in this chapter. Despite statements of the value of traditional taxonomic work for conservation (Wilson, 1988; 1992b), approaches such as developmental morphology remain chronically underfunded. The specter of a molecules versus morphology dichotomy remains a conceptual stumbling block, notwithstanding repeated claims that the distinct approaches are complementary. Few individuals have the opportunity, support, and time to learn the skills needed for all of high-throughput phylogenomic data acquisition, computational phylogenetics, developmental morphology, anatomy, and organismal fieldwork. But, a new and potentially revolutionary source of Big Data has appeared, one that promises to provide biogeographic and ecological data at an unprecedented scale.

At the time of this writing (August 2020), the online platform iNaturalist supported 47,578,993 observations of organisms posted by 1,257,585 observers, with identifications provided by 152,767 users (iNaturalist, 2020). Launched in 2008, iNat's growth has accelerated in recent years: in August 2018, there were 12 million observations, and in June 2019, 20 million observations; in July and August 2020, 5 million observations were added (pers. obs.). Another platform, eBird, hosts 737 million observations by half a million observers (eBird Team, 2019). Citizen scientists providing photographs, descriptions, and locality information in a casual manner can rapidly expand the scale of data far beyond what an individual researcher could achieve in a lifetime of work.

Elsewhere (Quinn, 2021), I have commented on some challenges raised by these massive online platforms, some (though not all) of which can be ameliorated via the involvement of professional taxonomists, who should be appropriately compensated for this labor. The hope is that the current popularity of citizen science at some institutions is not a flash-in-the-pan form of outreach that resembles advertising but a meaningful investment in an emerging source of Big Data.

Scientists have used these data in a diverse range of studies, but the potential of these platforms to support species delimitation lies largely in the future (c.f. Rovito, 2017). Documenting what organisms are present where, what they look like, how they behave, and how they interact has always been critical to species delimitation, but it is difficult to estimate the potential of multiplying such observations to unprecedented scale. The possibility of working with billions of observations across decades radically changes not only what evidence can be produced but also what questions can reasonably be asked. Big Data came to systematics in the form of methods to obtain genome-scale data and computational power to analyze this data, but a new dimension is on the horizon: harnessing crowd sourcing to collect millions, and potentially billions, of natural history observations.

Future work may thus incorporate both molecular and phenotypic data at massive scales. It is sometimes thought that the sheer scale of molecular datasets constitutes an advantage (perhaps a decisive advantage) for phylogenomic approaches versus approaches that rely mostly or entirely on data about phenotypes. The apparent divide between molecule-based phylogenomic approaches to species delimitation and morphology-based 'traditional' approaches is historically contingent, and use of the distinct types of data does not entail significantly different conceptual or methodological frameworks (Huelsenbeck et al., 1996; de Queiroz, 2000; Wiens, 2004). Perhaps, the arrival of the Big Data age with respect to phenotypic data will lend support to calls to more meaningfully incorporate morphological, ecological, and behavioral studies together with genomic work on species delimitation (e.g., Hillis, 2019).

ACKNOWLEDGEMENTS

For comments on an earlier version of this manuscript, I thank Matt Haber, David Hillis, Kevin de Queiroz, Sam Sweet, and the phylogenetics reading group of the University of Idaho. For discussion, I thank Fiona Cowie, Tony Gill, Andy Gottscho, Brian Hinds, Scott Lidgard, Rudolf Meier, Bruce Rannala, and Olivier Rieppel.

I owe a great debt of gratitude to Kevin de Queiroz for discussing with me the concept of species, and many other aspects of systematics, for very many hours over several years.

The views and any errors in this chapter are entirely my own.

REFERENCES

Abadi, S., Azouri, D., Pupko, T., and Mayrose, I. (2019). Model selection may not be a mandatory step for phylogeny reconstruction. *Nature Communications*, 10(934):1–11.

Agar, J. E. (2018). What counts as threatened? Science and the sixth extinction. In Manning, P. and Savelli, M., editors, *Global Transformations in the Life Sciences, 1945–1980*, pages 180–194. University of Pittsburgh Press.

Apetrei, C., Robertson, D. L., and Marx, P. A. (2004). The history of SIVS and AIDS: Epidemiology, phylogeny and biology of isolates from naturally SIV infected non-human primates (NHP) in Africa. *Frontiers of Bioscience*, 9:225–254.

Avise, J. C., Arnold, J., Ball, R. M., Bermingham, E., Lamb, T., Neigel, J. E., Reeb, C. A., and Saunders, N. C. (1987). Intraspecific phylogeography: The mitochondrial DNA bridge between population genetics and systematics. *Annual Review of Ecology and Systematics*, 18(1):489–522.

Box, G. (1976). Science and statistics. *Journal of the American Statistical Association*, 71(356):791–799.

Brigandt, I. (2003). Species pluralism does not imply species eliminativism. *Philosophy of Science*, 70(5):1305–1316.

Carstens, B. C. and Knowles, L. L. (2007). Estimating species phylogeny from gene-tree probabilities despite incomplete lineage sorting: An example from *Melanoplus* grasshoppers. *Systematic Biology*, 56(3):400–411.

Carstens, B. C., Pelletier, T. A., Reid, N. M., and Satler, J. D. (2013). How to fail at species delimitation. *Molecular Ecology*, 22(17):4369–4383.

Chan, K. O., Alexander, A. M., Grismer, L. L., Su, Y. C., Grismer, J. L., Quah, E. S. H., and Brown, R. M. (2017). Species delimitation with gene flow: A methodological comparison and population genomics approach to elucidate cryptic species boundaries in Malaysian torrent frogs. *Molecular Ecology*, 26(20):5434–5450.

Chifman, J. and Kubatko, L. (2014). Quartet inference from SNP data under the coalescent model. *Bioinformatics*, 30(23):3317–3324.

Curtis, H. and Barnes, N. S. (1989). *Biology*, 5th ed. Worth Publishers.

Cuvier, G. (1817). *Le Règne Animal Distribuée d'après son Organisation*. Deterville.

Darwin, C. (1859). *On the Origin of Species by Means of Natural Selection, or the Preservation of Favoured Races in the Struggle for Life*. John Murray.

Dayrat, B. (2005). Towards integrative taxonomy. *Biological Journal of the Linnaean Society*, 85(3):407–415.

de Queiroz, K. (1998). The general lineage concept of species, species criteria, and the process of speciation: A conceptual unification and terminological recommendations. In Howard, D. J. and Berlocher, S. H., editors, *Endless Forms: Species and Speciation*, pages 57–75. Oxford University Press.

de Queiroz, K. (1999). The general lineage concept of species and the defining properties of the species category. In Wilson, R. A., editor, *Species: New Interdisciplinary Essays*, pages 49–89. MIT Press.

de Queiroz, K. (2000). Logical problems associated with including and excluding characters during tree reconstruction and their implications for the study of morphological character evolution. In Wiens, J. J., editor, *Phylogenetic Analysis of Morphological Data*, pages 192–212. Smithsonian Books.

de Queiroz, K. (2005a). Different species problems and their resolution. *Bioessays*, 27:1263–1269.

de Queiroz, K. (2005b). Ernst Mayr and the modern concept of species. *Proceedings of the National Academy of Sciences*, 102(S1):6600–6607.

de Queiroz, K. (2005c). A unified concept of species and its consequences for the future of taxonomy. *Proceedings of the California Academy of Sciences*, 56(18):196–215.

de Queiroz, K. (2007). Species concepts and species delimitation. *Systematic Biology*, 56(6):879–886.

de Queiroz, K. (2011). Branches in the lines of descent: Charles Darwin and the evolution of the species concept. *Biological Journal of the Linnaean Society*, 103:19–35.

eBird Team. (2019). *eBird 2019: Year in Review*. Available at: https://ebird.org/news/ebird-2019-year-in-review

Edwards, S. V. (2009). Is a new and general theory of molecular systematics emerging? *Evolution*, 63(1):1–19.

Edwards, S. V., Xi, Z., Janke, A., Faircloth, B. C., McCormack, J., Glenn, T. C., Zhong, B., Wu, S., Lemmon, E. M., Lemmon, A. R., Leaché, A. E., Liu, L., and Davis, C. C. (2016). Implementing and testing the multispecies coalescent model: A valuable paradigm for phylogenomics. *Molecular Phylogenetics and Evolution*, 94(Part A):447–462.

Endler, J. A. (1992). Natural selection: Current usages. In Keller, E. F. and Lloyd, E. A., editors, *Keywords in Evolutionary Biology*, pages 220–224. Harvard University Press.

Ereshefsky, M. (2001). *The Poverty of the Linnean Hierarchy: A Philosophical Study of Biological Taxonomy*. Cambridge University Press.

Ereshefsky, M. (2010). What's wrong with the new biological essentialism. *Philosophy of Science*, 77(5):674–685.

Etienne, R. S., Morlon, H., and Lambert, A. (2014). Estimating the duration of speciation from phylogenies. *Evolution*, 68(8):2430–2440.

Etienne, R. S. and Rosindell, J. (2012). Prolonging the past counteracts the pull of the present: Protracted speciation can explain observed slowdowns in diversification. *Systematic Biology*, 61(2):204–213.

Fisch, M. (1985). Whewell's consilience of induction: An evaluation. *Philosophy of Science*, 52:239–255.

Fitzhugh, K. (2006). *The Abduction of Phylogenetic Hypotheses*. Magnolia Press.

Flouri, T., Jiao, X., Rannala, B., and Yang, Z. (2018). Species tree inference with BPP using genomic sequences and the multispecies coalescent. *Molecular Biology and Evolution*, 35(10):2585–2593.

Freudenstein, J. V., Broe, M. B., Folk, R. A., and Sinn, B. T. (2017). Biodiversity and the species concept – lineages are not enough. *Systematic Biology*, 66(4):644–656.

Gottscho, A. D., Wood, D. A., Vandergast, A. G., Lemos-Espinal, J., Gatesy, J., and Reeder, T. W. (2017). Lineage diversification of fringe-toed lizards (Phrynosomatidae: *Uma notata* complex) in the Colorado desert: Delimiting species in the presence of gene flow. *Molecular Phylogenetics and Evolution*, 106:103–117.

Haber, M. (2012). Multilevel lineages and multidimensional trees: The levels of lineage and phylogeny reconstruction. *Philosophy of Science*, 79(5):609–623.

Haber, M. (2019). Species in the age of discordance. *Philosophy, Theory, and Practice in Biology*, 11:21.

Hartl, D. L. (1988). *A Primer of Population Genetics*. Sinauer Associates.

Helgen, K. M., Maldonado, J. E., Wilson, D. E., and Buckner, S. D. (2008). Molecular confirmation of the origin and invasive status of West Indian raccoons. *Journal of Mammalogy*, 89(2):282–291.

Helgen, K. M. and Wilson, D. E. (2003). Taxonomic status and conservation relevance of the raccoons (*Procyon* spp.) of the West Indies. *Journal of Zoology (London)*, 259:69–76.

Hennig, W. (1966). *Phylogenetic Systematics*. University of Illinois Press.

Hesse, M. B. (1968). Consilience of inductions. In Lakatos, I., editor, *The Problem of Inductive Logic*, pages 232–247. North Holland Publication Co.

Hillis, D. M. (1994). Application and accuracy of molecular phylogenies. *Science*, 264(5159):671–677.

Hillis, D. M. (2019). Species delimitation in herpetology. *Journal of Herpetology*, 53(1):3–12.

Huelsenbeck, J. P., Bull, J. J., and Cunningham, C. W. (1996). Combining data in phylogenetic analyses. *Trends in Ecology & Evolution*, 11:152–158.

Hull, D. (1992). Individual. In Keller, E. F. and Lloyd, E. A., editors, *Keywords in Evolutionary Biology*, pages 180–187. Harvard University Press.

iNaturalist. (2020). *Observations*. Available at: https://www.inaturalist.org/observations

Jamaluddin, Jamsari Amirul Firdaus, Tan Min Rau, and Mohd Nor Siti-Azizah. 2011. Genetic structure of the snakehead murrel, *Channa striata* (channidae) based on the cytochrome c oxidase subunit I gene: Influence of historical and geomorphological factors. *Genetics and Molecular Biology* 34(1): 152–160.

Jennings, W. B. (2016). *Phylogenomic Data Acquisition: Principles and Practice.* CRC Press.

Kelly, B. T., Beyer, A., and Phillips, M. K. (2008). *Canis rufus. The IUCN Red List of Threatened Species 2008*: e.T3747A10057394. http://dx.doi.org/10.2305/IUCN.UK .2008.RLTS.T3747A10057394.en

Lambert, A., Morlon, H., and Etienne, R. S. (2015). The reconstructed tree in the lineage-based model of protracted speciation. *Journal of Mathematical Biology*, 70(1–2):367–397.

Lennox, J. G. (2001). History and philosophy of science: A phylogenetic approach. *História, Ciências, Saúde-Manguinhos*, 8(3):655–669.

Mayr, E. (1942). *Systematics and the Origin of Species, from the Viewpoint of a Zoologist.* Harvard University Press.

McCormack, J., Huang, H., and Knowles, L. L. (2012). Ultraconserved elements are novel phylogenomic markers that resolve placental mammal phylogeny when combined with species tree analysis. *Genome Research*, 22:746–754.

McMullin, E. (1992). *The Inference that Makes Science.* Marquette University Press.

Monzón-Argüello, C., Patiño-Martínez, C., Christiansen, F., Gallo-Barneto, R., Cabrera-Pérez, M., Peña-Estévez, M., and Lee, P. (2015). Snakes on an island: Independent introductions have different potentials for invasion. *Conservation Genetics*, 16(5):1225–1241.

O'Hara, R. J. (1988). Homage to Clio, or, toward an historical philosophy for evolutionary biology. *Systematic Zoology*, 37(2):142–155.

O'Hara, R. J. (1992). Telling the tree: Narrative representation and the study of evolutionary history. *Biology and Philosophy*, 7(2):135–160.

O'Hara, R. J. (1993). Systematic generalization, historical fate, and the species problem. *Systematic Biology*, 42(3):231–246.

O'Hara, R. J. (1997). Population-thinking and tree-thinking in systematics. *Zoologica Sripta*, 26(4):323–329.

Papa, A. (2012). Dobrava-Belgrade virus: Phylogeny, epidemiology, disease. *Antiviral Research*, 95(2):104–117.

Paradiso, J. L. and Nowak, R. M. (1972). *Canis rufus. Mammalian Species*, 22:1–4.

Paterson, H. E. H. (1985). The recognition concept of species. In Vrba, E. S., editor, *Species and Speciation*, volume 4. Transvaal Museum Monograph 4: 21–29.

Pigliucci, M. (2003). Species as family resemblance concepts: The (dis)solution of the species problem? *Bioessays*, 25(6):596–602.

Pongracz, J. D., Paetkau, D., Branigan, M., and Richardson, E. (2017). Recent hybridization between a Polar Bear and Grizzly Bears in the Canadian arctic. *Arctic*, 70(2):151–160.

Quinn, A. (2016). Phylogenetic inference to the best explanation and the bad lot argument. *Synthese*, 193(9):3025–3039.

Quinn, A. (2021). Transparency and secrecy in citizen science: Lessons from herping. *Studies in History and Philosophy of Science* 85:208–217.

Rieppel, O. (2009). Species as a process. *Acta Biotheoretica*, 57(1–2):33–49.

Rieppel, O. (2010). The series, the network, and the tree: Changing metaphors of order in nature. *Biology and Philosophy*, 25(4):475–496.

Rovito, S. M. (2017). The geography of speciation in neotropical salamanders. *Herpetologica*, 73(3):229–241.

Scaduto, D. I., Brown, J. M., Haaland, W. C., Zwickl, D. J., Hillis, D. M., and Metzker, M. L. (2010). Source identification in two criminal cases using phylogentic analysis of HIV-1 DNA sequences. *Proceedings of the National Academy of Sciences*, 107(50):21242–21247.

Simpson, G. G. (1951). The species concept. *Evolution*, 5(4):285–298.

Simpson, G. G. (1961). *Principles of Animal Taxonomy*. Columbia University press.

Snyder, L. J. (2006). *Reforming Philosophy: A Victorian Debate on Science and Society*. University of Chicago Press.

Sober, E. (1984). Sets, species, and evolution: Comments on Philip Kitcher's 'species'. *Philosophy of Science*, 51(2):334–341.

Sober, E. (1988). *Reconstructing the Past: Parsimony, Evolution, and Inference*, 1st ed. MIT Press.

Sober, E. (1991). *Reconstructing the Past: Parsimony, Evolution, and Inference*, 2nd ed. MIT Press.

Sober, E. (2002). Bayesianism – Its scope and limits. In Swinburne, R., editor, *Bayes' Theorem*, pages 21–38. Oxford University Press.

Song, S., Liu, L., Edwards, S. V., and Wu, S. (2012). Resolving conflict in Eutherian mammal phylogeny using phylogenomics and the multispecies coalescent model. *Proceedings of the National Academy of Sciences*, 109(37):14942–14947.

Sukumaran, J. and Knowles, L. L. (2017). Multispecies coalescent delimits structure, not species. *Proceedings of the National Academy of Sciences*, 114(7):1607–1612.

Van Fraassen, B. C. (1980). *The Scientific Image*. Oxford University Press.

Velasco, J. (2019). Foundations of concordance views of phylogeny. *Philosophy, Theory, and Practice in Biology*, 11:20.

vonHoldt, B. M., Pollinger, J. P., Earl, D. A., Knowles, J. C., Boyko, A. R., Parker, H., Geffen, E., Pilot, M., Jedrzejewsky, W., Jedrzejewska, B., Sidorovish, V., Greco, C., Randi, E., Musiani, M., Kays, R., Bustamante, C. D., Ostrander, E. A., Novembre, J., and Wayne, R. K. (2011). A genome-wide perspective on the evolutionary history of enigmatic wolf-like canids. *Genome Research*, 21(8):1294–1305.

Whewell, W. (1840). *Philosophy of the Inductive Sciences, Founded upon Their History, Volume 1*. John W. Parker.

Whewell, W. (1858). *Novum Organon Renovatum*. John W. Parker.

Wiens, J. J. (2004). The role of morphological data in phylogeny reconstruction. *Systematic Biology*, 53(4):653–661.

Wiley, E. O. (1978). The evolutionary species concept reconsidered. *Systematic Zoology*, 27(1):17–26.

Wilkins, J. (2001). Defining evolution. *Reports of the National Centre for Science Education*, 21(1–2):29–37.

Wilkins, J. (2003). How to be a chaste species pluralist-realist. *Biology and Philosophy*, 18(5):621–638.

Wilkins, J. (2009). *Species: A History of the Idea*. University of California Press.

Will, K. W., Mishler, B. D., and Wheeler, Q. D. (2005). The perils of DNA barcoding and the need for integrative taxonomy. *Systematic Biology*, 54(5):844–851.

Wilson, D. E. and Reeder, D. M., editors. (2005). *Mammal Species of the World*, 3rd ed. Johns Hopkins University Press.

Wilson, E. O. (1988). The current state of biological diversity. In Wilson, E. O., editor, *Biodiversity*, pages 3–18. National Academies Press.

Wilson, E. O. (1992a). *The Diversity of Life*. Penguin.

Wilson, E. O. (1992b). Introduction. In Reaka-Kudla, M. L. and Wilson, D. E., editors, *Biodiversity II: Understanding and Protecting our Biological Resources*, pages 1–3. National Academies Press.

Wilson, P. J., Grewal, S., Lawford, I. D., Heal, J. N. M., Granacki, A. G., Pennock, D., Theberge, J. B., Theberge, M. T., Voigt, D. R., Waddell, W., Chambers, W. E., Paquet, P. C., Goulet, G., Cluff, D., and White, B. N. (2000). DNA profiles of the eastern Canadian wolf and the red wolf provide evidence for a common evolutionary history independent of the gray wolf. *Canadian Journal of Zoology*, 78(12):2156–2166.

Xi, Z., Liu, L., Rest, J., and Davis, C. C. (2014). Coalescent versus concatenation methods and the placement of *Amborella* as sister to water lilies. *Systematic Biology*, 63:919–932.

Yang, Z. (2015). The BPP program for species tree estimation and species delimitation. *Current Zoology*, 61(5):854–865.

Yang, Z. and Rannala, B. (2010). Bayesian species delimitation using multilocus sequence data. *Proceedings of the National Academy of Sciences*, 107(20):9264–9269.

Zeveloff, S. I. (2003). A review of the taxonomic and conservation statuses of the island raccoons. *Small Carnivore Conservation*, 29:10–12.

Zhang, C., Zhang, D.-X., Zhu, T., and Yang, Z. (2011). Evaluation of a Bayesian coalescent method of species delimitation. *Systematic Biology*, 60(6):747–761.

Zhang, J., Kapli, P., Pavlidis, P., and Stamatakis, A. (2013). A general species delimitation method with applications to phylogenetic placements. *Bioinformatics*, 29(22):2869–2876.

7 Species Delimitation Using Molecular Data

Megan L. Smith and Bryan C. Carstens

CONTENTS

7.1 INTRODUCTION

The Linnean shortfall, which describes the fact that only a small portion (1–10%) of extant species have been formally described (Brown and Lomolino 1998, but see Mora et al. 2011), is one of the most pressing challenges faced by the biological sciences. A lack of formal species description is likely to complicate conservation assessments (Beheregaray and Caccone 2007), bias evolutionary (Hortal et al. 2015), biogeographical (Whittaker et al. 2005), and ecological (e.g., Prada et al. 2014) studies, and have practical implications for disease ecology (e.g., Byrne et al. 2019), invasive species (Bickford et al. 2007), and wildlife management (Bickford et al. 2007). Amplifying this challenge is the ongoing loss of biodiversity (Costello et al. 2013), which makes addressing the Linnean shortfall a challenge with an inherent expiration date. For several decades, molecular data have been viewed as having the potential to address the Linnean shortfall (e.g., Hebert et al. 2003). However, despite their promise for this application, molecular data have a turbulent history of application in species delimitation, one that is complicated by researcher biases, a clear lack of best practices, and the varying information content of the data itself. While we do not hope to solve these problems in this chapter, we do hope that our discussion of the challenges inherent in delimiting species with genetic data will help researchers adopt useful strategies for practice.

7.2 MOLECULAR DATA AND THEIR INFLUENCE ON SPECIES DELIMITATION

Molecular data are now ubiquitous in the biological sciences. While they are easy to collect at the species level (McCormack et al. 2013) and have become central to

DOI: 10.1201/9780367855604-9

many evolutionary and ecological applications, including species delimitation, the widespread adoption and application of these data required biologists to adjust their thinking in various disciplines. For example, phylogeny inference has become far more quantitative and statistical since molecular data became common; a change prompted both by the increasing size of phylogenetic datasets and the change in the nature of the characters that form the basis of phylogenetic inference (Scornavacca et al. 2020). Similarly, taxonomists have been required to adopt both conceptual and practical changes in their approach to data analysis once massive amounts of molecular data became available to augment the trait data that were traditionally used to delimit species. Perhaps the most important of these was related to perspective. While investigations into the species level necessarily occur at the interface between phylogenetics and population genetics, initial attempts to apply molecular data to the question of species boundaries came primarily from systematists who were trained in phylogenetic biology. Influential papers encouraged researchers to apply phylogenetic thinking to intraspecific variation in a geographic context (e.g., Avise et al. 1987); a suggestion that found a receptive audience in researchers with a background in systematic biology and led to the exploding popularity of phylogeography. Phylogenetic species concepts *sensu lato*, such as genealogical species concepts (e.g., Baum and Shaw 1995) or criteria based on fixed allelic differences (e.g., population aggregation analysis; Nixon and Wheeler 1992), may have been natural outcomes of early efforts to apply phylogeographic data to detect species limits, albeit outcomes that proved difficult to apply in practice (e.g., Palumbi et al. 2001). Two developments from different disciplines, coalescent theory (Kingman 1982) and conceptual work on species concepts (e.g., Mayden 1997; de Queiroz 1998), led to a remarkable shift in how phylogeographic data were applied to the question of species limits.

Once multilocus sequence data became widely available in the early 2000s, researchers began to observe substantial incongruence in the inferred gene trees across sequenced loci (e.g., Funk and Omland 2003). After researchers had been encouraged to conceptualise intraspecific variation as the end point of phylogeny, the many discordant trees that they observed prompted new ways of thinking about the phylogenies that were inferred from sequence data collected in empirical systems. For example, the concept of the species tree was introduced (i.e., gene trees in species trees; Maddison 1997) to differentiate the phylogeny that can be estimated using individual genes' sequence data from the history of organismal diversification. Ultimately, it became more useful to think about phylogeny as a property that emerges from population-level processes because this enables incongruent empirical data to be modeled using coalescent theory (Kingman 1982). This radical shift, which began when the expectations of taxonomists met the realities of phylogeographic data, has resulted in the most substantive shift in systematic biology since the introduction of cladistic analysis.

Coalescent theory describes a stochastic model of the loss of alleles in a population via genetic drift. The broader implications of coalescent theory are relevant to species delimitation, although they were underappreciated until Hudson and Coyne (2002) described in detail the mathematical consequences of using a genealogical species concept. Their argument is as follows: at neutral loci, allelic variation that is present in a lineage at the time of speciation will gradually sort into monophyletic clades in the daughter lineages, but the rate at which this occurs is a property of the effective population size (Ne) of the parent lineage. While the expectation of the time required for this lineage sorting to occur is $4Ne^*$generations (Kingman 1982), Hudson and Coyne demonstrate that there is considerable variance around this expectation; for example, it would take 9–12 Ne^*generations for 95% of sampled loci to be reciprocally monophyletic. Even for species with modest effective population sizes (say 50,000 individuals), taxonomists would not be able to delimit species using a phylogenetic or genealogical species concept that uses monophyly as a criterion for hundreds of thousands of generations after the speciation event has occurred, even in simple cases where a single ancestor forms two new species with no further diversification. Given that many species have larger effective population sizes and complex patterns of diversification that may include introgression, the implication of coalescent theory to species delimitation is clear: genealogical and phylogenetic species concepts are difficult to apply near the species level because genealogies may not reflect the actual species phylogeny. Unless taxonomists are willing to accept that evolutionary lineages that are effectively independent of one another (and may have been for a million years!) do not obtain species status until all of this ancestral variation has sorted via genetic drift, the stochastic realities of lineage sorting require population-level thinking. Coalescent theory presently serves as the statistical foundation of modern phylogeographic inference, but another conceptual development was needed for the potential applicability of coalescent theory to the question of species boundaries to become clear.

Mayden (1997) and de Queiroz (1998) introduced a fundamental shift in how biologists thought about species concepts. They argued that while species concepts disagreed about the criteria used to recognise species (i.e., morphological distinctiveness, reproductive isolation, monophyly), all concepts fundamentally envisioned species as independent evolutionary lineages at the population or metapopulation level. The general lineage concept, proposed by de Queiroz (2005), encouraged researchers to equate species to independently evolutionary lineages regardless of the method used to identify them as such. This outlook on species fits nicely with modern coalescent-based methods for delimiting species. The conceptual unification of this concept with coalescent theory began during a symposium on species delimitation organised by the Society of Systematic Biologists at the 2006 Evolution Annual Meeting in Stoneybrook, New York. Kevin de Queiroz presented a lecture on species concepts, outlined his general lineage concept, and mentioned how coalescent theory makes it possible to extend the general lineage concept into a unified species concept, where independent lineages can be recognised as species (de Queiroz 2007). In the same symposium, Lacey Knowles presented work that described a likelihood ratio test of lineage independence that enabled researchers to delimit species without relying on monophyletic gene trees (Knowles and Carstens 2007). This test utilised data simulated under the coalescent model where two lineages were independent and compared these data with those simulated under a model where the lineages were not independent. The lasting influence of this test has been felt in the general framework of the proposed statistical comparison (i.e., modeling the statistical fit of the data given two models, one where lineages are independent and one where lineages are combined), as many newer methods are based on statistical comparisons of species trees that include different groupings of putative species (Yang and Rannala 2010; Ence and Carstens 2011; Grummer et al. 2014; Leaché et al. 2014a).

Under the unified species concept, independently evolving lineages can be delimited as distinct species. While on a superficial level, this may appear to eliminate subjective decisions from the process of species delimitation, this definition of species is likely to result in over-splitting under some models of speciation when population genetic structure is present (Sukumaran and Knowles 2017). As more genomic data are gathered, many algorithms become more effective at identifying population genetic structure, highlighting the need for sanity checks in species delimitation (e.g., Jackson et al. 2017b), where the intuition of the taxonomist is considered. Given that taxonomists generally do not wish to name all populations as species due to practical considerations (Zachos et al. 2020), additional considerations may be required. For example, Zachos et al. (2020) distinguish between the process of grouping organisms into 'species taxa' and making the subjective decision of whether these taxa should be ranked as species in the Linnaean classification system. To the extent that researchers do not view each independently evolving population as warranting species recognition, the coalescent and related models and methods cannot address this second aspect of species delineation, which requires taxonomic expertise and subjective thought. Regardless, coalescent-based approaches to species delimitation provide valuable information about the status and history of species taxa, and this information can serve as the basis for integrative taxonomic efforts.

7.3 PRACTICAL CONSIDERATIONS IN SPECIES DELIMITATION

Genomes accumulate nucleotide substitutions at a rate that is influenced by demographic processes (i.e., gene flow, population size change), natural selection, and recombination as the population evolves over time. The pattern of nucleotide variation across individuals sampled from multiple lineages within a species complex will retain information about the recent history of that complex, and any method used to delimit species with molecular data will attempt to access this information. However, decisions made by researchers can potentially influence the results of a species delimitation analyses. Perhaps the most important factor to consider at the outset of an investigation is the sample design. As with any source of inference, the strength of the signal is likely to be positively correlated with the size of the dataset, although comprehensive evaluations of this relationship have not been conducted for all methods. Note that the size of the dataset is best measured on two axes, the number of loci and the number of samples, as the former determines how many independent realizations of the coalescent process are sampled, and the latter determines how well the allelic and/or genotype frequencies of the sample match the actual values from the empirical system. An equally important consideration is to document what information exists about potential division of individuals within a nominal taxon. For example, are there described subspecies? Allopatric populations? Evident environmental gradients that could serve to divide a population? Any of these factors could serve to guide researchers as they acquire samples and choose individuals for sequencing. They can also influence the types of analyses that are chosen by researchers once the genetic data are collected. Related to each of these is the question of what type of genetic data to collect. Data can be collected on a locus-by-locus basis using polymerase chain reaction and Sanger sequencing methods, but this can be tedious work. Next-generation-sequencing technologies enable research-ers to collect data from thousands of loci, either in the form of sequence capture techniques (e.g., Faircloth et al. 2012) or using restriction-digest approaches (Miller et al. 2007). Notably, the technology used for sequencing also affects downstream methodological choices, as some methods are designed for use with single nucleotide polymorphism (SNP) data while others are designed for sequence data.

Carstens et al. (2013) proposed that researchers conceptualise species delimi-tation as a two-step process. Since many taxa lack obvious partitions, such as described subspecies or populations that are clearly allopatric, the first analyses for many investigations should be discovery approaches that do not require samples to be partitioned prior to analysis. Discovery approaches include methods such as STRUCTURE (Pritchard et al. 2000) or ADMIXTURE (Alexander and Lange 2011), which implement algorithms that cluster samples into groups based on some criterion, such as minimizing Hardy–Weinberg disequilibrium, as well as methods based on genetic distances (e.g., Automatic Barcode Gap Discovery [ABGD]) and those based on gene tree diversification (e.g., Generalized Mixed Yule Coalescent [GMYC)]; Pons et al. 2006). The key information obtained via the use of these methods is a division of the samples into two or more groups that can serve as the basis for the next step of species delimitation. Methods that require samples to be partitioned prior to analy-sis, such as species-tree-based programs (e.g., BPP (Yang and Rannala 2010), BFD* (Leaché et al. 2014a) and those based on demographic models (e.g., delimitR (Smith and Carstens 2020), PHRAPL (Jackson et al. 2017a)), work on some level by comparing the probability of the data given the model where a key component of the model is the assignment of samples to each putative lineage. See Box 7.1 for additional exam-ples of species discovery and species validation approaches, and Rannala and Yang (2020) for a recent review of several approaches. Note that some investigations omit the first step (i.e., discovery) because there are *a priori* groupings of samples (e.g., Morales and Carstens 2018). Others conduct both steps sequentially, with sample partitions in the validation stage informed by the clustering of samples from the discovery phase (e.g., Leaché and Fujita 2010). One challenge to this approach is how to treat samples where there is evidence of admixture (i.e., genetic ancestry in an individual sample that can be traced to two or more populations). Some research-ers remove these samples from the validation analysis, since we know that gene flow can interfere with species tree estimation (Eckert and Carstens 2008; Leaché et al. 2014b). However, this should not be done if divergence with gene flow models are included, because it could presumably bias the validation analysis. Notably, some discovery and most validation approaches rely on particular models of the specia-tion process, and the choice and application of such models can greatly impact the results of species delimitation analyses. In the following, we discuss popular models employed in species delimitation and their potential shortfalls.

7.4   THE IMPORTANCE OF MODELS

The Multispecies Coalescent Model (MSCM) addresses the difficulties of applying genealogical and phylogenetic species concepts near the species level by directly modeling the coalescent process (Knowles and Carstens 2007). By modeling the causes of incomplete lineage sorting, methods based on the MSCM allow researchers to go beyond a monophyly criterion and address whether observed genealogies are consistent with different numbers of species. Species delimitation methods based on the MSCM have proliferated since its development (e.g., Yang and Rannala 2010; Ence and Carstens 2011; Leaché et al. 2014a), and many allow researchers to use genetic data to assess the probability of different numbers of species.

While the MSCM is undoubtedly a powerful approach to delimiting species with genetic data, it is not without its limitations. As with any model-based approach, the MSCM makes certain assumptions, which if violated, may render the results of species delimitation under the model unreliable. For example, MSCM methods rely on *a priori* definitions of populations or putative species (i.e., they are validation approaches). When populations are estimated using genetic data from sparse sampling, geographic clines can be mistaken for discrete populations, and this can lead to over-splitting under the MSCM (Chambers and Hillis 2020). The MSCM also assumes that speciation is an instantaneous process, and recent results demonstrate that when this is not the case, but rather, speciation is protracted, the MSCM will over-split, delimiting population structure as distinct species (Sukumaran and Knowles 2017).

Perhaps the best-known violation of the MSCM is the presence of gene flow between populations or species. The MSCM models only genetic divergence and does not consider the possibility of post-divergence gene flow between lineages. However, gene flow is thought to be important in speciation, and is implicated in many empirical systems, including *Myotis* bats (Morales et al. 2017) and flowering plants on Lord Howe Island (Papadopulos et al. 2011). Simulation studies demonstrate that ignoring gene flow causes overestimates of population sizes and underestimates of divergence times under the MSCM (Leaché et al. 2014b), and BPP (Yang and Rannala 2010) may delimit populations as species even when levels of gene flow between populations are high (Jackson et al. 2017b; Leaché et al. 2019). However, recent attempts to use more appropriate models that consider gene flow, for example, have improved error rates and led to more meaningful species delimitation (Jackson et al. 2017b; Leaché et al. 2019; Smith and Carstens 2020).

Considering these results, it is clear that the choice of appropriate models is essential for species delimitation using genetic data. While choosing an appropriate model is not always straightforward, recent advances in simulation-based approaches provide a promising avenue for species delimitation. Software for simulating large genomic datasets under models including divergence, gene flow, and population size changes has improved vastly in speed and computational efficiency in recent years (e.g., Excoffier et al. 2013). More recently, the development of tree-sequence recording has permitted simulating tens of thousands of replicates of genomic datasets under models that include selection as well as demographic processes (Kelleher et al. 2016; Haller et al. 2019). The ability to simulate many replicates of large genomic datasets under various models permits researchers to then use either Approximate Bayesian Computation (e.g., Camargo et al. 2012) or machine learning approaches (e.g., Pei et al. 2018; da Fonseca et al. 2020; Smith and Carstens 2020) to find the model that generates data most similar to the observed data. By combining new powerful simulation approaches with machine learning, researchers are effectively limited only by their creativity and the computational resources available when designing a model set. It should be noted that larger model sets inevitably lead to increased difficulties in differentiating among models (Pelletier and Carstens 2014), even when machine learning approaches are employed (Smith and Carstens 2020), because the distance in model space between these models decreases. While choosing models to compare *a priori* requires researchers to make decisions about which processes are likely to be important in their focal system, researchers are implicitly making such decisions when they utilise tools that explore a limited number of processes, like methods based on the MSCM. By using approaches that ignore processes like gene flow, researchers assume that those processes are not important. Leaving the power to determine which models to test to researchers who are experts in their study system takes advantage of their knowledge of the taxa, similarly to traditional taxonomic investigations. We view this aspect of defining a model set as a positive aspect of species delimitation, but it could lead to biases when all models considered are a poor fit to the data, or when researchers limit their model set too strictly to match misleading *a priori* knowledge of the study system. Tools to directly assess model fit, like posterior predictive simulations (e.g., Fonseca et al. 2021) or composite likelihood ratio tests (e.g., Excoffier et al. 2013), may help researchers to diagnose such situations.

Evaluating a broader array of models not only prevents erroneous inference due to model violations but also may provide novel insights into the processes driving speciation. Different modes of speciation involve different demographic and selective processes, and by modeling these processes directly, researchers may be able to address not only how many species are present but also the processes that gave rise to these species. For example, gene flow and directional selection may play an important role when divergent ecological selection drives speciation in sympatry (Coyne and Orr 2004) or when reinforcement drives speciation between once isolated populations (e.g., in Phlox; Hopkins and Rausher 2011). On the other hand, when speciation occurs in allopatry, genetic drift, natural selection, or some combination of the two may drive divergence (Coyne and Orr 2004). By designing models based on predictions about the mode of speciation and then using machine learning or other approaches to determine which of those models best reproduces the observed data, researchers can identify the most likely mode of speciation in their system as well as the number of species present. Of course, doing so requires researchers to explicitly state an operational species concept.

While model selection itself provides insights into the number of species and the process of speciation, it also permits more accurate parameter estimation (Thomé and Carstens 2016). When parameters are accurately estimated, they provide insight into the magnitude of divergence and gene flow between populations, essential parameters for determining whether populations represent independent evolutionary units (Rannala and Yang 2020). Additionally, parameter estimates may lend insight into the correspondence of speciation events with geologic and climatic processes.

For example, more precise parameter estimates might provide resolution on how the Pleistocene glaciations impacted speciation, or the extent to which divergence can occur with gene flow.

7.5 PROSPECTS FOR THE FUTURE

As genomic data become increasingly available near the species level, opportunities to connect process to pattern in taxonomy are ripe. Already, with the rise in popularity of the Multispecies Coalescent Model in species delimitation, taxonomists have begun to embrace the link between population-level and species-level processes and to use models based explicitly on these processes to evaluate species delimitation hypotheses. With further advances in the nature of genetic data and in the models and computational tools available to taxonomists, we believe that the field of molecular-based species delimitation will rely increasingly on evolutionary genetics, linking genetic variation to the specific evolutionary processes that drove speciation. As our understanding of the importance of selective processes on structuring genetic variation increases, future developments may take advantage of this and model speciation as the complex interplay of neutral and selective processes that it is. This should shed additional light on the history of populations and prove invaluable to taxonomists when evaluating the species status of lineages.

Although molecular data held (and continue to hold) great promise for species delimitation, the importance of other data sources, including morphological and ecological data, should not be overlooked. The call for so-called integrative taxonomy (Weins and Penkrot 2002; Sites and Marshall 2004; Dayrat 2005; Winker 2009; Padial et al. 2010; Schlick-Steiner et al. 2010; Yeates et al. 2011) highlights the potential benefits of combining data types when inferring species boundaries. Phenotypic and ecological data have further power to illuminate the process of speciation and to allow researchers to distinguish among population- and species-level variation (Cadena and Zapata 2021). Further, following up molecular studies with phenotypic and ecological investigations may provide diagnostic characters, without which the recognition of distinct species in the field by conservation biologists and ecologists is impossible. In short, the availability of molecular data does not eliminate the need for phenotypic and ecological data. Rather, by combining molecular, phenotypic, and ecological data, researchers can better understand how genetic divergence, phenotypic divergence, and ecological divergence differ across putative species, which should not only inform taxonomic efforts but also shed light on the processes of speciation and diversification in a way that either data type on its own could not (Winker 2009; Cadena and Zapata 2021).

7.6 SPECIES DESCRIPTION IS A NECESSARY LAST STEP IN A DELIMITATION ANALYSIS

Although species delimitation studies have flourished in recent years, a remarkably small number of those studies follow up with the description of delimited species. Pante et al. (2015) found that ~47% of integrative taxonomy studies published between 2008 and 2013 did not describe newly delimited species. Without following up with taxonomic revisions, species delimitation studies hardly address the Linnean shortfall that we often claim as the motivation for the field. A variety of factors likely contribute to the failure of many studies to describe the species that are inferred by species delimitation investigations. First, the lack of species description might signal a lack of confidence in the results – an unwillingness to commit to the delimited species (Pante et al. 2015). As mentioned earlier, taxonomists may not view all independently evolving populations as warranting formal species recognition (Zachos et al. 2020), and thus, some lack of species description may only reflect that delimited entities do not meet a particular taxonomist's criteria for describing a new species. Second, it may be that researchers plan to follow up the delimitation results with further morphological, behavioral, or other types of taxonomic investigation (Pante et al. 2015), particularly since describing species with non-traditional characters (e.g., molecular characters) remains difficult (Satler et al. 2013). Due to the ease of collecting molecular data, it could be that these investigations are more easily completed and published than integrative work that incorporates multiple data types. Third, researchers could feel inhibited by the formal rules associated with taxonomic description in the Zoological or Botanical Codes, potentially due to a lack of training (Pante et al. 2015; Pearson et al. 2011). Finally, there are likely to be fewer professional rewards for publishing in the taxonomic literature, for while these papers have a long potential history of citation, they likely will receive less notice in the immediate future than works published in the general interest literature. The competition for space is fierce for journals in the latter category, and editors might balk at devoting several pages to species description; thus, general interest journals rarely publish taxonomic revisions (Pante et al. 2015; Agnarsson and Kunter 2007). Pressure to publish in journals with high impact factors may therefore discourage authors from including species descriptions in their work. For species delimitation to address the Linnean shortfall, these issues must be addressed, so that discovered species are subsequently described. We urge funding panels to demand that proposals which include species delimitation also include species description. Furthermore, senior scientists need to be more vocal to their administration in highlighting the importance of species description, particularly when conducted by early career researchers.

BOX 7.1   Examples of Popular Species Discovery Approaches

Population Genetic Structure. STRUCTURE uses a Bayesian clustering approach to assign individuals to populations and estimate population allele frequencies (Pritchard et al. 2000). STRUCTURE assumes that markers are unlinked and that each population is in Hardy–Weinberg equilibrium (Pritchard et al. 2000). Recent advances have improved the computational efficiency of the approach used in STRUCTURE (Raj et al. 2014). STRUCTURE is often combined with ad-hoc methods (e.g., Earl 2012; Evanno et al. 2005) to estimate the number of populations. Like STRUCTURE, structurama (Huelsenbeck et al. 2011) assumes

Hardy–Weinberg equilibrium, but it also uses a Dirichlet-process prior on the number of populations and reversible jump MCMC to allow simultaneous inference of population assignments and the number of populations.

Generalized Mixed Yule Coalescent (GMYC). The GMYC (Pons et al. 2006) takes ultrametric gene trees (i.e., rooted trees where a molecular clock has been enforced) as input. It then infers the transition point between branching events corresponding speciation events (the Yule process), and branching events corresponding to allele coalescence within species (the coalescent process). Reid and Carstens (2012) introduced a Bayesian implementation of the GMYC (**bGMYC**), which takes as input a posterior distribution of gene trees and outputs posterior distributions of the number and composition of species.

Automatic Barcode Gap Discovery (ABGD). ABGD (Puillandre et al. 2012) takes as input short sequences and searches for a gap in the distribution of pairwise differences between sequences. ABGD requires that the user supply a prior maximum divergence of intraspecific diversity, and this value determines how finely ABGD divides individuals into species. Like the GMYC, ABGD is limited to single-locus data.

Multivariate Methods. Multivariate methods (e.g., methods based on Principal Components Analysis [PCA]) are powerful because they are fast and do not make assumptions about evolutionary models generating population and species divergence. Adegenet is a popular software package that combines PCA and Discriminant Analysis of Principal Components to assign individuals to populations (Jombart et al. 2010).

Machine Learning. Recently, Derkarabetian et al. (2019) applied a suite of machine learning approaches to perform species discovery analysis. They applied Random Forests, Variational Autoencoders, and t-Distributed Stochastic Neighbor Embedding to assign individuals to populations (or species) and found that they have high power to identify population structure. As with the multivariate methods described earlier, these approaches do not make assumptions about the evolutionary models generating population and species divergence.

BOX 7.2   Examples of Popular Species Validation Approaches

BPP. BPP (Yang and Rannala 2010) is a fully Bayesian approach to multilocus species delimitation. BPP takes as input DNA sequence data and uses reversible jump MCMC to estimate the species tree topology, the number of populations, genetic diversity, and population divergence. Recently, the model underlying BPP was extended to allow estimation of introgression probabilities (Flouris et al. 2019).

Bayes Factor Delimitation. Bayes Factors can be used to compare models when marginal likelihoods of the competing models are available and were initially used in the context of species delimitation by Carstens and Dewey (2010) and Grummer et al. (2014). The popular species delimitation software BFD* (Leaché et al. 2014a) applies Bayes Factor Delimitation to genome-wide SNP data by using SNAPP to calculate marginal likelihoods for all hierarchical arrangements of individuals into predefined populations and then comparing models using Bayes Factors.

PHRAPL. PHRAPL (Jackson et al. 2017a) takes as input gene trees and population assignments and can be applied to species delimitation (Jackson et al. 2017b). To compare different delimitation hypotheses, PHRAPL approximates the likelihood of models that differ in the number of species and the species history and compares models using information theory.

CLADES. CLADES (Pei et al. 2018) simulates data belonging to either the same or different species and then trains a support vector machine (SVM) to recognise whether two samples come from the same or different species. Using this SVM, CLADES can classify DNA sequence data sampled from populations as belonging to the same or different species. Finally, if there are more than two putative species, CLADES maximises the likelihood of the species status of all populations.

delimitR. delimitR (Smith and Carstens 2020) takes as input a multidimensional Site Frequency Spectrum (mSFS) and allows users to specify models including divergence, migration (primary or upon secondary contact), and population size changes. Then, fastsimcoal2 (Excoffier et al. 2013) is used to simulate mSFS under each model. Finally, a Random Forest classifier is constructed using the R package 'abcrf' (Pudlo et al. 2016) and used to select the best model and to estimate classification error rates under each model.

SODA. SODA (Rabiee and Mirarab 2021) uses the algorithm of the popular species tree inference software ASTRAL (Mirarab et al. 2014; Zhang et al. 2018) and expected patterns of quartet frequencies to infer species boundaries.

REFERENCES

Agnarsson, I., and M. Kunter. 2007. Taxonomy in a changing world: Seeking solutions for a science in crisis. *Systematic Biology* 56:531–539.

Alexander, D. H., and K. Lange. 2011. Enhancements to the ADMIXTURE algorithm for individual ancestry estimation. *BMC Bioinformatics* 12:246.

Avise, J. C., J. Arnold, R. M. Ball, E. Bermingham, T. Lamb, J. E. Neigel, C. A. Reeb, and N. C. Saunders. 1987. Intraspecific phylogeography: The mitochondrial DNA bridge between population genetics and systematics. *Annual Review of Ecology and Systematics* 18:489–522.

Baum, D. A., and K. L. Shaw. 1995. Genealogical perspectives on the species problem. *Experimental and Molecular Approaches to Plant Biosystematics* 53:123–124.

Beheregaray, L. B., and A. Caccone. 2007. Cryptic biodiversity in a changing world. *Journal of Biology* 6:9.

Bickford, D., D. J. Lohman, N. S. Sodhi, P. K. L. Ng, R. Meier, K. Winker, K. K. Ingram, and I. Das. 2007. Cryptic species as a window on diversity and conservation. *Trends in Ecology and Evolution* 22:148–155.

Brown, J. H., and M. V. Lomolino. 1998. *Bioegeography* (2nd ed.). Sunderland: Sinauer.

Byrne, A. Q., V. T. Vredenburg, A. Martel, F. Pasmans, R. C. Bell, D. C. Blackburn, M. C. Bletz, J. Bosch, C. J. Briggs, R. M. Brown, A. Catenazzi, M. Familiar López, R. Figueroa-Valenzuela, S. L. Ghose, J. R. Jaeger, A. J. Jani, M. Jirku, R. A. Knapp, A. Muñoz, D. M. Portik, C. L. Richards-Zawacki, H. Rockney, S. M. Rovito, T. Stark, H. Sulaeman, N. T. Tao, J. Voyles, A. W. Waddle, Z. Yuan, and E. B. Rosenblum. 2019. Cryptic diversity of a widespread global pathogen reveals expanded threats to amphibian conservation. *Proceedings of the National Academy of Sciences* 116:20382–20387.

Cadena, C. D., and F. Zapata. 2021. The genomic revolution and species delimitation in birds (and other organisms): Why phenotypes should not be overlooked. *Ornithology* 138:ukaa069.

Camargo, A., M. Morando, L. J. Avila, and J. W. Sites Jr. 2012. Species delimitation with ABC and other coalescent-based methods: A test of accuracy with simulations and an empirical example with lizards of the *Liolaemus darwinii* complex (Squamata: Liolaemidae). *Evolution: International Journal of Organic Evolution* 66:2834–2849.

Carstens, B. C., and T. A. Dewey. 2010. Species delimitation using a combined coalescent and information theoretic approach: An example from North American *Myotis* bats. *Systematic Biology* 59:400–414.

Carstens, B. C., T. A. Pelletier, N. M. Reid, and J. D. Satler. 2013. How to fail at species delimitation. *Molecular Ecology* 22:4369–4383.

Chambers, E. A., and D. M. Hillis. 2020. The multispecies coalescent over-splits species in the case of geographically widespread taxa. *Systematic Biology* 69:184–193.

Costello, M. J., R. M. May, and N. E. Stork. 2013. Can we name Earth's species before they go extinct? *Science* 339:413–416.

Coyne, J., and H. Orr. 2004. *Speciation.* Sunderland: Sinauer.

da Fonseca, E. M., G. R. Colli, F. P. Werneck, and B. C. Carstens. 2020, Phylogeographic model selection using convolutional neural networks. *Molecular Ecology* 21:2661–2675.

Dayrat, B. 2005. Towards integrative taxonomy. *Biological Journal of the Linnean Society* 85:407–417.

de Queiroz, K. 1998. The general lineage concept of species, species criteria, and the process of speciation: A conceptual unification and terminological recommendations. In *Endless forms: Species and speciation*, eds. D. J. Howard and S. H. Berlocher, 57–75. Oxford: Oxford University Press.

de Queiroz, K. 2005. Different species problems and their solutions. *Bioessays* 26:67–70.

de Queiroz, K. 2007. Species Concepts and Species Delimitation. *Systematic Biology* 56:879–886.

Derkarabetian, S., S. Castillo, P. K. Koo, S. Ovchinnikov, and M. Hedin. 2019. A demonstration of unsupervised machine learning in species delimitation. *Molecular Phylogenetics and Evolution* 139:106562.

Earl, D. A. 2012. STRUCTURE HARVESTER: A website and program for visualizing STRUCTURE output and implementing the Evanno method. *Conservation Genetics Resources* 4:359–361.

Eckert, A. J., and B. C. Carstens. 2008. Does gene flow destroy phylogenetic signal? The performance of three methods for estimating species phylogenies in the presence of gene flow. *Molecular Phylogenetics and Evolution* 49:832–842.

Ence, D. D., and B. C. Carstens. 2011. SpedeSTEM: A rapid and accurate method for species delimitation. *Molecular Ecology Resources* 11:473–480.

Evanno, G., S. Regnaut, and J. Goudet. 2005. Detecting the number of clusters of individuals using the software STRUCTURE: A simulation study. *Molecular Ecology* 14:2611–2620.

Excoffier, L., I. Dupanloup, E. Huerta-Sánchez, V. C. Sousa, and M. Foll. 2013. Robust demographic inference from genomic and SNP data. *PLoS Genetics* 9:10.

Faircloth, B. C., J. E. McCormack, N. G. Crawford, M. G. Harvey, R. T. Brumfield, and T. C. Glenn. 2012. Ultraconserved elements anchor thousands of genetic markers spanning multiple evolutionary timescales. *Systematic Biology* 61:717–726.

Flouris, T., X. Jiao, B. Rannala, and Z. Yang. 2019. A Bayesian implementation of the multispecies coalescent model with introgression for phylogenomic analysis. *Molecular Biology and Evolution* 37:1211–1223.

Fonseca, E. M., D. J. Duckett, and B. C. Carstens. 2021. P2C2M. GMYC: An R package for assessing the utility of the Generalized Mixed Yule Coalescent model. *Methods in Ecology and Evolution* 12:487–493.

Funk, D. J., and K. E. Omland. 2003. Species-level paraphyly and polyphyly: Frequency, causes, and consequences, with insights from animal mitochondrial DNA. *Annual Review of Ecology, Evolution, and Systematics* 34:397–423.

Grummer, J. A., R. W. Bryson, and T. W. Reeder. 2014. Species delimitation using Bayes factors: Simulations and application to the *Sceloporus scalaris* species group. *Systematic Biology* 63:119–133.

Haller, B. C., J. Galloway, J. Kelleher, P. W. Messer, and P. L. Ralph. 2019. Tree-sequence recording in SLiM opens new horizons for forward-time simulation of whole genomes. *Molecular Ecology Resources* 19:552–566.

Hebert, P. D. N., A. Cywinska, S. L. Ball, and J. R. deWaard. 2003. Biological identifications through DNA barcodes. *Proceedings of the Royal Society B* 270:313–321.

Hopkins, R., and M. D. Rausher. 2011. Identification of two genes causing reinforcement in the Texas wildflower *Phlox drummondii*. *Nature* 469:411–414.

Hortal, J., F. de Bello, J. A. F. Diniz-Filho, T. M. Lewinsohn, J. M. Lobo, and R. J. Ladle. 2015. Seven shortfalls that beset large-scale knowledge of biodiversity. *Annual Review of Ecology, Evolution, and Systematics* 46:523–549.

Hudson, R. R., and J. A. Coyne. 2002. Mathematical consequences of the genealogical species concept. *Evolution* 56:1557–1565.

Huelsenbeck, J. P., P. Andolfatto, and E. T. Huelsenbeck. 2011. Structurama: Bayesian inference of population structure. *Evolutionary Bioinformatics* 7:EBO-S6761.

Jackson, N. D., A. E. Morales, B. C. Carstens, and B. C. O'Meara. 2017a. PHRAPL: Phylogeographic inference using approximate likelihoods. *Systematic Biology* 66:1045–1053.

Jackson, N. D., B. C. Carstens, A. E. Morales, and B. C. O'Meara. 2017b. Species delimitation with gene flow. *Systematic Biology* 66:799–812.

Jombart, T., S. Devillard, and F. Balloux. 2010. Discriminant analysis of principal components: A new method for the analysis of genetically structured populations. *BMC Genetics* 11:1–15.

Kelleher, J., A. M. Etheridge, and G. McVean. 2016. Efficient coalescent simulation and genealogical analysis for large sample sizes. *PLOS Computational Biology* 12:e1004842.

Kingman, J. F. C. 1982. The coalescent. *Stochastic Processes and Their Applications* 13:235–248.

Knowles, L. L., and B. C. Carstens. 2007. Delimiting species without monophyletic gene trees. *Systematic Biology* 56:887–895.

Leaché, A. D., and M. K. Fujita. 2010. Bayesian species delimitation in West African forest geckos (*Hemidactylus fasciatus*). *Proceedings of the Royal Society B* 277:3071–3077.

Leaché, A. D., M. K. Fujita, V. N. Minin, and R. R. Bouckaert. 2014a. Species delimitation using genome-wide SNP data. *Systematic Biology* 63:534–542.

Leaché, A. D., R. B. Harris, B. Rannala, and Z. Yang. 2014b. The influence of gene flow on species tree estimation: A simulation study. *Systematic Biology* 63:17–30.

Leaché, A. D., T. Zhu, B. Rannala, and Z. Yang. 2019. The spectre of too many species. *Systematic Biology* 68:168–181.

Maddison, W. P. 1997. Gene trees in species trees. *Systematic Biology* 46:523–536.

Mayden, R. L. 1997. A hierarchy of species concepts: The denouement in the saga of the species problem. In *Species: The units of diversity*, eds. M. F. Claridge, H. A. Dawah, and M. R. Wilson, 381–423. London: Chapman & Hall.

McCormack, J. E., S. M. Hird, A. J. Zellmer, B. C. Carstens, and R. T. Brumfield. 2013. Applications of next-generation sequencing to phylogeography and phylogenetics. *Molecular Phylogenetics and Evolution* 66:526–538.

Miller, M. R., J. P. Dunham, A. Amores, W. A. Cresko, and E. A. Johnson. 2007. Rapid and cost-effective polymorphism identification and genotyping using restriction site associated DNA (RAD) markers. *Genome Research* 17:240–248.

Mirarab, S., R. Reaz, Md. S. Bayzid, T. Zimmermann, M. S. Swenson, and T. Warnow. 2014. ASTRAL: Genome-scale coalescent-based species tree estimation. *Bioinformatics* 30:i541–i548.

Mora, C., D. P. Tittensor, S. Adl, A. G. B. Simpson, and B. Worm. 2011. How many species are there on earth and in the ocean? *PLoS Biology* 9:e1001127.

Morales, A. E., and B. C. Carstens. 2018. Evidence that *Myotic lucifigus* "subspecies" are five nonsister species, despite gene flow. *Systematic Biology* 67:756–769.

Morales, A. E., N. D. Jackson, T. A. Dewey, B. C. O'Meara, and B. C. Carstens. 2017. Speciation with gene flow in North American Myotis bats. *Systematic Biology* 66:440–452.

Nixon, K. C., and Q. D. Wheeler. 1992. Extinction and the origin of species. In *Extinction and phylogeny*, eds. M. J. Novacek and Q. D. Wheeler, 119–143. New York: Columbia University Press.

Padial, J. M., A. Miralles, I. De la Riva, and M. Vences. 2010. The integrative future of taxonomy. *Frontiers in Zoology* 7:6.

Palumbi, S. R., F. Cipriano, and M. P. Hare. 2001. Predicting nuclear gene coalescence from mitochondrial data: The three-times rule. *Evolution* 55:859–868.

Pante, E., C. Schoelinck, and N. Puillandre. 2015. From integrative taxonomy to species description: One step beyond. *Systematic Biology* 64:152–160.

Papadopulos, A. S., W. J. Baker, D. Crayn, R. K. Butlin, R. G. Kynast, I. Hutton, and V. Savolainen. 2011. Speciation with gene flow on Lord Howe Island. *Proceedings of the National Academy of Sciences* 108:13188–13193.

Pearson, D. L., A. L. Hamilton, and T. L. Erwin. 2011. Recovery plan for the endangered taxonomy profession. *BioScience* 61:58–63.

Pei, J., C. Chu, X. Li, B. Lu, and Y. Wu. 2018. CLADES: A classification-based machine learning method for species delimitation from population genetic data. *Molecular Ecology Resources* 18:1144–1156.

Pelletier, T. A., and B. C. Carstens. 2014. Model choice for phylogeographic inference using a large set of models. *Molecular Ecology* 23:3028–3043.

Pons, J., T. G. Barraclough, J. Gomez-Zurita, A. Cardoso, D. P. Duran, S. Hazell, S. Kamoun, W. D. Sumlin, and A. P. Vogler. 2006. Sequence-based species delimitation for the DNA taxonomy of undescribed insects. *Systematic Biology* 55:595–609.

Prada, C., S. E. McIlroy, D. M. Beltrán, D. J. Valint, S. A. Ford, M. E. Hellberg, and M. A. Coffroth. 2014. Cryptic diversity hides host and habitat specialization in a gorgonian-algal symbiosis. *Molecular Ecology* 23:3330–3340.

Pritchard, J. K., M. Stephens, and P. Donnelly. 2000. Inference of population structure using multilocus genotype data. *Genetics* 155:945–959.

Pudlo, P., J. -M. Marin, A. Estoup, J. -M. Cornuet, M. Gautier, and C. P. Robert. 2016. Reliable ABC model choice via random forests. *Bioinformatics* 32:859–866.

Puillandre, N., A. Lambert, S. Brouillet, and G. Achaz. 2012. ABGD, automatic barcode gap discovery for primary species delimitation. *Molecular Ecology* 21:1864–1877.

Rabiee, M., and S. Mirarab. 2021. SODA: Multi-locus species delimitation using quartet frequencies. *Bioinformatics* 36:5623–5631.

Raj, A., M. Stephens, and J. K. Pritchard. 2014. FastSTRUCTURE: Variational inference of population structure in large SNP data sets. *Genetics* 197:573–589.

Rannala, B., and Z. Yang. 2020. Species delimitation. In *Phylogenetics in the genomic era*. eds. C. Scornavacca, F. Delsuc, N. Galtier, 5.5:1–5.5:18. Self-published.

Reid, N. M., and B. C. Carstens. 2012. Phylogenetic estimation error can decrease the accuracy of species delimitation: A Bayesian implementation of the general mixed Yule-coalescent model. *BMC Evolutionary Biology* 12:196.

Satler, J. D., B. C. Carstens, and M. Hedin. 2013. Multilocus species delimitation in a complex of morphologically conserved trapdoor spiders (Mygalomorphae, Antrodiaetidae, *Aliatypus*). *Systematic Biology* 62:805–823.

Schlick-Steiner, B. C., F. M. Steiner, B. Seifert, C. Stauffer, E. Christian, and R. H. Crozier. 2010. Integrative taxonomy: A multisource approach to exploring biodiversity. *Annual Review of Entomology* 55:421–438.

Scornavacca, C., F. Delsuc, and N. Galtier. 2020. *Phylogenetics in the genomic era*. Open access book available from https://hal.inria.fr/PGE/.

Sites, J. W., and J. C. Marshall. 2004. Operational criteria for delimiting species. *Annual Review of Ecology, Evolution, and Systematics* 35:199–227.

Smith, M. L., and B. C. Carstens. 2020. Process-based species delimitation leads to identification of more biologically relevant species. *Evolution* 74:216–229.

Sukumaran, J., and L. L. Knowles. 2017. Multispecies coalescent delimits structure, not species. *Proceedings of the National Academy of Sciences* 114:1607–1612.

Thomé, M. T. C., and B. C. Carstens. 2016. Phylogeographic model selection leads to insight into the evolutionary history of four-eyed frogs. *Proceedings of the National Academy of Sciences* 113:8010–8017.

Weins, J. J., and T. A. Penkrot. 2002. Delimiting species using DNA and morphological variation and discordant species limits in spiny lizards (*Sceloporus*). *Systematic Biology* 51:69–91.

Whittaker, R. J., M. B. Araújo, P. Jepson, R. J. Ladle, J. E. Watson, and K. J. Willis. 2005. Conservation biogeography: Assessment and prospect. *Diversity and Distributions* 11:3–23.

Winker, K. 2009. Reuniting phenotype and genotype in biodiversity research. *BioScience* 59:657–665.

Yang, Z., and B. Rannala. 2010. Bayesian species delimitation using multilocus sequence data. *Proceedings of the National Academy of Sciences* 107:9264–9269.

Yeates, D. K., A. Seago, L. Nelson, S. L. Cameron, L. Joseph, and J. W. H. Trueman. 2011. Integrative taxonomy, or iterative taxonomy? *Systematic Entomology* 36:209–217.

Zachos, F. E., L. Christidis, and S. T. Garnett. 2020. Mammalian species and the twofold nature of taxonomy: A comment on Taylor et al. 2019. *Mammalia* 84:1–5. https://doi.org/10.1515/mammalia-2019-0009.

Zhang, C., M. Rabiee, E. Sayyari, and S. Mirarab. 2018. ASTRAL-III: Polynomial time species tree reconstruction from partially resolved gene trees. *BMC Bioinformatics* 19:153.

8 Taxonomic Order, Disorder, and Governance

Stijn Conix, Stephen T. Garnett,
Frank E. Zachos, and Les Christidis

CONTENTS

8.1 INTRODUCTION

The grey parrot (*Psittacus erithacus*), an endangered bird native to equatorial Africa, figures on all four major lists of the world's birds. However, this is not a sign of consensus among ornithologists, as two of the lists recognise one species, while the other two lists recognise two species. Similarly, all four lists include a babbler taxon *woodi*, which is treated as a subspecies of *lanceolatus* by one list and as a distinct species by the other three lists (del Hoyo 2020). There is also debate as to whether these two babbler taxa belong to the genus *Garrulax* or *Babax* (Dickinson and Christidis 2014). More generally, McClure et al. (2020) compared the four major bird lists and found that of the 665 raptor species recognised by at least one list, only 453, or 68%, were found in all four lists. Examples like these are also easy to find in other parts of the Tree of Life and illustrate what we will call 'taxonomic disorder': the classification of organisms into species is, at least for some groups, in a state of disarray. This disorder comes in a variety of forms. Groups with the same taxonomic rank are not always comparable, the same name is often used for multiple groups, some groups are known under a variety of names, and taxonomists frequently disagree about the appropriate rank and even identification of a group.

This disorder should not be taken as a failure of taxonomists. Instead, there are (at least) two broad reasons that make such disorder impossible to avoid and difficult to resolve. First, the organic world that taxonomists are trying to categorise is itself

DOI: 10.1201/9780367855604-10

disorderly (Dupré 1993; Kitcher 1984; Zachos 2016), and there are far more complex patterns in the organic world than a single taxonomy can capture. Taxonomies based on grounds as diverse as interbreeding, genealogy, and ecological interactions are all useful for different purposes but do not always coincide (Conix 2018; Ereshefsky 2002). In addition, many of the processes responsible for these patterns do not carve up the world into neat, discrete groups. Instead, evolutionary processes are continuous, leading to vague boundaries between some groups. As a consequence, taxonomists sometimes have to make partially subjective choices between various ways of classifying this disorderly world into a neat classification (Zachos et al. 2020). Most obvious here is the issue of ranking groups: whether a group is a subspecies or species can often be a matter of subjective interpretation, as is the designation of subspecies themselves (Donald 2020).

This disorder is further exacerbated by a second factor, namely, that there are simply very many species. Estimates of the total number of species on earth range between 5 million and several billion (if microbes are included), and only 2 million have currently been described (Costello et al. 2013; Larsen et al. 2017). Rough estimates of the number of taxonomists currently working are in the range of 40,000 (Costello et al. 2013), with a substantial number of those working on mammals and birds, and very few in some diverse groups like mites. This makes an additional layer of disorder hard to avoid. The project of describing all biodiversity is so big that efforts across the world, history, and taxonomic groups are not coordinated. It is inevitable, then, that some groups remain understudied and that in 'overstudied' groups, different taxonomists make the partially subjective choices in different ways. As a result, there are competing taxonomies for a few charismatic groups and only partial lists in lesser-known but highly diverse groups. In addition, because a name and type specimen do not fully determine the extension of taxon concepts, it is also inevitable that different interpretations of taxonomic descriptions proliferate (Sterner et al. 2020). While some of the disorder caused by these factors could in theory be resolved, taxonomists often lack the time to do this due to the large corpus of taxonomic names and concepts, and the large number of new species still awaiting description.

In short, there is no single best classification of organisms into species and higher ranks but a large body of taxonomic knowledge with competing hypotheses, duplicated information, understudied groups, and even various classifications fit for different purposes. Unfortunately, this is not merely an interesting theoretical phenomenon but has far-reaching practical consequences (Garnett and Christidis 2017). For example, the endangered grey parrot discussed earlier is in urgent need of protection against international trade and has for this reason been included by the Convention on the International Trade in Endangered Species (CITES) on the list of species that cannot be traded. Thus, it matters which group of organisms (out of the two alternatives proposed by experts) is connected to that name. More generally, biological taxonomy is one of biology's fundamental disciplines and plays a crucial role in identifying, naming, and classifying biodiversity. As such, its results are used, typically in the form of species lists, by both scientists and societal users in conservation policy, trade regulation, and biosecurity, among many other domains.

These users are faced with the challenge of selecting those parts that are up-to-date, reliable, and relevant for them. Because these users lack the expertise of specialised taxonomists – they may not even be aware that there are competing species lists for many taxa – this is no easy task, and they risk making a suboptimal choice even if they spend a lot of effort in choosing the list. These consequences are not confined to applied disciplines like conservation. For example, it has been shown that taxonomic variation has a worrying impact on the results of macro-ecological research and diversification analyses that use species lists as raw data (Faurby et al. 2016; Willis 2017).

This chapter considers the proposal to alleviate the problems that plague users of taxonomy by constructing a governed list of the world's recognised species. Such a list would be a go-to authority for users of taxonomy who lack the expertise to find and interpret the scientific taxonomic literature. In Section 8.2, we compare this solution with the main alternative and current system, namely, to make taxonomic information available in a variety of forms so that users can access the information that is most useful for them. We argue that while the current system has many merits, there are important benefits to having an authoritative governed list to supplement it. Section 8.3 surveys the current landscape of taxonomic services. We argue that most of the structure and materials for such a governed list are already available and give an outline of what is still lacking. Section 8.4 then argues that the short-comings of the current best list of the world's species can be overcome by adding a governance structure to it and provides a rough sketch of such a structure. Section 8.5 further strengthens this point by addressing two important challenges for the governance solution to the problems that users of taxonomy face. Section 8.6 concludes the chapter.

Before we turn to this, we want to emphasise that this paper focuses solely on *how taxonomic information should be communicated to policy-makers, taxonomists, and other scientists*. This question is distinct from deeper philosophical questions about the nature of species, the relations between different species concepts, or the suitability of species as a measure of biodiversity. Most importantly, the position we defend about the importance of a global list of the world's species is independent from views about the reality of species and the species category. It is perfectly possible to make a list of the world's accepted species even if those species, or the species category, are not 'real' (at least by many definitions in use). Thus, we believe that such a list is important regardless of whether there is taxonomic uncertainty about the entities listed. This also means that the solution for taxonomic disorder that we offer should not be taken as an attempt to resolve longstanding metaphysical questions about species. Instead, we propose a purely practical solution to resolve the practical problems caused by taxonomic disorder.

8.2 ONE LIST, MANY PERSPECTIVES, OR BOTH?

The organic world is multi-faceted and complex, and therefore, impossible to capture in all its detail by a single classification system. As a consequence, there is a rich diversity in folk-classifications that vary according to the needs of the users of those classifications.

For example, lichen naming and grouping practices vary substantially between the Samí of Northern Finnmark, the Sherpa, Limbu, Lama, and Rai of Nepal, and the Okanagan First Nations and reflect the different ways in which these people interact with lichen (Kendig 2020). The discussion in this chapter is restricted to scientific classifications within an evolutionary paradigm. However, as illustrated earlier, even within this narrow paradigm, similar problems of pluralism, diverging views, and disagreement arise.

The problem of taxonomic disorder, then, is one of making the multi-faceted, scattered, and sometimes contradictory findings of scientific taxonomy easy to interpret for societal and scientific users. There are two broad ways we could attempt to solve this problem: we could reduce this multi-faceted body of taxonomic knowledge to a single list of the world's species and other taxa, or we could organise and annotate the body of taxonomic knowledge so it is easier for users to navigate uncertainties. In the first option – call this the *guided approach* – the users' problem is solved by deciding for them which parts of taxonomic knowledge they can rely on. In the second option – call this the *facilitating approach* – the users' problem is solved by helping them choose the information they need.

The facilitating approach is the system that is currently in place: taxonomic information is offered on many platforms and in many shapes, and users can choose the lists, species hypotheses, and databases that work best for their purposes. This system is attractive for various reasons. First, it allows users to attune their selection of taxonomic knowledge to their particular needs. Taxonomy has many users with widely varying needs, and the flexibility of the facilitating approach allows each of them to work with taxonomic knowledge in the shape and format that are most useful for them. For example, some ecological, conservation, and evolutionary research may benefit from a rankless, purely phylogenetic classification with the smallest recognisable clades (so-called 'SNaRCs') at the lowest level (Mishler and Wilkins 2018), while other research requires species as independently evolving units (Crandall et al. 2000; Moritz 1994). In the facilitating approach, both SNaRCs and independently evolving units could be listed along with a system to map the two and understand the differences between these units. Similarly, biodiversity databases could benefit from different classification systems as their taxonomic backbone, depending on their characteristics and purposes. A specialised and fully manually curated database like Avibase may want to work with overlapping taxon concepts (Lepage et al. 2014), while other databases may prefer a single taxonomy of clades to enable users to retrieve various kinds of biodiversity data for those clades (Cellinese et al. 2021).

Second, the facilitating approach is attractive because it can reflect actual disagreements between taxonomists. Instead of giving the false impression that there is expert consensus about a taxon list, the facilitating approach can communicate disagreements to users in a way that allows them to take these disagreements into account when using taxonomic knowledge. For example, users could check whether their decisions are sensitive to the taxonomy they use and only investigate this decision in more detail if that is the case (Franz et al. 2016).

Finally, the facilitating approach is attractive because the level of detail of taxonomic information that is needed often depends on the spatial scale at which it is used. Often, the taxonomic resources available at the local or regional level are far more detailed than what could be included on the single global list of the guided approach. The facilitating approach aims to incorporate all this detailed taxonomic knowledge and enable users to include or exclude details depending on the context of their application. For example, conservation legislation in Belgium probably has no need of a taxonomy of bird populations – few, if any, bird species have multiple populations in a small country like Belgium – while this may be crucial for legislators at the European level. At the same time, such a population-level classification would probably be too fine-grained for a global list. Thus, one risk of the guided approach is that many users will not get taxonomic information at the level of detail that is suitable for their needs.

However, it is not clear that taxonomic flexibility – the main selling point of the facilitating approach – is always desirable. First, such flexibility may give users the opportunity to surreptitiously further their particular interests at the expense of the interests of others. For example, conservationists might choose a taxonomy that recognises more species as threatened to increase the funds they get. Conversely, real estate developers might choose a taxonomy that recognises fewer species in order to avoid building permits getting rejected because of conservation reasons. Indeed, various authors have claimed that such practices do occur (Isaac et al. 2004; Willan 2021).

Secondly, and more importantly, the facilitating approach only works well *if the entire body of taxonomic knowledge is fully integrated and well-organised.* Only where such well-organised information is available are users able to understand how different lists or hypotheses relate to each other, what the important differences are between competing lists or hypotheses, where experts disagree, and whether they don't comprehensively overlook important parts of the literature. This means that the facilitating approach is dependent on a well-functioning layer of metadata that enables the use and interpretation of taxonomic knowledge. If such a well-functioning layer of metadata does not exist, the simplicity and clarity of having a single authoritative list may well outweigh the benefits of flexibility for many users.

While substantial progress towards integration has already been made by taxonomists and biodiversity informaticians, such as in accessing and integrating legacy data (Agosti et al. 2019; Agosti and Egloff 2009; Penev et al. 2019), many longstanding challenges remain. Most important among these are resolving nomenclatural issues such as synonymy and homonymy (Boyle et al. 2013; Chamberlain and Szöcs 2013) and reconciling names with taxon concepts. Lacking solutions for these problems, taxonomic knowledge remains difficult to navigate, and users without expertise cannot fully benefit from the flexibility of the facilitating approach. This means that the current facilitating approach works well for experts (at least in theory) but not for the non-experts who make up a large proportion of all users of taxonomy.

The other option – the guided approach – bypasses the difficulty of integrating taxonomic knowledge by letting experts select the taxonomic knowledge for non-experts to use. In this way, users of taxonomic information do not have to delve into the taxonomic literature and databases themselves or have to make judgements about taxonomy with limited expertise. This guided approach has various advantages. First, it ensures that all relevant knowledge is taken into account and that this knowledge is interpreted adequately. This is by no means trivial, as taxonomic traditions in many taxa go back many decades, and it takes a long time to build familiarity with all relevant literature and names. Second, it saves users the costs and efforts of finding and processing the relevant information. Again, this gain in efficiency is far from trivial, because many organisations rely on taxonomists to provide them with advice, resulting in enormous duplication of effort. Finally, a single guided list makes it likely that various users will draw on the same taxonomic judgements. This is important, because national and international users are connected in various ways, and consistency in judgement about how to interpret the taxonomic information would streamline their interactions.

The guided approach also has an important downside: it lacks the flexibility of the facilitating approach. Some of the choices involved in selecting and interpreting taxonomic knowledge are subjective choices of the type discussed in the introduction. They are neither wrong nor right, with the best option depending on how the information will be used. By settling this choice for all users, it is unlikely that the taxonomic knowledge will be optimally attuned to each user. In addition, taxonomic experts are not infallible, and any authoritative list would be likely to contain mistakes as well as decisions with which many experts disagree. The risk, then, is that the guided approach entrenches the taxonomic perspectives that it endorses and sidelines the ones it rejects, even if these are all valid views held by some taxonomists.

In general, the strengths of each approach match the weaknesses of the other: the facilitating approach is attractive for its flexibility but undesirable because it only works for experts; the guided approach is attractive for its simplicity and user-friendliness but lacks the flexibility that some expert users require. Hence, we think that the solution is obvious: the best way forward is to pursue both approaches. In practice, this boils down to continued investment in biodiversity informatics to integrate taxonomic information as well as supplementing the current landscape of taxonomic services with an authoritative list of the world's taxa.

8.3   THE GUIDED APPROACH: WHERE WE ARE AND WHERE WE NEED TO GET

The previous sections argued that we need a taxon list that users can rely on as the standard, go-to source for taxonomic information. Garnett et al. (2020) propose a list of ten general principles for creating such a list (Box 8.1), including transparency, inclusiveness, and independence. In addition, Pyle et al. (2021) propose a set of items that this list should provide for each taxon it lists (Box 8.2), such as the scientific name and taxonomic status. Creating a list that meets all these principles and requirements is an ambitious aim, but fortunately, we do not have to start from scratch. An example of a list that meets many of the principles and minimal requirements listed in Boxes 8.1 and 8.2 is Catalogue of Life (COL henceforth).* The COL list contains nearly all of the world's described species (estimated to be 92% complete, with 1.8 million species listed at the time of writing) (cf. principle 10). These names have been gleaned from 168 databases and lists that it aggregates

---

* www.catalogueoflife.org/

and references (cf. principle 8). COL releases both monthly checklists and an annual checklist, and previous versions of most of these lists can be consulted (cf. principle 7). After a recent overhaul of its website and infrastructure for managing datasets, the names on the selected checklists of taxa and associated metadata are also easy to access for users.

BOX 8.1   Garnett et al.'s (2020) Principles for the Creation of a Global Species List. The principles shaded in grey are the ones COL already largely meets.

1. The species list must be based on science and free from non-taxonomic considerations and interference.
2. Governance of the species list must aim for community support and use.
3. All decisions about list composition must be transparent.
4. The governance of validated lists of species is separate from the governance of the naming of species.
5. Governance of lists of accepted species must not constrain academic freedom.
6. The set of criteria considered sufficient to recognise species boundaries may appropriately vary between different taxonomic groups but should be consistent when possible.
7. A global list must balance conflicting needs for currency and stability by having archived versions.
8. Contributors need appropriate recognition.
9. List content should be traceable.
10. A global listing process needs both to encompass global diversity and to accommodate local knowledge of that diversity.

BOX 8.2   Pyle et al.'s Minimal Requirements for Items on a Global Species List. Requirements shaded in grey are the ones COL already largely meets.

1. Each item on the global list must have a globally unique identifier.
2. Each item on the global list must have a version history that shows all changes to associated properties and metadata.
3. Each item on the global list must have a unique, standardised, widely recognised label such as a code compliant scientific name.
4. Each item should include nomenclatural authorship and date of publication.
5. Each item should include a source, such as a link to the database that provided the entry to the global list.

In other words, a list of the world's taxa is available. However, there are two important ways in which COL is limited in fully achieving the guided approach discussed earlier. First, while COL serves as (an important part of) the taxonomic backbone of major databases such as the Global Biodiversity Information Facility (GBIF) and Encyclopedia of Life (EoL), it has not been adopted by all major users of taxonomic information (cf. principle 2) (Bingham et al. 2017). International organisations and agreements like CITES, International Union for Conservation of Nature (IUCN), and the Convention on the Conservation of Migratory Species (CMS), as well as most countries, use their own classifications or rely on other resources. Because of this, COL does not yet come with the efficiency gains that make the guided approach so attractive.

Second, and most importantly, COL does not have fully transparent and well-justified procedures for selecting which lists to include and which to reject (cf. principle 3) for all groups of organisms. Such procedures are important, as there are many taxa for which there are multiple, incompatible checklists, particularly among taxa that have high salience to human society. With the guided approach, users have to trust the global listing organisation to choose between such lists for them. Hence, it is crucial that the procedures and justifications for these choices are both transparent and well-justified (cf. principles 1 and 3). Moreover, such procedures are important to ensure the quality of the global checklist, which depends directly on the quality of its component lists (cf. principle 6).

Note that these two issues with COL are closely entangled: transparent and well-justified procedures for compiling a list of the world's taxa are necessary to build users' trust in this list. In turn, such trust is necessary to get community-wide acceptance of the list and benefit fully from the efficiency gains of the guided approach. The consequence of these two problems is that COL does not yet solve the problems faced by users of taxonomic information.

8.4   THE GOVERNANCE OF A GLOBAL LIST OF THE WORLD'S TAXA

The two issues discussed earlier in relation to COL – a lack of transparency and incomplete global adoption – suggest that what is lacking in the current landscape of taxonomic information services is not a global list *per se* but a transparent governance structure. Such a governance structure could screen the component lists that make up the global list and in this way, serve as a warrant for users that the global list is authoritative and reliable. Consensus about the need for such a governance structure also emerged from a debate – initially spirited – on taxonomic disorder (Garnett and Christidis 2017; Thomson et al. 2018). This led to the formation of a Global Species List Working Group, under the auspices of the International Union of Biological Sciences (IUBS), that includes participants from both sides of the debate. The aim of this working group is to facilitate the design and implementation of a governance structure to resolve the problems that users of taxonomic information face in interpreting taxonomic disorder.

Note that there are two levels of lists at play here: a *global list*, such as that produced by COL, and many *component lists*, typically compiled by taxonomic communities, which together make up the global list. It follows that there are also two levels of governance to consider. On the level of the component lists, governance refers to the mechanisms and procedures used to compile a single list from all scientific hypotheses and research that exist about the taxon in question. These are likely to include mechanisms to deal with disagreement between experts as well as ways of involving as many relevant experts as possible in the creation of the list. For example, a working group of the International Ornithologists' Union is currently trying to create a single consensus list of the world's bird species from the four major competing lists available at the moment.*

The governance of component lists – which already exists in a variety of formal and less formal shapes – is strictly independent from the governance of the global list. The latter does not concern the decision about which species to include in a list but rather, which component lists to endorse and include in the global list. The details of such an endorsement body and the governance of the global list are beyond the scope of this chapter and would have to be filled in gradually in a conversation among all stakeholders. Instead, we propose some general requirements that the governance structure should meet in order to fulfill its purposes.

First, and most importantly, the endorsement body of the global list should evaluate not the contents of component lists *but the processes used to compile these component lists.* This is not merely motivated by the practical consideration that it is virtually impossible to evaluate the status of 1.8 million species twice (once when the component list is compiled and once when the list is included in the global list). The deeper reason is that only the taxonomic communities working on a taxon group are qualified to compile and evaluate a list of the entities in that group. Hence, discussion and decisions about the content of a component list should be reserved for the experts working on that taxon. The challenge for the global list is to evaluate whether the taxonomic communities that compile component lists do this well, and, if there are multiple lists, which list is best. This can be done by looking at the processes involved in compiling the component list: Did all experts get the chance to be involved? Were there appropriate mechanisms to resolve disagreement? Were there appropriate mechanisms to deal with conflict of interest? The idea is that only component lists compiled by an appropriate set of transparent processes can be fully endorsed as part of the authoritative global list. Lists meeting only some of the requirements could get different levels of endorsement in the case that there are no alternative lists available.

Secondly, we propose that non-taxonomic users of taxonomic information should be involved in the process of developing the governance mechanisms for the global list. More precisely, we argue that they should be involved in setting the list of criteria of adequacy for the component lists. This involvement makes it more likely that these users will adopt the resulting global list. This is not only because it allows these

---

* See www.internationalornithology.org/working-group-avian-checklists. Interestingly, the list used by COL is yet another one, not involved in this working group.

users to communicate what features a list needs to have to be useful for them, but also because involvement in the design makes it more likely that users will trust the list. User involvement could also incentivise taxonomic communities to participate in the global list, as it is likely to increase the impact of their efforts. At the same time, restricting the influence of users to the governance of processes rather than the actual contents of the list ensures that non-taxonomic considerations do not inappropriately shape these lists (Conix et al. 2021). For example, involvement in the design of criteria of adequacy would not enable users to exert pressure to include a particular species in order to increase its chances of getting protected.

Note that these proposals are compatible with a wide variety of governance models, including both the polycentric and monocentric approaches to list governance described by Lien et al. (2021). With a polycentric approach, taxonomic communities do not cede authority over component lists to the endorsement body. Instead, they cooperate voluntarily and on their own terms, and are incentivised to do so if important users adopt the global list. With a monocentric approach, taxonomic communities have to reach a consensus about a set of rules and principles to follow in compiling a global list and subsequently cede some authority to the endorsement body in their promise to follow these rules. With either approach, broad buy-in of, and cooperation with, taxonomic communities is crucial.

8.5 OBJECTIONS AND LIMITATIONS

We have sketched a rough outline of how the governance of a global species list could resolve some of the problems that users face in using taxonomic information. As the process of creating such a governance structure is only just starting, it is too early to evaluate the practical and organisational problems it will inevitably face. Still, there are at least two important problems that opponents of a guided approach might raise.

First, one might fear that the governance structure we propose is yet another grand initiative in a long list of similar 'integrating' initiatives that already populate the landscape of taxonomic information, such as Species 2000, ITIS, COL, CETAF and PESI (see also Bingham et al. 2017).[*] In response to this, it is worth noting that almost none of these initiatives tried to create a governance structure for taxonomic lists. One exception here is COL, where a governance model that can accommodate different taxonomic communities has always been part of the aim. However, because in practice, COL mostly tried to serve as a backbone for other biodiversity informatics activities, its governance mechanisms are not well attuned to the needs of international biodiversity conservation and management. In addition, our proposal is not to add a new initiative to an already overpopulated landscape but to develop or enhance a governance structure that will make these initiatives more effective.

Secondly, one may argue that investing in a governance solution is unwise when we are on the brink of further advances in biodiversity informatics. The idea is that

---

[*] For more information on these initiatives, see www.itis.gov/, www.eu-nomen.eu/portal/, www.sp2000 .org/home, https://cetaf.org/.

many of the current limitations of the facilitating approach can be overcome by progress in biodiversity informatics. If such progress were to make taxonomic information easier to find, interpret, and use without reducing it to a single list, this might be preferable to a well-governed list. Indeed, in an ideal world, users could perhaps even use biodiversity data without making a detour by semi-arbitrary Linnaean ranks and species descriptions. In that sense, investing in biodiversity informatics could be a better long-term solution than investing in a single list of the world's taxa.

While we appreciate this objection, we are not very optimistic that such bioinformatics solutions to taxonomic disorder will be ready in the near future or, even when ready, that they will rapidly be adopted by users. The idea of organising biodiversity policy around species lists, developed over many decades, has enormous institutional inertia that will not easily be disrupted by improvements in informatics. Eventually, a global list of the world's taxa may indeed become superfluous, but in the meantime, many users need clear and as-simple-as-possible taxonomic information to deal with pressing societal issues such as the biodiversity crisis and climate change. Because various initiatives have been working towards an authoritative global list for three decades and in many senses are nearly there, we believe it sensible to continue building on these efforts rather than abandon them for solutions that seem relatively far away.

8.6 CONCLUSIONS

This chapter has argued that the problems that users have in choosing the appropriate taxonomic information can be resolved by adopting a guided approach to disseminating taxonomic information through species lists. In practice, this guided approach would ideally boil down to improving the governance structure to COL, currently the most comprehensive list of the world's species, such that COL can transparently choose between competing taxonomies. The purpose of this governance structure would be to evaluate the processes used to compile component lists and to choose between component lists on the basis of these processes whenever there are competing alternatives. By involving both taxonomic communities and users in the design of this governance structure, we can ensure that the resulting list meets the requirements of as many stakeholders as possible. This, in turn, makes it more likely that the list will gradually be adopted as the authoritative and standard species list for all users who prefer a simple list over taxonomic information that caters to their needs.

BIBLIOGRAPHY

Agosti, D., Catapano, T., Sautter, G., Kishor, P., Nielsen, L., Ioannidis-Pantopikos, A., et al. (2019). Biodiversity Literature Repository (BLR), a repository for FAIR data and publications. *Biodiversity Information Science and Standards*, 3, e37197. https://doi.org/10.3897/biss.3.37197.

Agosti, D., & Egloff, W. (2009). Taxonomic information exchange and copyright: The Plazi approach. *BMC Research Notes*, 2(1), 53. https://doi.org/10.1186/1756-0500-2-53.

Bingham, H., Doudin, M., Weatherdon, L., Despot-Belmonte, K., Wetzel, F., Groom, Q., et al. (2017). The biodiversity informatics landscape: Elements, connections and opportunities. *Research Ideas and Outcomes*, 3, e14059. https://doi.org/10.3897/rio.3.e14059.

Boyle, B., Hopkins, N., Lu, Z., Raygoza Garay, J. A., Mozzherin, D., Rees, T., et al. (2013). The taxonomic name resolution service: an online tool for automated standardization of plant names. *BMC Bioinformatics*, 14(1), 16. https://doi.org/10.1186/1471-2105-14-16.

Cellinese, N., Conix, S., & Lapp, H. (2021). Phyloreferences: Tree-Native, reproducible, and machine-interpretable taxon concepts. *Philosophy, Theory, and Practice in Biology*, 14, 7. https://doi.org/10.3998/ptpbio.2101.

Chamberlain, S. A., & Szöcs, E. (2013). Taxize: Taxonomic search and retrieval in R. *F1000Research*, 2. https://doi.org/10.12688/f1000research.2-191.v2.

Conix, S. (2018). Radical pluralism, classificatory norms and the legitimacy of species classifications. *Studies in History and Philosophy of Science Part C: Studies in History and Philosophy of Biological and Biomedical Sciences*, 73, 27–43. https://doi.org/10.1016/j.shpsc.2018.11.002.

Conix, S., Christidis, L., Pyle, R. L., Costello, M. J., Zachos, F. E., Bánki, O. S., et al. (2021). Towards a global list of accepted species III. Independence and stakeholder inclusion. *Organisms Diversity & Evolution*, 21, 631–643. https://doi.org/10.1007/s13127-021-00496-x.

Costello, M. J., May, R. M., & Stork, N. E. (2013). Can we name Earth's species before they go extinct? *Science*, 339(6118), 413–416. https://doi.org/10.1126/science.1230318.

Crandall, K. A., Bininda-Emonds, O. R. P., Mace, G. M., & Wayne, R. K. (2000). Considering evolutionary processes in conservation biology. *Trends in Ecology & Evolution*, 15(7), 290–295. https://doi.org/10.1016/S0169-5347(00)01876-0.

del Hoyo, J. (2020). *All the Birds of the World*. http://imis.nioz.nl/imis.php?module=ref&refid=330632. Accessed 30 March 2021.

Dickinson, E., & Christidis, L. (2014). *The Howard and Moore Complete Checklist of the Birds of the World*. (4th ed., Vol. 2). Eastbourne, UK: Aves Press.

Donald, P. F. (2020). Accounting for clinal variation and covariation in the assessment of taxonomic limits: Why we should remember the 'rules.' *Ibis*. https://doi.org/10.1111/ibi.12908.

Dupré, J. (1993). *The Disorder of Things: Metaphysical Foundations of the Disunity of Science*. Cambridge, MA: Harvard University Press.

Ereshefsky, M. (2002). Linnaean ranks: Vestiges of a Bygone era. *Philosophy of Science*, 69(S3), S305–S315. https://doi.org/10.1086/341854.

Faurby, S., Eiserhardt, W. L., & Svenning, J.-C. (2016). Strong effects of variation in taxonomic opinion on diversification analyses. *Methods in Ecology and Evolution*, 7(1), 4–13. https://doi.org/10.1111/2041-210X.12449.

Franz, N., Gilbert, E., Ludäscher, B., & Weakley, A. (2016). Controlling the taxonomic variable: Taxonomic concept resolution for a southeastern United States herbarium portal. *Research Ideas and Outcomes*, 2, e10610. https://doi.org/10.3897/rio.2.e10610.

Garnett, S. T., & Christidis, L. (2017). Taxonomy anarchy hampers conservation. *Nature*, 546(7656), 25–27. https://doi.org/10.1038/546025a.

Garnett, S. T., Christidis, L., Conix, S., Costello, M. J., Zachos, F. E., Bánki, O. S., et al. (2020). Principles for creating a single authoritative list of the world's species. *PLOS Biology*, 18(7), e3000736. https://doi.org/10.1371/journal.pbio.3000736.

Isaac, N. J. B., Mallet, J., & Mace, G. M. (2004). Taxonomic inflation: Its influence on macroecology and conservation. *Trends in Ecology & Evolution*, 19(9), 464–469. https://doi.org/10.1016/j.tree.2004.06.004.

Kendig, C. (2020). Ontology and values anchor indigenous and grey nomenclatures: a case study in lichen naming practices among the Samí, Sherpa, Scots, and Okanagan. *Studies in History and Philosophy of Science Part C: Studies in History and Philosophy of Biological and Biomedical Sciences*, 84, 101340. https://doi.org/10.1016/j.shpsc.2020.101340.

Kitcher, P. (1984). Species. *Philosophy of Science*, 51(2), 308–333.

Larsen, B. B., Miller, E. C., Rhodes, M. K., & Wiens, J. J. (2017). Inordinate fondness multiplied and redistributed: The number of species on Earth and the new pie of life. *The Quarterly Review of Biology*, 92(3), 229–265. https://doi.org/10.1086/693564.

Lepage, D., Vaidya, G., & Guralnick, R. (2014). Avibase – A database system for managing and organizing taxonomic concepts. *ZooKeys*, 420, 117–135. https://doi.org/10.3897/zookeys.420.7089.

Lien, A., Christidis, L., Conix, S., Costello, M. J., Zachos, F. E., Bánki, O. S., et al. (2021). Towards a global list of accepted species IV: Overcoming fragmentation in the governance of taxonomic lists. *Organisms Diversity & Evolution*, 21, 645–655. https://doi.org/10.1007/s13127-021-00499-8.

McClure, C. J. W., Lepage, D., Dunn, L., Anderson, D. L., Schulwitz, S. E., Camacho, L., et al. (2020). Towards reconciliation of the four world bird lists: hotspots of disagreement in taxonomy of raptors. *Proceedings of the Royal Society B: Biological Sciences*, 287(1929), 20200683. https://doi.org/10.1098/rspb.2020.0683.

Mishler, B., & Wilkins, J. S. (2018). The hunting of the SNaRC: A snarky solution to the species problem. *Philosophy, Theory, and Practice in Biology*, 10. https://doi.org/10.3998/ptpbio.16039257.0010.001.

Moritz, C. (1994). Defining 'Evolutionarily Significant Units' for conservation. *Trends in Ecology & Evolution*, 9(10), 373–375. https://doi.org/10.1016/0169-5347(94)90057-4.

Penev, L., Dimitrova, M., Senderov, V., Zhelezov, G., Georgiev, T., Stoev, P., & Simov, K. (2019). OpenBiodiv: A knowledge graph for literature-extracted linked open data in biodiversity science. *Publications*, 7(2), 38. https://doi.org/10.3390/publications7020038.

Pyle, R. L., Christidis, L., Conix, S., Costello, M. J., Zachos, F. E., Bánki, O. S., et al. (2021). Towards a global list of accepted species V. The devil is in the detail. *Organisms Diversity & Evolution*, 21, 657–675. https://doi.org/10.1007/s13127-021-00504-0.

Sterner, B., Witteveen, J., & Franz, N. (2020). Coordinating dissent as an alternative to consensus classification: Insights from systematics for bio-ontologies. *History and Philosophy of the Life Sciences*, 42(1), 8. https://doi.org/10.1007/s40656-020-0300-z.

Thomson, S. A., Pyle, R. L., Ahyong, S. T., Alonso-Zarazaga, M., Ammirati, J., Araya, J. F., et al. (2018). Taxonomy based on science is necessary for global conservation. *PLOS Biology*, 16(3), e2005075. https://doi.org/10.1371/journal.pbio.2005075.

Willan, R. C. (2021). Magallana or mayhem? *Molluscan Research*, 1–5. https://doi.org/10.1080/13235818.2020.1865514.

Willis, S. C. (2017). One species or four? Yes!...and, no. Or, arbitrary assignment of lineages to species obscures the diversification processes of Neotropical fishes. *PLOS ONE*, 12(2), e0172349. https://doi.org/10.1371/journal.pone.0172349.

Zachos, F. E. (2016). *Species Concepts in Biology: Historical Development, Theoretical Foundations and Practical Relevance*. Basel: Springer.

Zachos, F. E., Christidis, L., & Garnett, S. T. (2020). Mammalian species and the twofold nature of taxonomy: A comment on Taylor et al. 2019. *Mammalia*, 84(1), 1–5. https://doi.org/10.1515/mammalia-2019-0009.

Section 3

Ranks and Trees and Names

9 Ecology, Evolution, and Systematics in a Post-Species World

Brent D. Mishler

CONTENTS

9.1 INTRODUCTION

The species rank needs to be abandoned, along with all other taxonomic ranks, for a number of reasons that I have written about extensively elsewhere (Mishler 1999; 2010; Mishler and Wilkins 2018; Mishler 2021). So, in this chapter I will not repeat those arguments in detail, I just summarise them in the following two paragraphs and go on from there to address how we can proceed with biological research, biodiversity studies, and conservation biology under a fully rankless world view. The argument I am making is an empirical one. I acknowledge that the Biological Species Concept (BSC) and most other published species concepts are logical and internally consistent theories about how biodiversity *could* be organised. My argument is simply that *all* published species concepts (including my own Phylogenetic Species Concept; Mishler and Brandon 1987) do not fit nature as we are coming to know it. Species are not a special kind of taxon; they are taxa like those at other current ranks. There is no distinctive species rank to be found in nature. The seemingly endless and ever-expanding debate over species concepts is because we have all been trying to capture a will-o'-the-wisp. That way madness lies ….

The recent availability of extensive genetic and genomic data within and between named species has demonstrated that interbreeding (and other forms of horizontal transfer) among lineages occurs at multiple hierarchical levels, while at the same time, a lack of interbreeding among lineages occurs at multiple hierarchical levels

DOI: 10.1201/9780367855604-12

as well. Thus, phylogenetic structure (and relevance to ecological and evolutionary processes) emerges at many levels; there is no particular level where rampant inter-breeding abruptly transitions to zero interbreeding, despite the prevalent view that just such a magical distinction exists (e.g., Hennig 1966; Nixon and Wheeler 1990). There is no empirical justification for treating the species level as unique. It is just one level on the tree of life among many other levels, as first recognised by Darwin (1859) in probably the most novel idea in his seminal book (see Mishler 2010 for discussion).

To say that the species *rank* is artificial does not mean that currently named species *taxa* are artificial – on the contrary, many of them are real entities in nature. The important point is that their reality is due to the same processes acting above and below the traditional 'species' level (Mishler 1999; 2010). The reality of lineages and clades (see Mishler and Wilkins 2018 for the important distinction between these concepts) is due to the same suite of processes acting at multiple levels of the tree of life, albeit with the balance of processes shifting in interesting ways. The actual situation is richer and much more interesting than the Mayrian BSC allows – there are nested entities smaller and larger than traditional species that play important roles in ecology and evolution and deserve equal attention.

In making the transition to fully rankless classification, it is of course important to ensure that the community can still do everything it is accustomed to doing with species, both practically and theoretically. The remainder of this chapter assumes we are going to abandon the species rank, explores theoretical and practical implications of a rankless phylogenetic approach to terminal taxa, and makes the argument that is it not only *possible* to use rankless classification for all purposes across biology, but *better* (Mishler 2021). A multi-level approach to systematics, ecology, and evolution is more accurate for studying the origin, maintenance, and conservation of biodiversity. A revolution in many topics, including systematics, diversification (formerly known as 'speciation'), niche evolution, biogeography, coevolution, and conservation prioritisation, will follow once a rigid focus on the species level is replaced by this multi-level view.

9.2   THE GENERAL EPISTEMOLOGICAL APPROACH NEEDED IN THE POST-SPECIES WORLD

The *multi-level approach* I am advocating is fundamentally the same for all the topics considered in the following sections. The first step recommended in all cases is to build a phylogeny down to the finest scale that is feasible with current data. This is not the place to go into the vast topic of building phylogenetic trees, but it is important to note three aspects in the present context:

(1) The terminals in *all* phylogenetic trees are (and must be) *semaphoronts* (signal bearers in Hennig's sense; 1966), i.e., specimens from which data have been obtained ranging from DNA sequences to morphology to behavior. Groups of semaphoronts that are identical with current data might be lumped into OTUs for analysis, but we must be prepared to separate them if new data shows them to be heterogeneous (Mishler 2005). In practice, we sometimes put species names (or other taxonomic

ranks) at the tips of phylogenies as a short cut for illustration purposes, but fundamentally, the terminals on a phylogenetic tree are always semaphoronts – sampled to represent previously named taxa perhaps, but *never* actually the whole taxon.

(2) Given recent knowledge coming from whole genome sequencing of how extensive horizontal transfer, incomplete lineage sorting, and convergence are, monophyly can no longer mean monophyly in every single gene tree in the genome (we would have no monophyletic groups that way). There are bound to be gene trees that are incongruent with organism-level phylogenies. As discussed by Baum (2007; 2009); Degnan and Rosenberg (2009), monophyly of clades is an ensemble property based on the preponderance of congruent gene trees. Relationships of higher-level lineages are an emergent property that is not expected to be matched by the history of all contained lineages; e.g., someone is still the same person following a liver transplant, and a clade still represents a cross-section of a lineage even if some horizontal transfer is represented in its genome.

(3) The phylogenies do not have to be based on DNA sequence data – that is really helpful to have, but initial phylogenies can be made from morphological data analysed properly (Mishler 2005). So, a phylogenetic approach can still be used for relatively poorly known groups. The SNaRCs (i.e., smallest named and registered clades, Mishler and Wilkins 2018; see later for discussion) found using only morphological data might be relatively coarse-scale, and finer SNaRCs will likely be found in the future once DNA sequence data are added, but a good start on all the topics covered here can be made using the morphological SNaRCs and the phylogenies connecting them.

Given a tree, the next steps that I recommend involve using the full tree to build classifications, assess biodiversity, and address theoretical and applied questions about ecological and evolutionary processes. Specific methods can include comparing sister groups at all levels as well as analysing reconstructed changes in traits along all the branches. Using phylogenies, one can achieve a much greater understanding of biogeography, ecology, and evolution than without them. One can also better carry out practical objectives in assessing biodiversity and developing conservation plans using a phylogeny. Using the full phylogenetic framework contrasts favorably in all ways to the old approach of just considering species, their ranges, ecology, and traits, which only allowed (at best) a limited, one-dimensional, non-evolutionary view of biodiversity. For example, conservation actions that consider only the species level are non-evolutionary, because variation within and among species is ignored.

9.3 SYSTEMATICS

One important area of study that is much better done using a phylogeny than without is classification (Hennig 1966; Nelson 1973; Farris 1983). In general, a classification that is based on the primary process that is producing the entities being classified (evolution in the case of biology) has the greatest information content for what is currently known and the greatest predictive value for unknown information (discussed in detail in Mishler 2000; 2009). Phylogenetic classification has taken over the field of taxonomy almost completely, except at the species level, where debate continues to rage over whether species taxa are the same as taxa at other ranks or are special in some way.

In my view, the only logical approach is for naming of any taxon (including at the species level) to be done *after* the phylogeny is built using semaphoronts, as discussed earlier. The approach advocated by some to name species *first* using non-phylogenetic clustering criteria, whether using morphological or molecular data (e.g., Nixon and Wheeler 1990; Mallet 1995), and then use species as units to build phylogenies is chaotic. You can't have a phylogenetic system where one level of taxa is non-phylogenetic. It's like having a classification of apples where the basic level is oranges!

Not all levels of the tree are (or should be) named. Taxa are named for the levels of the tree that are well supported by data and thus, hypothesised to be clades that will be confirmed by future studies and therefore stable. They are named because the author concludes that they will be useful clades for people to talk about in the future, perhaps because of interesting evolutionary novelties they display. Formal international codes of nomenclature are needed to guide naming of taxa so that the application of a name is clear in the future as more semaphoronts are added to phylogenetic analyses.

Traditional classification systems as codified in the International Code of Botanical Nomenclature (ICBN) and International Commission on Zoological Nomenclature (ICZN) were begun under a non-evolutionary worldview and employ ranks in their rules for naming, which has led to numerous problems as applied to the modern world with our expanding knowledge of the tree of life with its thousands of known levels. To put the problems into two categories: (1) *Ranks* – there are not nearly enough ranks, but more importantly, the clades named at a particular rank, say genus or family, are not comparable in any way (e.g., age, included diversity, amount of morphological or genetic difference with their sister group, etc.) even though scientists often act as though they are comparable. Furthermore, needless taxonomic instability is caused when taxa at the same rank are found to be nested inside each other – this is particularly the case using binomial species names, because the name of a species is often forced to change when nothing whatsoever has changed about its circumscription, which is very frustrating to users of classifications, not to mention confusing in biodiversity informatics; (2) *Precision for naming clades* – the current codes of nomenclature use only one type, which makes it impossible to triangulate to a specific node on a tree in such a way that the name's application is clear and stable in future studies (de Queiroz and Gauthier 1994).

The PhyloCode has been developing for a couple of decades and was just published (Cantino and de Queiroz 2020). It solves the two main categories of problems indicated earlier by getting rid of the ranks in taxonomy and by using two or more 'specifiers' (instead of single type specimens) to clearly define what clade is being named so that it is still clear after future sampling. As published, the PhyloCode is a major step forward for phylogenetic classification, but it is still not logically complete. One rank still remains embedded in the PhyloCode – you guessed it – the always troublesome and divisive species level! The community of PhyloCode supporters is still split on this, with zoologists largely wanting to hold on to the species rank and botanists largely wanting to get rid of that rank with all the rest (see extensive discussion in Cellinese, Baum, and Mishler (2012).

It is not clear how this debate will resolve in the PhyloCode community, but for my purposes in this chapter, I adopt a completely rank-free approach that views all taxa as hypothesised clades to be named with a rankless uninomial. The clade names are clearly connected to the phylogeny by using multiple 'specifiers' (i.e., semaphoronts!) to triangulate to a particular node, instead of the single type used by the current codes. This includes the finest-level clades that an author chooses to name; these are the Smallest Named and Registered Clades or SNaRCs (Mishler and Wilkins 2018). There is no ontological implication that SNaRCs *are* the smallest clades that exist. Instead, the naming of SNaRCs, as for taxa at all levels, is epistemological – they are the smallest clades that can be justified to the author's satisfaction by current data. Other clades may be discovered inside a SNaRC in the future, in which case its clade name remains unchanged, but it is no longer a SNaRC. Thus, the SNaRC is not a rank, and clades so named are not necessarily comparable with each other biologically, so counting them remains as problematical a measure of biodiversity as counting species or other taxa (more on that in following sections).

While PhyloCode-style taxa have no ranks, the hierarchy remains. Each named clade gets only a uninomial, but all the more inclusive named clades it belongs to are available in the database *RegNum* (www.phyloregnum.org/). So, the nestedness of names is easily machine-readable, and some of the higher clade names for a particular uninomial (SNaRC or otherwise) could be listed in parentheses in text as needed to orient human readers as to the position of that clade. The node on the tree of life where the clade name was bestowed remains stable into the future if other clades come to be named above it (or inside it) or even if its position on the tree changes. So, the name is stable while progress in systematics continues without the frustrating name changes caused by the ranks in current codes of nomenclature; e.g., when one genus is discovered to nest inside another, and many species names must be changed when nothing has changed about their circumscription.

Therefore, biodiversity informatics can be much improved with a rankless approach. Clades have a stable name, and information can be securely attached to those names, such as their inferred age, geographic distribution, traits, etc. This allows legitimate comparisons of taxa to be made, comparing like with like, enabling true comparability, which is impossible using taxon ranks. For example, one could pull from the database all the clades of plants in California that originated about 5 million years ago and compare them for their apomorphic traits or diversity relative to their sister group. Or, one could pull out all the clades (at whatever depth in time) where wings first evolved and compare them with their closest relatives to look for genes, morphological traits, or ecological situations that are associated with the evolution of flight.

9.4   ECOLOGY

The species rank is highly entrenched in ecology, where a particularly Aristotelian view of species has played a major role in the development of both theory and practice. Species are generally treated in ecology as homogeneous entities equally related to all others and without internal structure. For example, it is commonly assumed that the nature of inter- and intra-specific competition is fundamentally different (e.g., MacArthur and Levins 1967). Niche theories have always been linked to species (Hutchinson 1965), which is manifested most recently in the burgeoning species distribution modeling approach. Current climate change studies virtually always frame their questions pretty much as a creationist would, asking 'how will this species respond to climate change?' with very little consideration of the effects of within-species variation or between-species relationships – very little consideration of evolution, period. Ecologists have trusted what taxonomists have called species; risky behavior if you examine how taxonomists fight over what species are in general and in particular cases (Agapow et al., 2004).

The history of this rigid, non-evolutionary viewpoint on species in ecology would make an interesting study in itself. Ecologists seem to have been more sure of the uniqueness of species than the systematists, who had to grapple with actually trying to decide what is a species and what isn't. The Aristotelian use of species in ecology may partly be due to a sociological disconnect between systematics and ecology that continues today. But for the purposes of this chapter, it is more relevant to ask whether this entrenchment of a rigid species concept was ever necessary.

Clearly, clades at different levels can play an important role in ecological networks. The ecological interchangeability of two organisms – i.e., whether or not they are identical with regard to their performance of some ecological function – is a contingent question that depends on the function and depends on the organism. Some ecological roles can be quite broad: all plants might be validly lumped into the 'primary producer' role in a food web. On the other hand, other ecological roles can be quite narrow: perhaps just one genetic variety of a fungus attacks a particular tree in a forest. So, very often, a node in a network of ecological interactions is most accurately represented not as a species-rank taxon but as something more or less inclusive.

Likewise, clades at different levels can have a distinct ecological niche. There are very large clades that fill a niche in a global sense (e.g., sharks as ocean top predators, large clades of fungi as wood decomposers, blue-green bacteria as nitrogen fixers). On the other hand, different parts of a named species can play a different ecological role at different locations. For just a few examples: one plant species that occurs on an elevational gradient can play different roles in pollinator communities at different elevations (Lefebvre et al. 2018), males and females can have a different niche in both plants and animals (Dawson and Geber 1999), young and old individuals can have a very different niche (perhaps most striking in marine animals that have planktonic larvae yet sessile adults; Pechenik 1999).

In short, ecology needs to be studied in a multi-level fashion. Taxa at the current species level may of course be of relevance, but that is not nearly the only level of relevance. Thus, when building environmental niche models, one should not focus only on species. Phylogenetically distinct subgroups within named species should be modeled separately. And when examining niches for clades larger than the named species, one should model their ranges directly, using their own distributional data, rather than the prevailing approach that models species ranges and then uses ancestral state reconstruction to posit niches for higher clades (e.g., Smith and Donoghue 2010).

9.5 EVOLUTION

'Speciation' is widely held to be an important process in evolution; some go as far as to say that it is the *only* process producing new biodiversity (Schluter and Pennell 2017). Speciation is usually taken to mean the cessation of interbreeding between lineages, but as has become abundantly clear, the likelihood of interbreeding and other forms of horizontal transmission among lineages does not usually go from 100% (panmixia) to 0% (total reproductive isolation) at a certain moment in time as two lineages diverge. Divergence in reproductive compatibility usually happens gradually as lineages diverge, and different mechanisms might be involved at different levels in this process, thus, the origin of reproductive isolation is another topic needing to be studied in a multi-level fashion.

The whole idea of looking for species 'boundaries', while a popular pursuit these days (e.g., Harrison and Larson 2014), employs a metaphor that is nonsensical under an evolutionary world view. Genealogical relationships among organisms go back in time, so a tree is a far better metaphor for biodiversity than a map (Stevens 1994). Taxa including species do not have 'boundaries' in the present day – their defining point is back in time if they are monophyletic. No one would talk about looking for the 'boundary' of the Mammalia; instead, they would look to see whether there is an inferred common ancestor. It makes no more sense to talk about species boundaries – the map analogy is simply the wrong one to fit biological process.

Population geneticists nearly universally believe that different questions and methods apply within species as compared with between species. For example, the field called 'phylogeography' assumes that distance and clustering methods apply within species, while phylogenetic methods apply only among species (Avise 1989). But if the species level is nothing special, then drawing such a line at the 'speciation' point is problematic. Many studies in population genetics take the previously named species in a group as given *a priori*; i.e., they trust what taxonomists have decided are species (risky behavior, as discussed in the previous section). For example, Doyle et al. (2004) address genetic structure and hybridisation among taxonomic species of *Glycine* without ever addressing what sort of species concept has been applied or whether those species are well supported by data. Maybe, the currently accepted species of *Glycine* are all 'bad', and thus, comparing gene trees within and among them is meaningless!

Building phylogenies using semaphoronts, as described earlier, should always be the first step in an evolutionary study. Once this is done, it can be useful to use clustering methods to look for finer-scale structure among organisms and populations within SNaRCs. Such structure might be important for the detection of fine-scale evolutionary processes that could influence the origin of distinct clades. This relates to the distinction most evolutionary biologists see between 'microevolutionary' and 'macroevolutionary' processes. That distinction may be a good one, but the named species level is not the best dividing line between the two – the SNaRC level is a more logical dividing line because of the explicit criteria used for naming them.

Diversification studies likewise usually assume that species are a unique and comparable level on the tree of life (e.g., Nürk et al. 2020). In this manner, the number of described species in a clade is most often used to define its diversity. There are so many things wrong with this that it is hard to know where to start. Taxonomic study is not evenly spread among taxonomic groups or geographic regions; some groups (particularly those that are useful to humans or visually showy) or regions (particularly those with many local taxonomists) are way over-studied as compared with others. Ranking traditions vary among different communities of taxonomist specialists – some communities are dominated by splitters, some by lumpers. For all these reasons, the number of named species in a group is a terribly inadequate measure of diversity in either an absolute or a comparative sense (Faurby et al. 2016; Willis 2017).

What should replace it? Phylogenetic diversity (PD; Faith 1992) is a much better method for comparing diversity between clades (Miller et al. 2018). There are still issues of concern when using PD, of course. Different clades may have been phylogenetically studied more intensively than others, and the SNaRCs are therefore finer, and more phylogenetic branches are known. So, care needs to be taken to examine sampling efforts and intensity of study within clades being compared for possible normalisation or other adjustments to the data. For example, good variables to include for normalising biodiversity comparisons with PD might be the number of collections that have been made, the number and distribution of accessions that have been sampled for DNA sequencing, or the type of markers used.

To understand the 'cutting edge' of divergence – the process formerly known as 'speciation' but generalised to include processes in addition to the cessation of gene flow – one should compare the terminal-most sister clades in one's study group with each other to see which traits (i.e., genomic, morphological, physiological, geographic, ecological, developmental, reproductive, etc.) are reconstructed to change on the branches separating them. Those characters are the candidates for possible causes or constraints affecting divergence in that particular sister pair and can be tested via field and lab studies.

To summarise this section, if we accept that there is no magical dividing line separating within- from among-species processes, we can proceed to develop a multi-level approach to evolutionary studies just as argued earlier for ecological studies. Divergence (splitting of lineages), reticulation (lineages coming back together again), and extinction (lineages ceasing their existence) all can and do happen at several levels in the same groups of organisms. Thus, selection and drift can and do happen at multiple levels at once (Brandon 1982). We need to take off our species blinders and try to find out what is going on at each level. The best approach is building gene trees connecting semaphoronts, looking for congruence and conflict among them, and adding the context of geography, ecology, and phenotypic traits.

9.6   CONSERVATION BIOLOGY

Biodiversity is commonly mapped and assessed on the landscape using species distributions. Conservation prioritisation is likewise universally considered to be based on an examination of species richness, rarity, etc. (Agapow et al., 2004). The assumption is made that species are *the* units of biodiversity and therefore, the entities upon which conservation should focus (Myers et al. 2000). As discussed in other contexts earlier, this assumption seems like one only a creationist would make. In an evolutionary worldview, it should be clear that named species are all different from each other. Even if we assume species are clades (which they very often are not), they are not named at the same depth on the tree of life. Due to different customs in different scientific communities, some are shallow (named by 'splitters'), and some are more inclusive (named by 'lumpers'). Thus, named species are not comparable units, nor can they be made comparable, because evolution is a continuous process, and thus, the level where a species name is applied is intrinsically arbitrary, as Darwin (1859) himself saw clearly.

An additional issue is that conservation has tended to focus on one species at a time, through Red List efforts around the world and laws such as the Endangered Species Act in the United States. While these efforts have had their successes, I would argue that we do not have nearly enough time in the current world to do conservation in this manner. The pressures of human population growth and concomitant land use changes and climate change are just too overwhelming. We need to immediately (i.e., in the next decade) save what natural lands we can, and we need to be brutally efficient in how we spend money and time on lands to conserve as much of biodiversity as we can. Triage is needed, with priorities set by rigorous quantitative methods that move beyond consideration of only species occurrences (e.g., Myers et al. 2000).

Therefore, instead of the traditional approach that focuses on a single species at a time, biodiversity should be mapped and conservation priorities assessed on the landscape using the distributions of *all* clades on the tree of life. The new field of *spatial phylogenetics* (Mishler et al. 2014; Thornhill et al. 2016; 2017) is designed to do just this by converting the tree of life into geographic information system (GIS) layers that can be used along with other kinds of GIS layers (such as climate, protection status, and threats from climate change and changes in land use) to give an evolutionary view of conservation priorities. This approach combines a phylogeny and a spatial dataset containing the geographic distribution of the tips of the phylogeny. Diversity and endemism are measured based on phylogenetic relatedness, branch length, and range size of all lineages on the phylogeny (Thornhill et al. 2017; Kling et al, 2018).

PD (Faith 1992) is the core metric used in spatial phylogenetics. It sums the branch lengths leading to all tips of the phylogeny that are present in a given location (usually including the path to the base of the phylogeny). There are different PD values for the same tree topology, given different ways of representing branch lengths, called the *facets of phylodiversity* by Kling et al. (2018); see Figure 9.1 for illustration. The branch lengths can be represented by the reconstructed number of

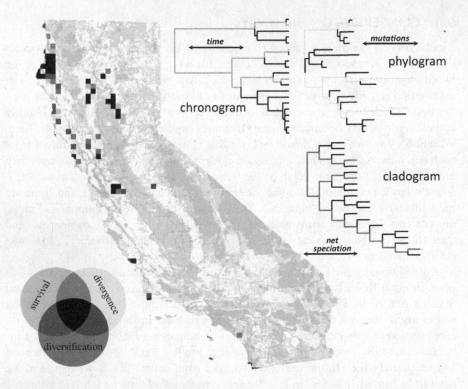

FIGURE 9.1 The top 50 highest-priority sites for preserving California's native plants for each of three facets of phylodiversity (see text for more discussion): genetic uniqueness (divergence among lineages represented), historic speciation rate (diversification), and independent evolutionary history (survival). Areas high in all three measures are in black. The algorithm used starts with GIS layers for current protection status of land and intactness of natural land cover (shown in background on the map, where grey indicates areas that are already protected or degraded). The algorithm is applied iteratively; everything is recalculated after each highest priority is chosen, and thus, it is applying a phylogenetic complementarity criterion. In this manner, priority is given to poorly protected locations that have high intactness of natural land cover and high biodiversity value (i.e., that contain many lineages that have small ranges and are not protected well elsewhere on the map). (Modified from Kling, M.M., et al., *Philosophical Transactions Royal Society B*, 374, 20170397, 2018.) **This Figure is available to download in color from the Support Materials tab on www.routledge.com/9780367425371**

mutations along each branch, resulting in a *phylogram*. The branch lengths can be represented by the estimated time they each existed, using fossil-derived calibrations, resulting in a *chronogram*. The branch lengths can each be represented as equal, resulting in a *cladogram*. These different facets have different meanings for assessing conservation priorities: a location with high PD on the phylogram has high genetic diversity; a location with high PD on the chronogram has high lineage survival time; and a location with high PD on the cladogram has high net diversification (speciation–extinction) represented (Kling et al. 2018). All facets are of interest, although one may be of greater interest for particular conservation goals.

In addition to these PD patterns, it is also important for most conservation goals to take into account the geographic rarity of lineages. For this purpose, a new metric was proposed called phylogenetic endemism (PE; Rosauer et al. 2009). This metric is like PD except calculated on a topology with branch lengths that are negatively range-weighted; i.e., each branch is divided by its range size (with the range of branches defined as the union of the ranges of all descendants of the branch). In this way, widespread branches shrink down, and PE is influenced primarily by the branches with relatively small ranges. Locations with high PE represent a concentration of lineages that have small ranges, likely of high conservation priority and also indicative of potential evolutionary and ecological processes. Figure 9.1 shows a conservation prioritisation for locations in California (Kling et al. 2018) using all three facets of PD and PE within a sophisticated algorithm that applies an important concept in conservation biology called *complementarity* (i.e., the best decision for the next step in conservation should be based on what has already been conserved; Justus and Sarkar 2002).

The initial formulation of PE confounded the original branch length with range size to some extent in that PE is high only for long-branch range-restricted taxa (often termed 'paleoendemics'). To examine the distribution of branch lengths on a map and in particular, to find concentrations of recently evolved range-restricted taxa (often termed 'neoendemics'), Mishler et al. 2014 added two more metrics, both of which are ratios. Relative phylogenetic diversity (RPD) is a ratio of PD measured on the phylogram or chronogram (in the numerator) to PD measured on the cladogram (in the denominator). Relative phylogenetic endemism (RPE) is a ratio of PE measured on the phylogram or chronogram (in the numerator) to PE measured on the cladogram (in the denominator). Mishler et al. 2014 also developed a spatial randomisation to test the statistical significance of all these phylodiversity metrics. Significantly high RPD or RPE indicates an unexpected concentration of long branches (range-restricted ones in the case of RPE), and conversely, significantly low RPD or RPE indicates an unexpected concentration of short branches. The two-step procedure they developed, called Categorical Analysis of Neo- And Paleo-Endemism (CANAPE), first identifies centers of significantly high PE and then classifies them into centers dominated by neoendemism (rare short branches), paleoendemism (rare long branches), or a mixture of both.

9.7 SUMMARY AND CONCLUSION

Using PD-based methods, we can discover patterns of biodiversity, classify it, develop hypotheses about ecological and evolutionary processes affecting it, and proceed with urgent conservation assessment – all without using ranks and in particular, without using species. Instead of counting taxa at some rank, we can use the phylogeny itself as a measuring device. The three facets of phylodiversity discussed earlier illustrate the versatility of PD applied to different representations of branch lengths, e.g., phylograms (inferred amount of trait changes), chronograms (inferred amount of time), or cladograms (with all branches considered the same length). All of those topology types can be range-weighted as well, yielding different PE measures. They can be compared with each other (e.g., the RPD and RPE metrics) to gain additional insights. In addition to these various phylogenetic measures of alpha diversity, there is a full set of phylogenetic beta diversity measures for use in assessing turnover between different locations, including range-weighted versions (Laffan et al. 2016).

Thus, we can do everything we need to do in academic biology, and in applied conservation biology, using the phylogeny itself. We can do things better, because we are looking at pattern and process realistically at many nested evolutionary levels instead of pretending that everything important happens at a single depth (i.e., the level formerly known as species). It is urgent to make this conceptual change from a species-based view to a multi-level view to complete the Darwinian Revolution in systematics, ecology, and evolution, and even more importantly, to conserve as much biodiversity as we can in the face of human pressures on the world.

REFERENCES

Agapow, P.M., O. Bininda-Emonds, K.A. Crandall, J.L Gittleman, G.M. Mace, J.C. Marshall, and A. Purvis. 2004. The impact of species concept on biodiversity studies. *Quarterly Review of Biology* 79: 161–179.

Avise, J.C. 1989. Gene trees and organismal histories: A phylogenetic approach to population biology. *Evolution* 43: 1192–1208.

Baum, D. 2007. Concordance trees, concordance factors, and the exploration of reticulate genealogy. *Taxon* 56: 417–26.

Baum, D.A. 2009. Species as ranked taxa. *Systematic Biology* 58: 74–86.

Brandon, R. 1982. The levels of selection. *PSA: Proceedings of the Biennial Meeting of the Philosophy of Science Association* 1: 315–323.

Cantino, P.D., and K. de Queiroz. 2020. *International Code of Phylogenetic Nomenclature (PhyloCode)*. Boca Raton, FL: CRC Press.

Cellinese, N., D.A. Baum, and B.D. Mishler. 2012. Species and phylogenetic nomenclature. *Systematic Biology* 61: 885–891.

Darwin, C. 1859. *On the Origin of Species by Means of Natural Selection*. London: John Murray.

Dawson, T.E., and M.A. Geber. 1999. Sexual dimorphism in physiology and morphology. In Geber, M.A., Dawson, T.E., and Delph, L.F. (eds.), *Gender and Sexual Dimorphism in Flowering Plants*, pp. 175–215. Berlin, Heidelberg: Springer.

Degnan, J.H., and N.A. Rosenberg. 2009. Gene tree discordance, phylogenetic inference and the multispecies coalescent. *Trends in Ecology & Evolution* 24: 332–340.

Doyle, J.J., J.L. Doyle, J.T. Rauscher, and A.H.D. Brown. 2004. Diploid and polyploid reticulate evolution throughout the history of the perennial soybeans (*Glycine* subgenus *Glycine*). *New Phytologist* 161: 121–132.

de Queiroz, K., and J. Gauthier. 1994. Toward a phylogenetic system of biological nomenclature. *Trends in Ecology & Evolution* 9: 27–31.

Faith, D.P. 1992. Conservation evaluation and phylogenetic diversity. *Biological Conservation* 61: 1–10.

Farris, J.S. 1983. The logical basis of phylogenetic analysis. In Platnick, N. and Funk, V. (eds.), *Advances in Cladistics. Vol. 2*, pp. 7–36. New York, NY: Columbia University Press.

Faurby, S., W.L. Eiserhardt, and J.-C. Svenning. 2016. Strong effects of variation in taxonomic opinion on diversification analyses. *Methods in Ecology and Evolution* 7: 4–13.

Harrison, R.G., and E.L. Larson. 2014. Hybridisation, introgression, and the nature of species boundaries. *Journal of Heredity* 105 Supplement 1: 795–809.

Hennig, W. 1966. *Phylogenetic Systematics*. Urbana, IL: University of Illinois Press.

Hutchinson, G.E. 1965. *The Ecological Theater and the Evolutionary Play*. New Haven, CT: Yale University Press.

Justus, J., and S. Sarkar. 2002. The principle of complementarity in the design of reserve networks to conserve biodiversity: A preliminary history. *Journal of Biosciences* 27 Supplement 2: 421–435.

Kling, M.M., B.D. Mishler, A.H. Thornhill, B.G. Baldwin, and D.D. Ackerly. 2018. Facets of phylodiversity: Evolutionary diversification, divergence, and survival as conservation targets. *Philosophical Transactions Royal Society B* 374: 20170397.

Laffan, S.W., D.F. Rosauer, G. Di Virgilio, J.T. Miller, C.E. González-Orozco, N. Knerr, A.H. Thornhill, and B.D. Mishler. 2016. Range-weighted metrics of species and phylogenetic turnover can better resolve biogeographic transition zones. *Methods in Ecology and Evolution* 7: 580–588.

Lefebvre, V., C. Villemant, C. Fontaine, and C. Daugeron. 2018. Altitudinal, temporal and trophic partitioning of flower-visitors in Alpine communities. *Scientific Reports* 8: 4706.

Macarthur, R., and R. Levins. 1967. The limiting similarity, convergence, and divergence of coexisting species. *American Naturalist* 101: 377–385.

Mallet, J. 1995. A species definition for the modern synthesis. *Trends in Ecology & Evolution* 10: 294–299.

Miller, J.T., G. Jolley-Rogers, B.D. Mishler, and A.H. Thornhill. 2018. Phylogenetic diversity is a better measure of biodiversity than taxon counting. *Journal of Systematics and Evolution* 56: 663–667.

Mishler, B.D. 1999. Getting rid of species? In R. Wilson (ed.), *Species: New Interdisciplinary Essays*, pp. 307–315. Cambridge, MA: MIT Press.

Mishler, B.D. 2000. Deep phylogenetic relationships among 'plants' and their implications for classification. *Taxon* 49: 661–683.

Mishler, B.D. 2005. The logic of the data matrix in phylogenetic analysis. In V.A. Albert (ed.), *Parsimony, Phylogeny, and Genomics*, pp. 57–70. Oxford, UK: Oxford University Press.

Mishler, B.D. 2009. Three centuries of paradigm changes in biological classification: Is the end in sight? *Taxon* 58: 61–67.

Mishler, B.D. 2010. Species are not uniquely real biological entities. In F. Ayala and R. Arp (eds.), *Contemporary Debates in Philosophy of Biology*, pp. 110–122. Hoboken, NJ: Wiley-Blackwell.

Mishler, B.D. 2021. *What, if Anything, Are Species?* Taylor and Francis Group, Boca Raton, FL: CRC Press. https://www.taylorfrancis.com/books/oa-mono/10.1201/9781315119687/anything-species-brent-mishler

Mishler, B.D., and R.N. Brandon. 1987. Individuality, pluralism, and the phylogenetic species concept. *Biology and Philosophy* 2: 397–414.

Mishler, B.D., N. Knerr, C.E. González-Orozco, A.H. Thornhill, S.W. Laffan, and J.T. Miller. 2014. Phylogenetic measures of biodiversity and neo- and paleo-endemism in Australian *Acacia. Nature Communications* 5: 4473.

Mishler, B.D., and J.S. Wilkins. 2018. The hunting of the SNaRC: A snarky solution to the species problem. *Philosophy, Theory, and Practice in Biology* 10: 1–18. http://dx.doi.org/10.3998/ptpbio.16039257.0010.001.

Myers, N., R. Mittermeier, C. Mittermeier, G. da Fonseca, and J. Kent. 2000. Biodiversity hotspots for conservation priorities. *Nature* 403: 853–858.

Nelson, G. 1973. Classification as an expression of phylogenetic relationships. *Systematic Zoology* 22: 344–359.

Nixon, K.C., and Q.D. Wheeler. 1990. An amplification of the phylogenetic species concept. *Cladistics* 6: 211–223.

Nürk, N.M., H.P. Linder, R.E. Onstein, M.J. Larcombe, C.E. Hughes, L.P Fernández, P.M. Schlüter, L. Valente, C. Beierkuhnlein, V. Cutts, M.J. Donoghue, E.J. Edwards, R. Field S.G.A. Flantua, S.I. Higgins, A. Jentsch, S. Liede-Schumann, and M.D. Pirie. 2020. Diversification in evolutionary arenas—assessment and synthesis. *Ecology and Evolution* 10: 6163–6182.

Pechenik, J.A. 1999. On the advantages and disadvantages of larval stages in benthic marine invertebrate life cycles. *Marine Ecology Progress Series* 177: 269–297.

Rosauer, D.F., S.W. Laffan, M.D. Crisp, S.C. Donnellan, and L.G. Cook. 2009. Phylogenetic endemism: A new approach to identifying geographical concentrations of evolutionary history. *Molecular Ecology* 18: 4061–4072.

Schluter, D., and M. Pennell. 2017. Speciation gradients and the distribution of biodiversity. *Nature* 546: 48–55.

Smith, S.A., and M.J. Donoghue. 2010. Combining historical biogeography with niche modeling in the *Caprifolium* clade of *Lonicera* (Caprifoliaceae, Dipsacales). *Systematic Biology* 59: 322–341.

Stevens, P.F. 1994. *The development of biological systematics*. New York, NY: Columbia University Press.

Thornhill, A.H., B.G. Baldwin, W.A. Freyman, S. Nosratinia, M.M. Kling, N. Morueta-Holme, T.P. Madsen, D.D. Ackerly, and B.D. Mishler. 2017. Spatial phylogenetics of the native California flora. *BMC Biology* 15: 96.

Thornhill, A.H., B.D. Mishler, N. Knerr, C.E. Gonzalez-Orozco, C.M. Costion, D.M. Crayn, S.W. Laffan, and J.T. Miller. 2016. Continental-scale spatial phylogenetics of Australian angiosperms provides insights into ecology, evolution and conservation. *Journal of Biogeography* 43: 2085–2098.

Willis, S.C. 2017. One species or four? Yes!...and, no. Or, arbitrary assignment of lineages to species obscures the diversification processes of Neotropical fishes. *PLoS ONE* 12(2): e0172349.

10 The Species before and after Linnaeus

Tension between Disciplinary Nomadism and Conservative Nomenclature

Alessandro Minelli

CONTENTS

DOI: 10.1201/9780367855604-13

10.1 INTRODUCTION

The units of biodiversity recognised in many branches of biology are called species and are mostly named by using the binomials of Linnaean taxonomy, although in many circumstances, researchers use instead a diversity of undisciplined non-Linnaean formulas (Minelli 2019b).

Most biologists, and probably a number of philosophers of biology too, will construe this circumstance as implying that 'species taxa, as all other taxa in biological classification, serve as the foundation for all other biological analyses and hence should be as similar to one another as possible' (Bock 2004:183). However, general agreement on this issue has emerged as highly controversial and possibly beyond hope of definitive solution.

This 'species problem' probably troubles philosophers more than biologists, but to address it, we need their combined efforts as well as a refreshed and extended reading through the literature in which species, as a term applied to animal and plant forms, has been meandering, accompanying taxonomy from its uncertain beginnings until Linnaeus and beyond.

Despite a few comprehensive studies, such as those of Mayr (1982), Wilkins (2009a, 2009b, 2018), Richards (2010), and Zachos (2016), and pithy papers on selected old authors, of which I only mention Cain's (1958, 1994) classic studies of Linnaeus and Bauhin, a revisitation of early authors may prove to be a worthwhile exercise if based on a direct, critical examination of the original texts. Within the limits of a chapter, this cannot be accomplished with the deserved accuracy and completeness. Nevertheless, I think that the analysis presented in the following pages may help refresh the terms of the ongoing debate about the species in what we call today the biological sciences.

10.2 GENUS AND SPECIES: PHILOSOPHY VERSUS ZOOLOGY AND BOTANY

My analysis focusses on a selection of authors from the sixteenth to the early nineteenth century; nevertheless, it is worth starting with Aristotle. In his texts about animals, we find no explicit classification, but Aristotle's 'formalization of individual kinds (species) and of collective groups (genera) was the point of departure for the more perceptive and elaborate classifications of the later period' (Mayr 1982:153). How well-founded is actually this traditional opinion? As remarked by Reydon (2020), γένος (genos) in Aristotle is not a formal taxonomic category. A genus, in his 'biological' works, was a group of organisms on which he fixed attention, a set of fairly similar organisms among which, however, it is possible to identify subgroups on the basis of stable diagnostic characters (*differentiæ specificæ* of Aristotle and the Scholastics). Philosophers (from Neoplatonists to Scholastics) who picked up from Aristotle the categories of genus and species used these as two terms in a logical relationship without ever hooking them up to the natural world as levels of a taxonomy (see Pavlinov 2022). Similarly, Renaissance zoologists and botanists who used these terms (or at least one of them, either species or genus) do not seem to have been directly inspired by the Aristotelian-Scholastic tradition.

Eventually, an alignment between the formal structure of classification and the logical schemes of scholastic philosophy was proposed by Linnaeus (1751) by suggesting [aphorism 155]* an equivalence between the sequence of the botanical *membra* (*classis, ordo, genus, species, varietas*) and the philosophical categories *genus summum*, *intermedium* and *proximum*, *species* and *individuum*. But Linnaeus regarded this as nothing more than analogy, since he offered, in parallel, two further illustrative examples, with a geographical sequence (*Regnum, Provincia, Territorium, Parœcia, Pagus*) and a military one (*Legio, Cohors, Manipulus, Contubernium, Miles*), respectively. Linnaeus derived from the Scholastic tradition (indirectly, i.e., essentially through his predecessors in natural history) the names, rather than the meanings, of these categories; his actual interest was mereology. Linnaeus also used the word *species* in other contexts, e.g., to indicate different sorts of trunk, i.e., stem, culm, scape, peduncle, petiole, frond, and stipe [82], or different kinds of buds [85: *Species Gemmarum variae sunt*].†

The animals described by Renaissance scholars such as Leonard Fuchs and Pietro Andrea Mattioli, among the herbalists, and Conrad Gesner and Ulisse Aldrovandi, among the authors of zoological encyclopedic books, are – like those of folk botany and zoology (Atran 1990; Freudenstein et al. 2017) – simply recognisable, reidentifiable clusters of organisms, those that Sterelny (1999:119) has characterised as phenomenological species.

10.3 CATALOGUE VERSUS CLASSIFICATION

Irrespective of whether the term *genus* or *species* is employed, we shall not take as granted that the named units of biological diversity described by botanists and zoologists throughout the centuries derive from the application to natural history of the Aristotelian and Scholastic tradition in logics. Moreover, it is worth discussing whether, or in which respect, encyclopedic works as Fuchs' herbal (*De historia stirpium*, 1542) or species lists such as John Ray's (1660, 1677) catalogues of plants are organised according to a classification and thus represent, or not, early efforts in biological taxonomy.

Deciding either way depends on what we mean by classification and as a consequence, on whether we accept alphabetical ordering as a peculiar kind of classification. These issues are clearly discussed in Hjørland (2017), on which the following paragraph is based.

In a sense, a dictionary, e.g., of plant names could be described as a classification, where the ordering principle is the alphabetic sequence of the entries, and each entry is intended as the name of a class to which plant individuals are referred, based on more or less explicitly acknowledged criteria (mostly, but not necessarily,

---

* Linnaeus' *Philosophia Botanica* is articulated into 365 sections, or aphorisms. In this chapter, all numbers in square parentheses refer to the correspondingly numbered aphorisms in that book.
† See also Sober's (1984: 337) insightful remark that 'The kinds of structure that may be discovered in the search for Laws of Form might have been termed "species", if the term had not been preempted by another research program [...]. So whether something counts as a species concept is itself a historical question.'

morphological similarity). This may satisfy Bliss' (1929:143) definition: 'A classification is a series or system of classes arranged in some order according to some principles or conception, purpose or interest, or some combination of such'. However, to regard the simple alphabetical ordering of plant or animal names as a classification makes little sense to a systematist. The problem is not so much the implicit acceptance of a concept of species as class rather than as historical individual (an issue I will not discuss in this chapter; for its background, see Ghiselin 1997; Stamos 2003; Wilkins 2009a, 2009b; Richards 2010; Zachos 2016); rather, for the natural historian, this would correspond simply to the identification of phenomenological entities worth a name (Sterelny 1999), whereas a taxonomy would require their arrangement in a scheme in which at least two levels are recognized. This corresponds to 'the *systematic* classification involved in the design and utilization of taxonomic schemes such as the biological classification of animals and plants by genus and species' (Suppe 1989:292). The alphabetic ordering of entries, instead, fits into Suppe's broader definition of *conceptual* classification as 'intrinsic to the use of language, hence to most if not all communication. Whenever we use nominative phrases we are classifying the designated subject as being importantly similar to other entities bearing the same designation; that is, we classify them together' (Suppe 1989:292).

Ordering and classifying are two different operations, and to some extent, either of them can be present in the absence of the other. For example, entries can be retrieved from modern databases such as BOLD or GeneBank according to multiple ordering criteria, but these are embodied in the output routines rather than associated to the individual records. Instead, these retain variable amounts of taxonomic information, at least in the scientific name of the organism (binomial = genus name + specific epithet), which provides possible filters for a selective search of records.

When Lamarck (1815–22) in the *Histoire naturelle des animaux sans vertèbres* arranged the main groups of invertebrates from the simplest to the most complex, thus reversing the traditional order, he demonstrated that the link between classification and ordering is not indissoluble.

I will briefly present here examples of early encyclopedic treatments with no visible ordering, followed by works based (in principle) on alphabet only, to arrive at the first natural history works in which some degree of taxonomic arrangement is deliberately applied and sometimes also critically discussed.

No visible order is perceptible, for example, in Thomas Moffett's (1634) *Insectorum sive Minimorum Animalium Theatrum*. Here, Liber I includes among others a chapter on scorpion, ant and lice (chap. 28. *De Scorpio, Formica, & Pediculis*) followed by one on the winged bug of the woods (chap. 29. *De Cimice Sylvestri Alato*), but virtually the same animals are treated again in chapters of Liber II (chap. 10. *De Scorpijs terrestribus*; chap. 16. *Formicarum Encomium*; chap. 13. *De Pediculis Brutorum & Plantarum*; chap. 25. *De Cimice*).

Examples of works where entries are ordered alphabetically are Leonhard Fuchs' herbal *De historia stirpium* (1542) and John Ray's early botanical works, *Catalogus plantarum circa Cantabrigiam nascentium* (1660) and *Catalogus plantarum Angliae* (1677).

A transition from a roughly alphabetic ordering to a distribution of items ('species') into named groups ('orders') is found in Conrad Gesner's works (Enenkel

2014). While in his first zoological treatise, on quadrupeds (Gesner 1551), mammals are arranged alphabetically,[*] a taxonomic effort is found in his later book on fishes (1560a)[†] and more explicitly, in the second edition of the *Icones Avium* (Gesner 1560b). In the latter work, some of the bird orders are quite natural; e.g., the first (diurnal birds of prey) and the second (nocturnal birds of prey) broadly correspond to orders of modern ornithology. Others do not. For example, only size (large or medium vs. small) separates birds in order III from those in order IV. And a typical songbird genus like the wagtail (*Motacilla*) is classified in order VIII rather than IV only because it frequents river banks.

A first step beyond merely ordering by alphabet is offered by Andrea Cesalpino (1583), who, following a first division of plants into trees and herbs, opens book II *De arboribus* of his *De plantis libri XVI* with an explanation of his criteria in distributing plants[‡]:

> As in all scientific texts order makes science more accessible and also helps memory, in explaining trees I regarded advisable to begin with the simplest one. Namely, the trees that under each flower bear only one seed or one seed-containing vessel (*conceptaculum*) are simpler than those that bear more than one. I call here *conceptaculum* a vessel that may contain a number of seeds with their own barks, such as the chestnut.[§]

> (Caesalpinus 1583:31)

Therefore, with Cesalpino, the different kinds of plants are distributed along the linear sequence of the chapters according to a criterion based on traits of the plants themselves rather than on alphabet, medical properties, or other.

10.4 GENUS AND SPECIES IN EARLY TAXONOMIES

Linnaeus' works consecrate the use of the terms *genus* and *species* as designating two important levels of the classification: one of the two (which one, I will discuss later) is indeed the name given to the fundamental units of the classification.

The terms *genus* and *species* are quite frequent, however, also in the catalogues and taxonomic works of many pre-Linnaean authors. It is critically important to

---

[*] For example, *Alces, Asinus, Bos & Vacca, Cacus, Calopodes, Camelopardalis, Camelus*. However, the pages about *Bos & Vacca* are followed not only by others on *Taurus* and *Vitulus* but also by accounts on *Boves sylvestres, Bison, Bonasus, Catoblepa, Boves Indici, Tarandus, Urus*, i.e. on wild members of the bovine family, plus reindeer (*Tarandus*), which belongs to the deer family, and the mythological *Catoblepa*. The alphabetical ordering is resumed with *Cacus*, but further diversions are found at other places.

[†] The fishes are distributed into 19 ordines, 17 of marine fishes and 2 of freshwater fishes. Within each order, bits of alphabetic order are retained, but not systematically.

[‡] All translations in this article are mine, except for the excerpts from Ray's *Historia Plantarum* (1686), as noted later.

[§] Cum in omni tractatione, ordo partim ad faciliorem doctrinam faciat, partim ad memoriam: in explicandis arboribus visum est a simplicioribus incipere. Sunt autem simpliciores, quæ sub uno flore unicum semen ferunt, aut unum seminis conceptaculum, quam quæ plura. Voco autem seminis conceptaculum quod aptum est plura semina cum propriis corticibus continere, ut Castanea.

examine a sample of them in order to understand to what extent the intended meaning of the terms *genus* and *species* is comparable to the meaning they will take in the works of Linnaeus. Inconsistencies are frequent, indeed, even within one work and in authors, e.g., John Ray, who have nevertheless made important contributions to systematics.

Among the authors of the great herbals of the sixteenth century, the use and meaning of the terms *genus* and *species* are quite variable, but authors mainly use *genus* for plant kinds approximately equivalent to Linnaean species, as Theophrastus and, after him, Dioscorides and Pliny did in classical Antiquity. An example from Fuchs (1542: 3): of *Absinthium*, he writes, there are three genera, according to Galen and Dioscorides.[*]

Mattioli (1568), when discussing related kinds of plants or animals, uses *genere* (genus) and *spetie* or *specie* (species) mostly in agreement with the relationship between the two terms according to Aristotle. For example, 'Juniper is of two species, id est, larger and lesser'[†] (Mattioli 1568:134); 'he [Pliny] writes of the parsnip, then immediately next of the Sisaro, as of the same genus'[‡] (Mattioli 1568:471). However, this is not enough to signal the adoption of a consistent framework of taxonomic relationships.

Of Caspar Bauhin's monumental *Pinax*, Cain (1994) rightly observed that the author probably did not intend species and genus as ranks in the Linnaean sense, despite a typographical arrangement of the work that on a cursory browsing might suggest otherwise. This typographical character is more evident in the second edition (Bauhin 1623) examined by Cain, less so in the first one (Bauhin 1596), from which I take the following examples – clear enough to dispel any doubt about the interpretation.

Bauhin (1596) lists five species of *Phalangium*, but of one of these (*Phalangium non ramosum*), he recognises two kinds (*maior* and *minor*) that he also calls species (p. 58). Of *Valeriana*: *Genera Dioscorides duo proposuisse videtur* (p. 291); of *Mandragora*: *Genera duo fecit Dioscorides* (p. 303); of *Helleborus*: according to Pliny, *Genera prima duo sunt, candidum & nigrum* (p. 339); of *Rosmarinus*: *Theophrastus duo genera fecit* (p. 409).

Even where nomenclature might suggest otherwise, the *species* of these botanists of the sixteenth and early seventeenth centuries are not species in the Linnaean sense of the term. Cesalpino, for example, lists five *species* of pomegranate (*Punica*), i.e., *Dulcia, Acria, Mixta, Acida, Vinosa* (p. 142), which, from their qualification, must be interpreted as simple cultivars.

One hundred years later, Ray is still inconsistent, as shown by the following examples, drawn both from his pupil Francis Willughby's[§] books on birds (Willughby

---

[*] Generum sunt tria, Galeno & Dioscoride testibus.
[†] E' il Ginepro di due specie, maggiore ciò è, & minore.
[‡] scrive egli [Plinio] della pastinaca, scrive subito poi del Sisaro, come congenere di quella.
[§] On the zoological works of Francis Willughby and the role of his mentor John Ray as editor of these books still unpublished by Willughby's death at 36, see Birkhead (2018), who lists all these volumes with Ray as author. I follow here traditional usage, listing the books on fishes and birds as Willughby's and those on quadrupeds and insects as Ray's.

1676) and fishes (Willughby 1686), works Ray edited and expanded before publication, and from his own books on four-footed animals and snakes (Ray 1693) and on insects (Ray 1710). Inconsistencies are conspicuous even within a single work.

For example, in the monograph on fishes, Rondelet's (1554:7) partitioning into sea, river, lake, and swamp fishes (*marinos, fluviatiles, lacustres*, and *palustres*) is rejected especially because 'it keeps apart fishes of the same genus *(ejusdem generis)*, i.e., those that agree in body shape, fin number and position and other traits characterizing genera, such as salmons and trouts, eels and congers, etc.'* (p. 20). Here, genus is used, more or less, in the Linnaean sense, but elsewhere it is not. For example, *genus* is used (pp. 21–22) for three of the five groups representing the primary subdivisions of the spiny-rayed fishes (our Acanthomorpha), among them the *Pisces Aculeati*, but also for each of the two divisions (*membra*) of the latter, i.e., those with two dorsal fins, the anterior of which with spiny rays, and those with one dorsal fin, of which the anterior rays are spiny and the posterior ones soft. Moreover, elsewhere in the same book, *genus* is used as equivalent to a Linnaean species: of cod (as *Asellus major vulgaris*), it is said that this genus of fish is caught every year in nearly infinite number especially in the Northern Ocean[†] (Willughby 1686:166).

The following sentence from the Preface to Willughby's ornithology book would suggest that the Aristotelian genus/species relationship has been fully translated into a taxonomic relationship of class/member inclusion: 'As the aim [of this book] is to describe accurately, to distinguish and to put the species into their particular classes or genera, in order to remove all obscurity and confusion'[‡] (Willughby 1676: page not numbered). Nevertheless, in the same work, *species* is sometimes used as just a synonym of kind. For example, whereas the first section of chapter XV (*De Columbis in specie*) is dedicated to '*Columba domestica seu vulgaris*. The Common Pigeon or Dove. Περιςτερά', the next section describes *Columbarum Domesticarum variæ species*.

A step forward in the direction of a classification with formally recognised ranks had been made by Joachim Jungius several years before:

> Unless plants are firmly arranged into genera and species under a constant principle rather than according to one or another botanist's caprice, the study of botany (*Phytoscopia*) will continue forever. [...] order imposes a limit on the infinity of sets of genera and species.[§]
>
> **(Jungius 1662:69)**

---

* Hanc ego divisionem minus probo [q]uia separat eos qui ejusdem generis sunt, hoc est qui & figura corporis & pinnarum numero ac situ aliisque accidentibus generum characteristicis conveniunt, Salmones v. g. a Truttis, Anguillas a Congris.

† Infinita propemodum multitudo hoc genus piscium in Oceano praesertim Septentrionali ab Anglis & Batavis quotannis capitur.

‡ Cùm vero Species omnes accurate describeré, distinguere & ad suas singulas classes seu genera reducere ad omnem obscuritatem & confusionem tollendam, scopus sit.

§ Plantæ nisi in certa Genera et Species constanti ratione, non pro lubitu hujus vel illius, redigantur, infinitum, quasi reddetur Phytoscopiae studium [...]. Ordo autem classium Generum, Specierum terminum infinitis imponit.

But Jungius' message remained largely unnoticed, probably because his pages on botanical classification were literally buried in a voluminous work about logic and the universal ordering of knowledge, whose distribution was quite limited. A botanist would have hardly sought inspiration in a text written by a mathematician, logician, and philosopher of science *ante litteram*.

The merit of being the first to build a popular classification with a hierarchy articulated on four ranks (class, section, genus, species) rests with Tournefort (1694). In the introductory chapters of his *Élémens de Botanique*, the author does not provide any comment about the rank of section, extensively used in the descriptive part of the text, but explains that

> it is not enough [...] to relate plants to their true genera: these same genera must be reduced to certain classes, so that we can see at a glance, & as in a general map, all the material which is the object of this science.[*]

> **(Tournefort 1694:2)**

> The only thing that remains to be done after what we have said about the genera of plants is to arrange them in a manner suitable for drawing up a general history of plants, a regular and convenient one. It is thus necessary to distribute genera into a number of classes. I will therefore call a class of plants the set of genera sharing a number of common marks that must necessarily be found in them, which distinguish them from all other genera.[†]

> **(Tournefort 1694:16)**

For a while, this taxonomic structure is used for plants only, but half a century later, Jakob Theodor Klein will use a similar classification structure in his books on birds and quadrupeds. In the first work (Klein 1750), the order of birds is divided into families and each of these into genera; some genera, however, are further subdivided into tribes, each of which roughly corresponds to a Linnaean genus.[‡] For example, Klein divides his genus *Passer* into five tribes (*'Vulgaris'* – i.e., *Passer* in the strict sense, *Emberiza*, *Linaria*, *Coccothraustes*, and *Fringilla*). *Emberiza* and *Fringilla* are recognised as distinct genera by Linnaeus (1758), who distributes instead between *Fringilla* and *Loxia* the species placed by Klein in the other three tribes. In Klein's (1751) work on quadrupeds, the structure of classification is closer to that of Linnaeus, indeed even closer than that of Linnaeus to the structure adopted by post-Linnaean authors, due to the interposition of the rank family (of 'Tournefortian'

---

[*] il ne suffit pas [...] de raporter les plantes à leurs véritables genres: il faut réduire ces mêmes genres sous certaines classes, en sorte que l'on puisse voir d'un coup d'oeil, & comme dans une carte générale, toute la matière qui fait l'objet de cette sience.

[†] La seule chose qu'il reste à faire après ce que nous avons dit des genres des plantes est de les disposer d'une maniere propre à dresser une histoire générale des plantes, qui soit régulière, & commode. Il est necessaire pour cela de partager les genres en certaines classes. J'apellerai donc une classe de plantes l'amas de plusieurs genres entre lesquels se doivent necessairement trouver certaines marques communes, qui les distinguent de tous les autres genres.

[‡] In current taxonomic usage, the tribe is an optional rank between subfamily and genus in descending order.

flavor) between order and genus. Eventually, Linnaeus will extend this hierarchical organisation to the entire *Systema Naturae*, organised *secundum classes, ordines, genera et species*.

10.5 NATURAL VERSUS ARTIFICIAL TAXA

For Linnaeus (at least around 1751, the year he published the first edition of *Philosophia Botanica*), the classification included both natural and artificial taxa, but he was confident that the study of currently unknown plants would allow a progressive approach to the full understanding of the divine plan of creation of plant species* [aphorism 160]. In detail, however, Linnaeus was anything but consistent. At first [152], he distinguished in the classification a part derived from theory, which is within reach of the true systematist only, from a part derived from practice, to which even those who are unable to adopt a spirit of system can contribute. Fruit of the purely practical approach are species and varieties, while classes, orders, and genera are identified *theorice*.[†] One might believe that this distinction corresponds to the distinction between natural and artificial taxa, but this is not the case. In aphorism 162, Linnaeus states that both species and genera are always the work of Nature, varieties are mostly a consequence of cultivation, and classes and orders are partly the work of Nature, partly the product of the systematist's arbitrary choice.[‡] More precisely, genera correspond to groups of species sharing sets of the most proximate common traits, as originally fixed at creation[§] (Linnaeus 1737, *Ratio operis*, aphorism 6, pages not numbered). However, Linnaeus does not offer criteria for the recognition of genera other than a similarly constructed *fructificatio*[¶] [159]. In spite of the name, the latter term includes both flower and fruit: 'The parts of the flower are calyx, corolla, stamen, pistil. Those of the fruit are pericarp, seed, receptacle. Therefore, the parts of the fructification are the flower and the fruit'** [87]. Varieties are produced by accidental causes, either natural or due to cultivation, and are reversible[††] [158].

A class was defined by Linnaeus [160] as the agreement of several genera in the parts of the flower and fruit.[‡‡] Eventually, this is not that different from Tournefort's (1694) definition:

---

* Artificiales classes succedaneæ sunt naturalium, usque dum omnes naturales sint detectæ, quas plura genera, nondum detecta, revelabunt, & tum limites classium difficillimi evadant.
† Dispositio vegetabilium divisiones s. conjunctiones docet; estque vel *Theoretica*, quæ Classes, Ordines, Genera; vel *Practica*, quæ Species & Varietates instituit. Dispositio plantarum, ex fundamento fructificationis, recentiorum inventum est. Practica ab eo potest tractari, qui de Systemate nihil intelligit.
‡ Naturæ opus semper est Species & Genus; culturæ saepius Varietas, naturæ et artis Classis & Ordo.
§ Genera tot sunt, quot attributa communia proxima distinctarum specierum, secundum quae in primordio creata fuere.
¶ Genera tot dicimus, quot similes constructæ fructificationes proferunt diversæ species naturales.
** Partes Floris: Calyx, Corolla, Stamen, Pistillum. [Partes] Fructus. Pericarpium, Semen, Receptaculum. [Partes] Fructificationis, itaque Flos, Fructus, sunt.
†† Varietates tot sunt, quot differentes plantæ ex ejusdem speciei semine sunt productae. Varietas est Planta mutata a caussa accidentali: Climate, Solo, Calore, Ventis, &c. reducitur itaque in Solo mutato.
‡‡ Classis est Generum plurium convenientia in partibus fructificationis.

I will give the name of plant class the collection of genera sharing certain common characteristics that distinguish them from all other genera.*

(Tournefort 1694:16)

Linnaeus regarded several among the classes recognised by botanists, e.g., umbellifers, legumes, and grasses, as natural. The remaining, artificial classes

> are substitutes for natural ones, until the discovery is made of all the natural classes which more genera, that have not yet been discovered, will reveal; and then the most difficult distinctions between classes may become apparent.†

[160]

Ostensibly artificial, instead, are orders, regarded by Linnaeus as

> subdivision[s] of the classes, [created by the botanist] so that it does not turn out to be necessary at one and the same time to distinguish more genera than the mind can easily comprehend. Indeed, it is easier to distinguish 10 genera rather than 100.‡

[161]

But let's return to the species.

10.6   IS THE SPECIES THE BASIC LEVEL IN THE TAXONOMIC HIERARCHY?

It is generally taken for granted (but see Pavlinov 2022 for opting instead in favor of the genus) that the species category is the basic level in the taxonomic hierarchy of biological classifications. However, it would be more sensible to say that the basic taxonomic category in biological classifications is called species. As discussed in the last section of this chapter, this reversed reading will offer help in addressing the so-called species problem.

The fundamental unit in a classification is not necessarily the lowest one, especially if the lowest ranks are optional and perhaps not associated to a standardised nomenclature: this is the case of subspecies, varieties, etc. in a classification where the species is overtly chosen as the basic unit.§ Caution in this respect is necessary

---

* J'apellerai donc une classe de plantes l'amas de plusieurs genres entre lesquels se doivent necessairement trouver certaines marques communes, qui les distinguent de tous les autres genres.

† Artificiales classes succedanea sunt naturalium, usque dum omnes naturales sint detectæ, quas plura genera, nondum detecta, revelabunt, & tum limites classum difficillimi evadant.

‡ Ordo est Classium subdivisio, ne plura Genera distinguenda simul & semel evadant, quam animus facile assequatur. Ordo est Classium subdivisio; facilius enim distinguuntur genera 10, quam 100. On logic and memory in Linnaeus, see the classic paper of Cain (1958).

§ The International Code of Zoological Nomenclature (1999) recognizes only one intraspecific level (the subspecies) as worthy of recognition and naming, but no rule requires that subspecies are recognized. More flexible is the botanical Code (Turland et al. 2018), which accepts up to five (optional) infraspecific ranks: subspecies, variety, subvariety, form, and subform.

when examining taxonomies up to Linnaeus, where the author's choice between genus and species is not always easy to determine. Here are two examples:

Although botanists' opinions differ about genera, nevertheless they agree more about species.*

(Jungius 1662:76)

[a]s the plants of the same genus still differ from each other by some peculiarity, we will call species all those which besides the generic character will have something singular that we will not notice in other plants of the same genus. For example, all those we will call Ranunculus (buttercup) will have a common character drawn from the structure of some of their parts; this will establish their genus, and will be suitable for Ranunculus only. But as all buttercups are alike only in this common character, while they are different in some others of their parts; the difference of these parts will establish the different buttercup species. The characters of the genera must satisfy two conditions: 1. be as similar as possible in all species; 2. be sensible and easy to observe, without having to use the microscope to discover them.†

(Tournefort 1694:13)

Both genus and species were still mentioned as basic systematic units by Spring (1838):

As for the artificial system the higher systematic units, so are the lower systematic units, the species and the genus, the foundations for the natural system, on which the architectural art of the systematist erects its steps and pinnacles from which we can overlook the army of natural beings in gradually wider circles.‡

(Spring 1838:8)

However, later in the same book, Spring seems to opt for the species altogether:

The task of natural history is to seek out the substance of every particular natural body and to bring it to scientific knowledge, i.e., to abstract the essential, unchangeable from

---

* Quamvis Botanici in Generibus discrepent, tamen in Speciebus magis conveniunt.
† comme les plantes de même genre diferent encore entre elles par quelque particularité, nous apellerons especes toutes celles qui outre le caractère générique auront quelque chose de singulier que l'on ne remarquera pas dans les autres plantes du même genre: Par exemple toutes celles que nous apellerons des Renoncules auront un caractère commun tiré de la structure de quelques-unes de leurs parties qui établira leur genre, & qui ne conviendra qu'aux seules Renoncules. Mais comme toutes les Renoncules ne se ressemblent que dans ce caractère commun, & qu'elles sont diferentes dans quelques autres de leurs parties; la diference de ces parties établira les diferentes especes de Renoncule. Les caractères des genres doivent avoir deux conditions: 1. être aussi semblables qu'il se peut dans toutes les especes; 2. être sensibles & faciles à remarquer; sans qu'on soit obligé d'employer le microscope pour les découvrir.
‡ Wie für das künstliche System die höheren, so sind für das natürliche die niederen systematischen Einheiten: die Art und die Gattung, Grund und Boden, auf dem die architektonische Kunst des Systematikers ihre Stufen und Zinnen errichtet, von welchen aus man das Heer der Naturwesen in allmählig weiteren Kreisen übersehen kann.

the individual and to fix it in a unity of knowledge. [...] Natural history's subject is therefore only the species [...], never the individual; [...] the natural scientist himself has only individuals before him; because only these have reality on our level of knowledge. [...] Knowledge of the species is, however, the foundation of natural history as a science. This should precede all other research.*

(Spring 1838:55)

10.7 ANIMALS VERSUS PLANTS – TWO DISTINCT TAXONOMIC TRADITIONS

An aspect largely overlooked by historians of the life sciences who have dealt with the precursors of the Linnaean classification is that early taxonomies were not intended as a common scheme for classifying all kinds of living beings, and these only.

On the one hand, before finding a point of arrival in Linnaeus' encyclopedic work, taxonomy (and with it, the fixation of genus and/or species as unit(s) of the classification) evolves separately for animals and plants. Almost all the authors contributed either to botany but not to zoology (e.g., Jungius, Tournefort), or vice versa (e.g., Klein). Even Ray, whose interests extended over both fields, never dealt with plants and animals in the same work and adopted very different treatments for plant versus animal species, as we will see in Section 10.8.

Plant taxonomy matures more rapidly than animal taxonomy, and this is well reflected in Linnaeus' scientific production. His first attempts at a system of nature 'to help the curious reader marching, according to a Map, in the highest kingdoms [of Nature]'† (Linnaeus 1735:3) were followed, in chronological succession, by a work on plant genera (Linnaeus 1737) followed by one on plant species (Linnaeus 1753), the first treatise in which Linnaeus used binomial nomenclature, following a slightly earlier suggestion, also with botanical examples, given in the little-cited dissertation *Pan suecicus* defended in Uppsala on December 9, 1749, by his pupil Nicolaus L. Hesselgren (Linnaeus 1749).

His zoological classification changed less compared with the first editions of *Systema Naturae* until the tenth (Linnaeus 1758), where for the first time, he adopted binomial nomenclature for animals too.

On the other hand, Linnaeus extended his taxonomy, and the associated nomenclature, also to minerals, the third of the traditional kingdoms of nature. His taxonomy was a taxonomy for all natural productions. This is understandable as an

---

* Die Aufgabe der Naturgeschichte ist es, die Substanz jedes besondern Naturkörpers aufzusuchen, und zur wissenschaftlichen Erkenntniß zu bringen, d. i. das Wesentliche, Unveränderliche vom Individuum zu abstrahiren, und in einer Einheit der Erkenntniß zu fixiren [...]. Vorwurf der Naturgeschichte ist daher nur die Art und der Artbegriff, nie das Individuum; [...] der Naturforscher selbst hat gleichwohl nur Individuen vor sich; denn nur diese haben auf unserer Stufe der Erkenntniß Realität. [...] Kenntniß der Arten ist aber Grund und Boden der Naturgeschichte, als Wissenschaft. Diese soll allen anderen Forschungen vorausgehen.
† Exhibui heic Conspectum generalem Systematis corporum Naturalium, ut curiosus Lector, ope Tabulæ huius Geographicæ quasi, sciat, quo iter suum in amplissimis his Regnis dirigat.

expression of a time at which an autonomous science of the living did not yet exist, but it is also an aspect that must be highlighted in a historical reconstruction of the evolution of 'species' as a taxon and as a category of classification.

The choice to apply the species category also to minerals does not prevent Linnaeus from recognising, as Ray had done for plants, a biological criterion of belonging to the same species, which obviously cannot be applied to minerals. On the one hand,

> it is abundantly clear to everybody that the living beings are propagated from eggs and that all eggs produce offspring that closely resembles the parent. Accordingly, in our times no new species are produced. Individuals multiply by generation [...] If in whatever species we follow backwards in time the multiplication of individuals [...] the series will eventually end up in just one parent. [...] As there are no new species [and] as living beings always generate offspring similar to themselves [...] it is necessary to attribute that generative unit to an Omnipotent and Omniscient Being.[*]

> **(Linnaeus 1744b:1)**

On the other hand, however, in minerals,

> Species very often are hard to determine to a satisfactory degree, since in these things there is no generation from eggs, but the multifarious diversity of polymorphic nature that is merely playing is both the crux of the art [the science of minerals] and the foundation of Metallurgy.[†]

> **(Linnaeus 1768:11)**

It would not be sensible to dismiss this uniform taxonomic treatment of animals, plants, and minerals as an oddity of past centuries. As remarked by Gilmour (1940:465), the peculiarity of the classification of animals and plants is restricted to the different kind of data on which it is based: 'Broadly speaking, the purpose of all classification is to enable the classifier to make inductive generalizations concerning the sense-data he is classifying'.

A generally neglected circumstance that helps understand the delay with which a dedicated taxonomy for the living beings was taking shape (as said, it did not even exist in Linnaeus) is the link between zoology and mineralogy, to the exclusion of botany, as the study subject of natural history, be this intended as a discipline in

---

[*] omnibus satis superque patet, viventia singula ex ovo propagari, omneque ovum producere sobolem parenti simillimam. Hinc nullæ species novas hodienum producuntur. Ex generatione multiplicantur individua. [...] Si hanc individuorum multiplicationem in unaquaque specie retrograde numeremus [...], series tandem in unico parente desinet [...]. Cum nullæ dentur novæ species; cum simile semper pariat sui simile [...], necesse est ut unitatem illam progeneratricem, Enti cuidam Omnipotenti & Omniscio attribuamus.

[†] Species saepissime vix ax ne vix sufficienter determinari queunt, cum in his genesis ex ovo, omnino nulla unquam existat, sed multiplex mere ludentis naturae polymorphae varietas sit et calamitas ipsius artis et Metallurgiae fundamentum.

university teaching or as the domain of expertise of a scholar (Minelli 2019a, 2020a).[*] Among the circumstances that may have determined this disciplinary coupling is the separation between the botanical garden, the theater of the study of plants, and the museum hosting the collections on which a systematic study of animals and minerals is based. As for the herbaria, at the time in the form of books rather than collections of loose sheets, they found their place on the disciplinarily neutral shelves of the library. Biology, as a science with an autonomous research program, did not yet exist. This is not the place to add to the still open debate on the criteria according to which the origin of biology can be established (e.g., Caron 1988; McLaughlin 2002; Minelli 2019a, 2020a; Nyhart 2019); however, it is appropriate to mention here the resounding refutation of the traditional (and Linnaean) division of nature into three kingdoms that opens the *Elenchus Zoophytorum* of Pallas (1766:3):

> It is customary to divide all the objects that make up our globe, and those that it contains, into three kingdoms – animal, plant, and mineral – but this distinction accepted until now is arbitrary and imaginary; Nature is organized very differently, if we look at the system of Nature with a spirit free from preconceptions, we must instead recognize within it the primary distinction between inert and brute bodies and living and organic ones.[†]

Linnaeus was not only the first modern author to offer in a single comprehensive work a detailed classification of all living beings. He was also one of the last. Lamarck was a botanist before being put in charge of the invertebrates as *professeur* at the national museum of natural history in Paris, and in his ground-breaking *Philosophie zoologique* (Lamarck 1809), he referred here and there to plants, but the botanical segment of his production is neatly detached from his subsequent zoological works. Readers of the *Histoire naturelle des animaux sans vertèbres* (Lamarck 1815–22) were unlikely the same naturalists who had used the *Flore françoise* (Lamarck 1778) to identify plants. Therefore, in the absence of the strict bond offered by treatment in the same work, it is no wonder that taxonomic traditions in botany and zoology began to diverge soon after Linnaeus' death. This was

---

[*] Combining zoology with mineralogy rather than with botany was a consolidated academic tradition. An example is provided by the many editions of Blumenbach's successful handbook of natural history. In the first edition (Blumenbach 1779–80), next to an introduction of 32 pages mainly devoted to a characterization of the three kingdoms of nature, 443 pages are devoted to animals, 204 to minerals (including 21 on fossils), but only 35 to plants. The relative amount of text allocated to the same topics remains the same throughout the editions, up to the following figures in the 12th and last edition (Blumenbach 1830): 21 (introductory matters), 311 (animals), 23 (plants), 144 (minerals, 17 of which on fossils). Other works offer even more clear-cut examples of this plants versus animals + minerals splitting. In 1760, Gronovius (1760a,b) published two bibliographic repertoires, one of botanical literature, the other of works on zoology and earth sciences. Similarly, nearly 100 years later, Agassiz and Strickland (1848, 1850, 1852, 1854) published a *Bibliographia zoologiae et geologiae*. In 1847, Louis Agassiz became Professor of Geology and Zoology at Harvard. Alpheus Spring Packard Jr taught zoology and geology at Brown University in Providence, Rhode Island, from 1878 until his death in 1905 (Cockerell 1920). More recently, Stephen Jay Gould had the distinction to be, at the same time, Alexander Agassiz Professor of Zoology and Professor of Geology at Harvard University.

[†] Corpora, quae Globum nostrum constituunt, eoque continentur, universa in tria vulgo Regna dispesci solent: Animale, Vegetabile, Minerale. Hanc receptam distinctionem arbitrariam prorsus atque imaginariam esse, longeque diverso instituto Naturam bruta inertiaque Corpora, a vivis atque organicis segregasse liquebit, si ab omni praeconcepta opinione libero animo Naturae Systema perlustraverimus.

remarkable, and annoying, because nomenclature began to diverge too. By 1842, at the time a committee of the British Association for the Advancement of Science was about to publish a first set of rules intended to ensure stability and universality in the creation and usage of names for animal species and higher taxa (Strickland et al. 1842), the Italian zoologist Luciano Bonaparte (a nephew of Napoleon) realised how far this trend had progressed. As a consequence, he launched an effort towards the establishment of common rules for the nomenclature of animals and plants (Minelli 2008). His efforts, however, failed, and a similar initiative launched in the last years of the past century (Greuter et al. 1998) does not seem to have enjoyed better success.

10.8   FIXISM AND DEFINITION OF SPECIES TAXA

10.8.1   Species Fixism

Having affirmed that the number of species recognised by botanists is the same as the number of species created in the beginning* [157], and reiterated that the species are very constant, since their generation is faithful continuity through time† [162], Linnaeus was inclined to downsize and ultimately reject the concerns that had arisen from the occasional discovery of plants that seemed to represent new species of very recent origin. Two of these 'new' plants had been described by Marchant (1719); another, the famous *Peloria*, was first described in the dissertation defended in Uppsala on December 19, 1744, under the presidency of Linnaeus, by his pupil Daniel Rudberg. *Peloria* was clearly related to the toadflax (*Linaria vulgaris*), with which it agreed in all characters except for a major difference in flower symmetry. The mechanism by which *Peloria* might have originated was suspected to be hybridisation, but a potential partner to *Linaria* had not been identified, and the hypothesis itself did not seem very plausible, *Peloria* being fertile rather than sterile like mules and hybrids in general (Linné 1744a).‡

Ray is often credited (e.g., Wilkins 2009:93) with having introduced species fixism, but the idea that all species were created in the beginning had long been expressed before. Ray wrote that 'Nature always scrupulously preserves all genera and species' (see later, in the discussion on hybrids) but did not present that thought as novel, and instead he added that he fully agreed with Gesner that none of the animal species created at the beginning of time had been completely lost§ (Ray 1693:76). As remarked by Enenkel (2014), on this view of creation, there was a place even for the animals discovered in modern times in the most remote parts of the world, e.g., in South America. Nieremberg (1635), for example, regarded them as part of God's creation despite the distance from Eden's garden: angels had provided for their transfer to their final destination.

---

* Species tot numeramus quot diversæ formæ in principio sunt creatae.
† Species constantissimæ sunt, cum earum generatio est vera continuatio.
‡ *Peloria* was long regarded as a mutant phenotype due to a mutation of major effect. Eventually, in 1999, it was discovered that the unusual symmetry of *Peloria* flowers has an epigenetic cause: it depends on the silencing by hypermethylation of a gene whose expression is critically important for the flower's symmetry (Cubas et al. 1999).
§ Ego cum Gesnero plane sentio, nullam omnino Animalis (initio creati) speciem penitus interiisse:

10.8.2   DEFINITION OF SPECIES TAXA

According to Cain (1997:337), 'the excellent naturalist John Ray [...] published [...] the first attempt to define living species by their reproduction (*Historia plantarum* I, 1686), now accepted as the definition of the biospecies'.

I find it hard to construe Ray's actual words as a definition of living species rather than as an empirical (experimental) criterion to check the conspecificity of similar but not identical kinds of plants:

> So that the number of plants can be gone into and the division of these same plants set out, we must look for some signs or indications of their *specific distinction* [italics mine] (as they call it). But although I have searched long and hard nothing more definite occurs than distinct propagation from seed. Therefore whatever differences arise from a seed of a particular kind of plant either in an individual or in a species, they are accidental and not specific. For they do not propagate their species again from seed; if a comparison is made between two kinds of plant, those plants which do not arise from the seed of one or the other, nor when sown from seed are ever changed one into the other, these finally are distinct in species.*

Ray did not intend to offer a definition of species but a criterion for the delimitation between similar species, that is, the foundation (not the description) of an Aristotelian *differentia specifica*. (The title of chapter XX of the first tome of *Historia plantarum* is indeed *De specifica (ut vocant) Plantarum differentia*.) In Ray's fixist scenario, the problem of species delimitation is reduced to checking how much variation can be accommodated within the strict limits of genealogical continuity:

> those [plants] which differ in species keep their own species forever, and one does not arise from the seed of the other and vice versa.†

In Ray's mind, this biological criterion was necessarily associated with the metaphysical creed that all species derive from God's creative act:

> For since in nature the number of species is fixed and determined, since 'God on the sixth day rested from all his labor', that is, from the creation of new species, however infinite might be the number of plants varying in color and multiplicity of flower, with new ones arising each year, we properly reject and exclude them from the grade and dignity of species.‡

---

* Ut Plantarum numerus iniri possit, & earundem divisio recte institui, oportet ut notas aliquas seu indicia specificæ (ut vocant) distinctionis investigemus. Nobis autem diu multumque indagantibus nulla certior occurrit quam distincta propagatio ex semine. Quæcunque ergo Differentiæ ex eiusdem seu in individuo, seu specie plantæ semine oriuntur, accidentales sunt, non specificæ. Hæ enim speciem suam satione iterum non propagant [...] aut si inter duas aliquas comparatio instituatur, quæ plantæ ex alterutrius semine non proveniunt, nec unquam semine satæ transmutantur in se invicem, eæ demum specie distinctæ sunt. – This excerpt from Ray (1686) and the two following ones from the same book are given here in Lazenby's (1995) translation.
† Quæ specie differunt speciem suam perpetuo servant, neque hæc ab illius semine oritur, aut vice versa
‡ Cum enim specierum numerus in natura certus & determinatus sit: cum **Deus sexto die ab omni opere suo**, hoc est, & novarum specierum creatione, requievit: floris autem colore & multiplicitate variantium plantarum numerus, novis quotannis exorientibus, infinitus sit, merito eas a specierum gradu & dignitate dejicimus & excludimus.

So much for Ray's criteria to delimit plant species. But there is no evidence that he extended this view to animals too. None of the zoological works to which John Ray contributed as the main author or as curator of the unpublished manuscripts left by Francis Willughby contains comments on the nature of the species or on criteria for delimiting them. On the contrary, some passages seem to exclude, on Ray's part, the adoption of the 'true breeding' criterion he proposed for plants.

In the volume on fishes, we find only a critical evaluation of the morphological differences that can justify the distinction between two closely similar species. For example, this is a comment on a comparison between an allegedly distinct species of gurnard and the one described in the previous lines: 'I would not easily accept that this species is distinct from the fish described in the previous lines, because they agree in all traits except for the lateral lines'* (Willughby 1686:280–281).

Similarly, in the monograph on insects, the brief descriptions of three beetles are followed by this skeptical comment: 'Differences among these three species are modest, thus I am not sure that they are actually distinct'[†] (Ray 1710:77). In the same book, no critical comment accompanies the use of the term *species* by his correspondent William Derham, from whom a long letter is reproduced that deals with larvae, pupae, and chrysalides (*aureliæ*). For example, Derham, after saying that 'The other Species of Aureliæ have large Heads somewhat like a wooden Mallet', adds that 'This Species may be also divided into two sorts, either such as springs from Worms, or such as do not' (Ray 1710:263).

10.9 VARIATION WITHIN THE SPECIES

10.9.1 VARIATION, ACCIDENTAL OR NOT

To have been issued at the beginning of times from an act of divine creation does not exempt species from liability to some degree of change. But what are the origin, amount, and nature of the variations that a species can present? Linnaeus and other authors address this problem explicitly, and in some cases, exploring variability leads them to calling fixism into question.

Linnaeus stated that the botanist does not bother about minor varieties[‡] [310] and explained:

Sex is the foundation of natural varieties, while all others are monstrous. Dioecious plants offer a unique very natural kind of varieties, with the distinction between male

---

* Piscem hunc a superius proxime descripto specie diversum esse haud libenter concesserim, notae enim omnes dempta linearum lateralium divaricatione conveniunt.
† Hæ tres species non multum inter se differunt ut dubitem an distinctæ revera sint. The three species are listed as Scarabæus niger mediæ magnitud[inis] corpore oblongo & angusto, antennis longis nodosis; Scarabæus medius, antennis brevioribus, elytris paulo longioribus; Scarabæus medius niger, alarum elytris longis, thoracis etiam tegmine longo.
‡ Varietates levissimas non curat Botanicus.

and female specimens: it is absolutely necessary that the botanist knows them and adds this information to the list of specific differences.[*]

[308]

Some degree of variation within the species is accommodated within a substantially fixist notion of species. We will still find this attitude in many post-Linnaean authors, as the following example demonstrates:

> The species is the fundamental group given by nature. Everything comes from it or returns to it; like the variety which is an accidental derivation, and the race which is a derivation that has become permanent.[†]

> **(Geoffroy Saint-Hilaire 1859:365)**

The topic received monographic treatment by Spring (1838) in a book that has been largely ignored except for Rieppel's (2016) illuminating analysis. Spring's conception of species was singularly dynamic: in Rieppel's (2016:8) translation, 'species do not persist in a state of "being", but instead are continuously "becoming"'[‡] (Spring 1838:49). Change is therefore intrinsic to each species' *Begriff*, a term that in other contexts might be translated as notion or concept but here means something like the essence of a particular species. This must be taken into account when reading the following:

> we would be in error if we accept the common opinion that those deviations (the varieties) were deviations from the species concept (*Artbegriff*), not within the species concept. Each time they belong to the species concept within which they are expressed, in so far as it was allowed. They are therefore also nothing accidental, as others think; for where the possibility is given by the species concept, there certain external conditions (influences) must always and every time produce certain changes or rather modes of development.[§]

> **(Spring 1838:59-60)**

---

[*] Sexus Varietates naturales constituit; reliquae omnes monstrosae sunt. Dioicae plantae constituunt unicum modum varietatum vere naturalem, in Mares & Feminas distinctum, quas nosse & differentiis addere Botanicis perquam necessarium est.

[†] L'espèce est le groupe fondamental donné par la nature. Tout en part ou y aboutit; comme la variété qui en est une dérivation accidentelle, et la race une dérivation devenue permanente.

[‡] Die Arten sind nicht, sondern sie werden. – As kindly brought to my attention by Frank Zachos, this is indeed virtually the same as Dobzhansky's (1937:312) sentence, 'Species is a stage in a process, not a static unit'. Spring's book, however, was not cited by Dobzhansky.

[§] Man fehlt sehr, wenn man der gewöhnlichen Meinung anhängt, als seien jene Abweichungen (die Varietäten), Abweichungen von dem Artbegriffe, und nicht in dem Artbegriffe. Sie gehören jedesmal zum Artbegriffe, in welchem sie ihrer Möglichkeit nach ausgesprochen sind. Sie sind daher auch nichts Zufälliges, wie Andere meinen; denn wo die Möglichkeit durch den Artbegriff gegeben, da müssen gewisse äussere Bedingungen (Einflüsse) immer und jedesmal gewisse Veränderungen oder vielmehr Entwicklungsweisen hervorbringen.

Variation (*Abartung*, variatio) must be distinguished from degeneration (*Ausartung*, degeneratio). The latter is a deviation from and outside the concept of species, which arises when a natural product is torn out of the sphere of external influences absolutely assigned to it by the species law.[*]

(**Spring 1838:70**)

As remarked by Rieppel (2016), Spring's dynamic conception of species was not a Darwinian one, but it could be easily rewritten in Darwinian terms. In accepting that a species possesses the capacity of changing based upon laws of change intrinsic to its organic constitution, this was quite similar to Buffon's views in his latest account on this subject, *De la dégénération des animaux* (Buffon 1766), discussed below, but also to much later views rooted in a timid adhesion to evolution, such as Wasmann's (1910): 'The varying degrees of capacity of evolution possessed by the primitive forms of the different natural species depend primarily upon the interior laws of evolution impressed upon their organic constitution'.

10.9.2 VARIATION IN SPACE

That animals and plants of the same species may grow to different sizes, or exhibit regional peculiarities under different climates, is something that could be easily accommodated within the framework of pre-Linnaean zoology and botany. Nevertheless, geographical variation was hardly regarded as an interesting issue to be addressed. Linnaeus' dismissive attitude towards *varietates minores* apparently settled matters for a while.

Evidence that invited naturalists to pay attention to geographic variation, however, began to get their attention. Exemplary is the essay in which Peter Simon Pallas described with unprecedented accuracy numerous species of rodents from Europe and Northern Asia, based on collections he gathered over six years of exploration in the vast area between St. Petersburg and the borders between Russia and China. For the first time, a zoologist compared the populations of many localities and documented with great accuracy (also through precise measurements of numerous traits of external morphology and sometimes, also of skeletal parts) variability both within and between populations, and did not ignore the frequent difficulty in drawing with certainty the boundaries between one species and another. Here is just one example, from his description of *Mus amphibius*:

> this species [is] common in Europe, and the same abounds throughout the whole of Russia and Siberia, as far as the Eastern ocean, under every climate, from the Caspian Sea as far as the glacial ocean, and shows up in a diversity of varieties [...] In the temperate and southern regions of Russia, around Iaikum & Volga, I observed everywhere [specimens] quite similar in color and size to those of Europe. [...] An all-black variety is quite common in Siberia [...]. I got specimens of another variety from the Ob region

---

[*] Wohl zu unterscheiden von der Abartung (*variatio*) ist die Ausartung (*degeneratio*). Letztere ist eine Abweichung von und außer dem Artbegriff, welche entsteht, wenn ein Naturprodukt aus der ihm von dem Artgesetze absolut angewiesenen Sphäre äußerer Einflüsse herausgerissen wird.

[...], which is of darkish grey color like the Europeans, but is remarkable for a large white irregular spot in the middle of the back above the shoulders, and at the same time often with a small white mark on the sternum.[*]

(Pallas 1778:80–81)

10.9.3 VARIATION IN TIME

Further problems will emerge later with respect to variation of living organisms through time. This, however, presupposes that extinct species are accommodated in the classification not among minerals but among animals or plants, in a place that does not depend on their status as living or extinct but only on their presumed affinities.

But extinct organisms did not find a place in the Linnaean classification that easily. In the XII edition[†] of *Systema Naturae* (Linnaeus 1768), eight genera of *Petrificata* are listed (*Zoolithus, Ornitholithus, Amphibiolithus, Ichthyolithus, Entomolithus, Helmintholithus, Phytolithus, Graptolithus*), but among the minerals. Some of the 'species' attributed to these genera can be compared with one of the species described today by paleontologists, for example (p. 160) *Entomolithus paradoxus*, a trilobite described by Linnaeus under this name in a previous paper (Linnaeus 1759). Other species, however, were overtly artificial, e.g., *Entomolithus succineus* (p. 161), a name indiscriminately applied to the most diverse array of insect remains in amber (*habitat* (sic) *intra Succinum*). A no less impressive example of the artificial character of these taxa is *Anthropolithus*, one of the genera of fossils (here collectively called Larvata) recognised in the XIII and last (posthumous) edition of *Systema Naturae*. Here (1793:386), of the two 'species' ascribed to *Anthropolithus*, one (*A. totalis*) is for complete human remains (*A. totum corpus humanum referens*), the other (*A. partialis*) for the specimens represented only by the skull or other bones (*A. cranii vel aliorum ossium*).

Eventually, extinct species end up classified next to the living forms to which they are most similar. Just as in the drawers of his collection, fossil shells and the shells of living mollusks were arranged together based on their similarities, so Lamarck arranged them in the *Histoire naturelle des animaux sans vertèbres* (1815–22). For the bivalve genus *Venus*, for example, Lamarck (1818:583–608) described a first batch of 88 living species, followed by a set of six fossil species. In a historical period

---

[*] Quandoquidem hanc speciem, in Europa licet communem, præter Buffonium & Daubentonum vix quisquam illustravit, eademque per totam Russiam atque Sibiriam, usque ad orientalem oceanum, sub omni climate, a mari caspio usque ad oceanum glacialem, abundat, variasque patitur varietates [...]. In temperatioribus & australibus Russiae, circa Iaikum & Volgam, subsimilem europææ colore atque magnitudine ubique observavi. [...] Per Sibiriam non infrequens occurrit varietas [...] tota atra [...]. Aliam varietatem ex Obensi regione, ubi frequens esse dicitur, accepi & ad Ieniseam observavit Messerschmidius, quæ solito colore e gryseo fuscescente europæos referunt, sed macula magna alba, irregulari, in medio dorso supra scapulas, simulque litura sæpe parva alba sterni insigniuntur.

[†] We cannot use the ed. X of *Systema Naturae*, the first in which the binomial nomenclature is used for animals (following the previous usage for plants only in *Species Plantarum*, Linnaeus 1753), as the planned volume on the mineral kingdom, whose manuscript is preserved at the Linnean Society of London, was never published (Hulth 1907).

in which biology was at last taking shape as the science of all living things, there was no place anymore for fossils in any taxonomy of inorganic products. In no uncertain terms, Isidore Geoffroy Saint-Hilaire (1859:167) will say a couple of decades later that general biology is the comparative study of all beings that live on the surface of the globe or have lived there in the past.*

10.10   HYBRIDS AND THE NATURE OF SPECIES

10.10.1   KIRCHER AND RAY ON HYBRIDS

If some degree of variation in space and time can be accommodated within the fixed limits of species as created, what about hybrids? Do they have a place in the plan of Creation? Nieremberg (1635), among others, believed that they do, in so far as they are fertile. Kircher (1675) even believed that Noah exploited hybridisation to ensure optimal use of the limited space on the Ark: the species of hybrid origin could be formed anew provided that specimens of the two parental species were saved from the Flood:

> Of quadrupeds believed to have been saved on the Ark. – As there is a nearly infinite number of animal species produced by promiscuous mating between different species, it is not likely that all of them were introduced into the Ark, as they could be gener-ated later by spurious gathering starting with real good species [...] First among these is the mule, spurious animal generated by mare and female donkey, that all exegetes exclude from the Ark, as it could arise after the Flood by promiscuous mating between numerous horses and donkeys. Similarly, the giraffe (*Camelopardalis* animal), born of Panther (*Pardus*) and Camel (*Camelus*), could be excluded from the Ark.†

> **(Kircher 1675:67–68)**

However, if hybrids cannot generate offspring similar to themselves, they must be regarded as monsters rather than as new species. Interestingly, this argument is not found in Buffon, who is widely credited with having introduced reproductive isola-tion as a criterion for the distinction between species, but in Ray:

---

* Tous les résultats généraux, toutes les hautes vérités auxquels a conduit, dans les temps modernes, et de nos jours surtout, l'étude comparée des êtres qui vivent ou ont vécu à la surface du globe, sont donc du domaine de cette science supérieure [Biologie générale].

† *De Quadrupedibus qua in Arca conservata putantur.* – Cum infinitae propemodum animalium species sint ex promiscuo diversarum specierum coitu productæ, illas omnes intra Arcam introductas fuisse non est verisimile, cum ille postmodum ex genuinis & propriis speciebus spurio congressu generari potuerint [...] Et primo quidem loco se sistit Mulus, animal spurium, ex Equo & Asina generatum, quem omnes Interpretes ab Arca eximunt, cum post Diluvium multiplicibus jam Equis & Asinabus, ex promiscuo congressu nasci potuerit. Ab hac quoque eximi potest Camelopardalis animal ex Pardo, & Camelo natum.

The Mule is not a certain and constant animal species, but a hybrid and spurious one, generated by a donkey that has mated with a mare; it does not propagate its own species, so it would be better to regard it as a monster.*

(Ray 1693:64)

The Mouflon is an animal typical of Sardinia and Corsica, about which wrote the Ancients, but it is not certain that today it can be found in nature. […] I would say (Gesner affirms) that it would be absurd to suspect that it is completely extinct, as Pliny believed, since Nature always scrupulously preserves all genera and species. […] Perhaps the Mouflon, as Albert [= Albert the Great, *Albertus Magnus*] believed, was a particular genus born of a goat and the Aries, which could disappear.†

(Ray 1693:75–76)

10.10.2 BUFFON ON HYBRIDS

Like Ray, Buffon was not interested in reproductive isolation per se, as a criterion to separate species, but in the maintenance of genealogical continuity:

we must consider as the same species that which by means of copulation is perpetuated & maintains the similarity of the species, & as different species those which, by the same means, cannot produce anything together.‡

(Buffon 1749, Tome II:10–11)

this property common to animals and plants, this power to produce its like, this chain of successive existences of individuals, which constitutes the real existence of the species.§

(Buffon 1749, Tome II:18)

It is true that he added:

a fox will be a different species from a dog, if indeed by the copulation of a male and a female of these two species nothing results; & even if the result would be a kind of

---

* Non est certa & constans Animalis species, sed hybrida & spuria, ab Asino equam ineunte genita; quæ speciem suam non propagat, ideoque inter monstra potius reputanda.
† Musimon Animal Sardiniæ & Corsicæ peculiare, Veteribus memoratum, an hodie in rerum natura inveniatur dubium est. […] Interiisse quidem suspicari, ut Plinius, absurdum dixerim (inquit Gesnerus) cum Natura rerum genera & species omnes sedulo semper conservet. […] Verum Musimon fortasse (ut voluit Albertus) spurium quoddam genus fuit ex capra et Ariete natum, quod deficere potuit.
‡ on doit regarder comme la même espèce celle qui, au moyen de la copulation, se perpetue & conserve la similitude de cette espèce, & comme des espèces différentes celles qui, par les mêmes moyens, ne peuvent rien produire ensemble
§ Examinons de plus près cette propriété commune à l'animal & au végétal, cette puissance de produire son semblable, cette chaîne d'existences successives d'individus, qui constitue l'existence réelle de l'espèce.

mule [=hybrid]; as this mule would not produce anything, that would suffice to establish that the fox & the dog would not be of the same species.[*]

(Buffon 1749, Tome II:10–11)

but once again, he did not point to reproductive isolation per se, as the previous sentence continues:

since we have supposed that to constitute a species, it takes a continuous, perpetual, invariable production, that is, one like that of other animals.[†]

(Buffon 1749, Tome II:10–11)

Popular accounts of Buffon's thoughts on hybrids and speciation usually mention his deductions from the sterility of mules, but in pages where he discusses this case, he makes it clear that the nearly complete reproductive isolation between horse and donkey cannot be generalised as providing a biological criterion revealing that two animals belong to different species. To credit Buffon with introducing a biological notion of species based on reproductive isolation does not fit at all with his views on hybrids as expressed in detail in a long chapter of volume XIV (1766) of the *Histoire naturelle*:

Be that as it may, it is certain from all that we have just explained that mules [hybrids] in general, which have always been accused of impotence and sterility, are however neither really sterile, nor generally infertile; & that it is only in the particular species of the mule coming from the donkey & the horse, that this sterility is manifested, since the mule [=hybrid] which comes from the goat & of the ewe, is as fruitful as its mother or its father; since in birds most of the mules which come from different species are not infertile: it is therefore in the particular nature of the horse and the donkey that we must seek the causes of the infertility of mules which come from them; & instead of supposing sterility as a general & necessary defect in all mules, restrict it on the contrary only to the mule coming from the donkey & the horse, & again give great limits to this restriction, since these same mules can become fertile under certain circumstances, and especially by approaching to some degree their original species.[‡]

(Buffon 1766, Tome XIV:342–343)

---

[*] un renard sera une espèce différente d'un chien, si en effet par la copulation d'un mâle & d'une femelle de ces deux espèces il ne résulte rien, & quand même il en résulteroit un animal mi-parti, une espèce de mulet, comme ce mulet ne produiroit rien, cela suffiroit pour établir que le renard & le chien ne seroient pas de la même espèce

[†] puisque nous avons supposé que pour constituer une espèce, il falloit une production continue, perpétuelle, invariable, semblable, en un mot, à celle des autres animaux.

[‡] Quoi qu'il en soit, il est certain partout ce que nous venons d'exposer, que les mulets en général qu'on a toujours accusés d'impuissance & de stérilité, ne sont cependant ni réellement stériles, ni généralement inféconds; & que ce n'est que dans l'espèce particulière du mulet provenant de l'âne & du cheval, que cette stérilité se manifeste, puisque le mulet qui provient du bouc & de la brebis, est aussi fécond que sa mère ou son père; puisque dans les oiseaux la plupart des mulets qui proviennent d'espèces différentes, ne sont point inféconds: c'est donc dans la nature particulière du cheval & de l'âne, qu'il faut chercher les causes de l'infécondité des mulets qui en proviennent; & au lieu de supposer la stérilité comme un défaut général & nécessaire dans tous les mulets, la restreindre au contraire au seul mulet provenant de l'âne & du cheval, & encore donner de grandes limites à cette restriction, attendu que ces mêmes mulets peuvent devenir féconds dans de certaines circonstances, & surtout en se rapprochant d'un degré de leur espèce originaire.

This view of species combines with Buffon's idiosyncratic idea that only the female sex provides for species identity:

> In the common ordinance of Nature, it is not the males, but the females, who constitute the unit of the species: we know from the example of the sheep which can serve two different males & produce both from the goat and the ram, that the female influences the specificity of the product much more than the male, since from these two different males only lambs are born, that is to say, individuals specifically resembling the mother.[*]

> (Buffon 1766, Tome XIV:339)

10.11 THE NOMADIC BEHAVIOR OF SPECIES NAMES AND THE ORIGIN OF THE SPECIES PROBLEM

10.11.1 SAME NAMES IN DIFFERENT TAXONOMIES?

> Knowing plants is precisely knowing the names that have been given to them in relation to the structure of some of their parts. This structure makes up the character that essentially distinguishes plants from one another. The idea of this character must be inseparably united with the name of each plant; & without this precaution the language of Botany would be in a strange confusion.[†]

> (Tournefort 1694:1–2)

> Once the genera have been established, it is necessary to fix forever the names that must be used when referring to them.[‡]

> (Tournefort 1694:13)

Linnaeus was more explicit in stressing that knowledge is bound to names: 'If you do not know the names, your knowledge of things also dies'[§] [210].

The success of Linnaeus' encyclopedic work promoted to an unprecedented degree a strict association between taxa (above all, species taxa) and their names:

---

[*] Dans l'ordonnance commune de la Nature, ce ne sont pas les mâles, mais les femelles, qui constituent l'unité des espèces: nous savons par l'exemple de la brebis qui peut servir à deux mâles différens & produire également du bouc & du bélier, que la femelle influe beaucoup plus que le mâle sur le spécifique du produit, puisque de ces deux mâles différens il ne naît que des agneaux, c'est-à-dire, des individus spécifiquement ressemblans à la mère.

[†] Connoître les plantes, c'est précisément savoir les noms qu'on leur a donné par raport à la structure de quelques-unes de leurs parties. Cette structure fait le caractere qui distingue essentiellement les plantes les unes d'avec les autres. L'idée de ce caractere doit estre inseparablement unie au nom de chaque plante; & sans cette précaution le langage de la Botanique seroit dans une confusion étrange.

[‡] Les genres étant établis, il est necessaire de fixer pour toujours les noms dont il faut se servir pour les exprimer.

[§] Nomina si nescis, perit et cognitio rerum.

In order to obtain a satisfactory and comprehensive knowledge of the vast landscape of Nature, a methodical arrangement and a nomenclature become indispensable, if we do not want confusion to throw us into an inextricable labyrinth. [...] Therefore, in the need to follow a method of nomenclature, and in the scarcity of monographs about [the organic productions of] our sea, it is convenient to appeal to the Authors of general treatises: and it was very convenient to follow the most correct, the most concise, the most expressive of all languages and systems – that of the famous Cavalier Linnaeus.*

(Olivi 1792:4–5)

By using it today, however, we must not take for granted that a name introduced by Linnaeus has the same meaning as in Linnaeus' works. Until now, this potential conflict has only been discussed in terms of the relationship between species as recognised by taxonomy versus species of evolutionary biology or other disciplines, but we must first focus on the relationship between stages of an evolving taxonomy. Of course, it is not possible to identify a precise temporal divide between the 'orthodox' Linnaean tradition and a taxonomy in which the species taxa take on a different nature. However, an important step marking a conceptual change hidden behind the continued use of the same names was the introduction of types.

10.11.2 Types in Taxonomy and Nomenclature

Taxonomy requires solid criteria to delimit taxa and clear rules to ensure stability and universality of nomenclature. An important tool for achieving this goal is the designation of reference types.

In biology, this term has been used with different meanings (e.g., Simpson 1940; Farber 1976; Witteveen 2016). We can ignore here types in the sense of idealised, generalised morphological models, such as Owen's (1848) archetype of the vertebrate skeleton, or Turpin's (1837) *Urpflanze* inspired by Goethe's idealistic plant morphology, but we should pay attention to other kinds of types: the type specimen of a named species and the lower taxon selected as the type of an immediately higher taxon in the classificatory hierarchy.

The importance of the latter kind of type was acknowledged when the Strickland Committee drafted for the British Association for the Advancement of Sciences the document generally recognised as the first code of zoological nomenclature:

When a genus is subdivided into other genera, the original name should be retained for that portion of it which exhibits in the greatest degree its essential characters as at first defined. Authors frequently indicate this by selecting some one species as a fixed point of reference, which they term the 'type of the genus.' When they omit doing so,

---

* Per ben conoscere in tutta la sua estensione il vasto quadro della Natura, una sistemazione metodica, ed una nomenclatura diventano indispensabili, se non vogliamo che la confusione ci getti in un inestricabile laberinto. [...] Nella necessità per tanto di seguitare un metodo di nomenclatura e nella scarsezza degli Scrittori particolari del nostro mare, convenne ricorrere ai Trattatisti generali: ed era ben conveniente, che si adottasse il più giusto, il più conciso, il più espressivo di tutti i linguaggi, e sistemi, quello del cel. Cav. Linneo.

it may still in many cases be correctly inferred that the first species mentioned on their list, if found accurately to agree with their definition, was regarded by them as the type.

(Strickland et al. 1842:6)

Type specimens, instead, are not mentioned in the 'Strickland code', but probably, Hugh Strickland was again the first to stress their importance, just a couple of years later. Having remarked that at the British Museum

the scientific classification of bird specimens is making great progress, under the able superintendence of the two Messrs. Gray, [...] ornithologists will soon possess in this collection a standard model which may be applied with advantage to other museums. This latter object will be greatly aided by the recent publication of catalogues [...] of all the species contained in the museum. [In t]hese catalogues, [t]he classification and the scientific nomenclature are based on sound principles [...] and every specimen is separately enumerated, with its locality and the name of its donor, which is especially important in a collection containing the *type-specimen*s [italics as in the original], from which original descriptions have been made. The zoological catalogues of the British Museum will now become standard works of reference.

(Strickland 1845:215)

Thus, type specimens are firm references for taxonomy but also for nomenclature: that is, type specimens are onomatophores, i.e., name-bearers (Simpson 1940:421).

The codes of nomenclature do not contain a definition of the species category, but they set precisely the rules for the adoption and use of the names with which species taxa must be designated. Consequently, the use of a universal nomenclature that many would like to adopt in all biological disciplines sets precise limits on the way we shall circumscribe species taxa, both in taxonomy and in the other disciplines too.

The transition from Linnaeus' species taxa to the species taxa of the current taxonomy, with a nomenclature governed by the codes, went on almost unnoticed, because it was a long, progressive evolution within the same discipline (taxonomy) and therefore, within the same community of researchers. Within this, when voices are raised proclaiming the need for radical change, a need may also be felt to switch to a different system of nomenclature and to question the choice of the species as fundamental taxonomic unit. This happened with phylogenetic systematics, the development of which, starting from the conceptual and methodological renewal initiated by Hennig, questioned the maintenance of the Linnaean classification ranks (Griffiths 1976) and suggested launching a new 'phylogenetic' nomenclature (de Queiroz and Cantino 2020). This proposal, strongly defended by some authors (e.g., de Queiroz 1997; de Queiroz and Cantino 2001; Bryant and Cantino 2002; Pleijel and Rouse 2003; Cantino 2004; Pleijel and Härlin 2004; Laurin et al. 2006), has been strongly rejected by others (e.g., Lidén and Oxelman 1996; Dominguez and Wheeler 1997; Benton 2000; Forey 2002; Carpenter 2003; Keller et al. 2003; Nixon et al. 2003). According to some authors, in a consequent taxonomic system based on phylogenetic principles, there would not even be a place for species; the lowest unit recognised in the system should be instead an operationally defined LITU (least inclusive taxonomic unit) (Pleijel and Rouse 1999). The prospect of abandoning the species category has been discussed by LaPorte (2007), who concluded that no alternative to the use of 'species' among those proposed for a rank-free systematics has provided a proper replacement for the species taxon.

10.11.3 TAXONOMY VERSUS OTHER BIOLOGICAL DISCIPLINES

What are the services taxonomy can offer or those that other disciplines expect from it?

On the one hand, a widespread predictive value. If two species are classified in the same genus, they will likely share many traits or properties, so a good knowledge of one may allow predictions about the other. The expected similarities will gradually decrease if the two species, rather than being ascribed to the same genus, are classified into two different genera of the same family, or into two different families of the same order, but far from confidently, as the relationship between ranking and recency of common ancestry is inconsistent at least (Zachos 2011). This predictive value, however, can be read against a phylogenetic tree much better than it can be extracted from a classification, and anyway, with caution: Jenner (2006) has shown the logical flaws underlying the common assumption that model organisms can be expected to represent in an unambiguous way the clades of which they are a part.

The other service that taxonomy can offer to all life sciences is a reasonably stable nomenclature, not so much for higher taxa as for species. To speak of *Aves* rather than birds (or *oiseaux*, *Vögel*, *uccelli*, etc.) makes little difference; this is one of the reasons why the ICZN does not regulate the names of taxa of rank higher than those of family and optional related ranks such as the superfamily. On the other hand, stability and universality of names are crucially important at the level of species – or better, of what passes under the name of species. And this is where the so-called species problem arises.

The problem arises from the conflict between the unavoidable diversity of meanings the term *species* takes in disciplines other than taxonomy, such as evolutionary biology, ecology, etc., contrasting with the rigidity of a nomenclature that remains anchored to the species understood as the unit of classification in what is considered as 'the' taxonomy, i.e., the taxonomy of Linnaean tradition. However, the use of the same Linnaean binomials in taxonomy, evolutionary biology, ecology, etc. is a *consequence* of the fact that we call *species* all the *different* biodiversity units worth recognition in all these disciplines, but it does not attest that these units are, or can be, coextensive.

The changing meaning of the term *species*, moving from one discipline to another, is a typical case of a *nomadic concept*. This term was proposed (Stengers 1987) to describe concepts for which the meaning and domain of application change with the new contexts into which they migrate. *Nomadic* concepts take different meanings according to the different disciplines with which they are associated. Each discipline acts as a semantic *anchor context*.

I have recently suggested (Minelli 2020b) that in this association between concepts and disciplines, it may be worthwhile to perform an experiment by reversing their roles. If disciplines are traditionally taken for granted, as *anchor disciplines* around which concepts may move, a reversed relationship between disciplines and core concepts may prove useful as an epistemic tool to refresh the traditional divides between biological disciplines. It may be fruitful to fix a few *anchor concepts* and to explore the diverging consequences of their association with *nomadic disciplines*.

In the case of species, dozens of different species notions are found in the literature, although Wilkins (2011) collapses them to seven distinct definitions – all traceable to one species concept – with 27 variations and mixtures. In fact, most of these different definitions (or variations) refer to the different kinds of biological diversity relevant in the context of different biological disciplines. As units whose scope of application is to some (but variable) extent superimposable onto that of the phenomenological species of taxonomy, they almost always receive the name of species. From a lexical point of view, this is not objectionable. However, this generalised use of the term *species* entitles biologists of all disciplines to apply to them the Linnaean names of the species taxa. This choice is very often harmless, but not always.

More critical is another consequence of the generalised use of the term *species* for all basic kinds of biological diversity. This is the expected feedback from the different biological disciplines to taxonomy: if Linnaean names have to be useful in evolutionary biology, ecology, etc., they should refer to units with as much overlap as possible with the units in which this or that discipline is interested; ideally, to all of them, hence the species problem. This has been mostly addressed in either of two ways (Brigandt 2020).

The first way is to decide that the units of taxonomy should be, for example, those of evolutionary biology, or segments of the phylogenetic tree, to the exclusion of any other criterion. For example, according to Bock (2004:183), 'the value of the biological species concept or any other species concept can be decided only by a consideration of its role within evolutionary theory, not by a consideration of systematic practice'. More radically, according to Szalay and Bock (1991:10), the species concept is not a part of systematics but an integral part of evolutionary theory.

A second way is to look for a general species concept within which the largest number of independent and sometimes conflicting criteria may be accommodated (e.g., Mayden 1997; de Queiroz 1998).

However, there is a third, more convincing way out of the 'species problem'; that is, to accept a certain degree of taxonomic and nomenclatural pluralism to account for the different units of representation of biological diversity required in the different disciplinary or operational contexts, most of which were not foreseeable in the times of Linnaeus or Ray (Pavlinov 2020).

In the past, pluralism was advocated because no species concept apparently works throughout the whole diversity of living organisms (e.g., Mishler and Donoghue 1982): for example, the biological species concept cannot be used for partitioning the diversity of uniparental organisms into units comparable to those identified by reproductive isolation in the biparental ones. This is not the kind of pluralism advocated here. As noted quite long ago by Kitcher (1984), the description and naming of the different kinds of units relevant to the different biological disciplines require pluralism of species notions. This is not without consequences, however, because the resulting classifications are not the same: the number and circumscription of the units recognised following the adoption of different criteria may differ in an unpredictable way and to an unpredictable extent (Conix 2018; Brigandt 2020). Moreover, pluralistic taxonomy necessarily requires some degree of *nomenclatural pluralism*: a poorly recognised trend in this direction is still operating and may deserve more attention and eventually, ruling: for example, Molecular Operational Taxonomic Units (MOTUs) identified through comparison of DNA sequences do not correspond necessarily to conventional taxonomic species, and no simple or universal rule exists to 'translate' them into Linnaean species. Provisionally recognised MOTUs do not get a Linnaean name but are labelled with a formula (Minelli 2020c).

What can these different 'species'-level units have in common? What may play the role of anchor species concept in respect to the nomadism of biological disciplines? I think that this role can only be played by the traditional phenomenological species in the sense of Sterelny (1999). The 'general species concept' suggested by Pavlinov (2020) is similarly intended to unite and explain all particular ones rather than to replace them, as suggested by others as mentioned.

This does not mean preferring a phenotypic concept to one based on genealogical continuity, reproductive isolation, or the ecological role of species. Instead, it means embracing a certain degree of fuzziness, of imprecision. But, imprecise concepts promote integration and thus, eventually benefit science (Brigandt 2010; Brigandt and Love 2012; Waters 2014; Neto 2020).

10.12 CONCLUSIONS

In the literature on the historical roots of the usage in biology of fundamental taxonomic categories called *genus* and *species*, a number of circumstances are often taken for granted, which are not supported by a perusal of the pre-Linnaean literature. Among these are:

- That the early use of those terms, in the zoological or botanical literature, derives directly and consciously from the tradition of Aristotelian and Scholastic logic;
- That the use of these terms in texts about plants or animals always has a taxonomic meaning;
- That this meaning broadly corresponds to the usage of *genus* and *species* by Linnaeus;
- That in the natural sciences, the term *species* has always been restricted to plants and animals, while any use of the term in other areas, for example in mineralogy, is improper or abusive;
- That the species, rather than the genus, is the unit, or the only unit, of Linnaean taxonomy;
- That the relationship between species taxon and species name was the same in Linnaeus and in the whole of the Linnaean taxonomic tradition till our times.

A selective but reasonably dense sampling of texts suggests otherwise. The continuity of taxonomic usage through time is not provided by smooth and controlled change in the definition and circumscription of taxonomic units, at least since the recognition and naming of species taxa by Linnaeus, but by the uniform adoption of the binomial names of the Linnaean tradition.

This has continued up to the present, irrespective not only of the advent of an evolutionary reading of the living world but also of the fact that the biodiversity units we need to recognise in the different branches of biology, e.g., ecology and evolutionary biology, do not necessarily overlap with the phenomenological species of the tradition to which the names of the Linnaean nomenclature are still attached.

Eventually, this has caused a nomadic meandering of the meaning of the term *species*, whose consequences have finally emerged as the species problem. Tentative solutions have been advanced, either by arbitrarily restricting the meaning of *species* to a notion (unlikely to be applicable throughout the whole tree of life) based on the units of biodiversity emerging, for example, from evolutionary theory, or from a suitable 'dissection' of a phylogenetic tree. A less popular alternative suggests looking for a compromise, with a generalised species concept jointly inspired by many of the current definitions.

In my opinion, it is more sensible to accept that biological disciplines, including taxonomy, need a diversity of species-level units. The phenomenological species of the pre-Linnaean and Linnaean tradition can still function as an anchor concept and provide species names in so far as the so-called species recognised in other biological disciplines are largely coextensive with them; but, we must accept that the nomadic association of the different disciplines to the anchor of taxonomy necessitates both conceptual and nomenclatural pluralism.

ACKNOWLEDGEMENTS

I am grateful to Igor Pavlinov, John Wilkins, and Frank Zachos for their invitation to contribute a chapter to this book, and also for stimulating and enriching my mind on matters of species through their important publications on this topic.

REFERENCES

Agassiz, L. and Strickland, H. E. 1848, 1850, 1852, 1854. *Bibliographia zoologiae et geologiae. A general catalogue of all books, tracts, and memoirs on zoology and geology*, 4 vols. London: Ray Society.

Atran, S. 1990. *Cognitive foundations of natural history: Towards an anthropology of science*. Cambridge: Cambridge University Press.

Bauhin, C. 1596. *Phytopinax, seu, Enumeratio plantarum ab herbariis nostro seculo descriptarum*. Basileae: per Sebastianum Henricpetri.

Bauhin, C. 1623. *ΠΙΝΑΞ [Pinax] Theatri botanici Caspari Bauhini [...] sive Index in Theophrasti, Dioscoridis, Plinii et botanicorum qui à seculo scripserunt opera*. Basileae Helvet.: typis Ludovici Regii.

Benton, M. J. 2000. Stems, nodes, crown clades, and rank-free lists: Is Linnaeus dead? *Biological Reviews* 75:633–48.

Birkhead, T. 2018. *The wonderful Mr Willughby: The first true ornithologist*. London: Bloomsbury.

Bliss, H. E. 1929. *The organization of knowledge and the system of the sciences*. New York: Henry Holt and Company.

Blumenbach, J. F. 1779–1780. *Handbuch der Naturgeschichte*. Göttingen: Dieterich.

Blumenbach, J. F. 1830. *Handbuch der Naturgeschichte. Zwölfte rechtmäßige Ausgabe*. Göttingen: Dieterich.

Bock, W. J. 2004. Species: The concept, category and taxon. *Journal of Zoological Systematics and Evolutionary Research* 42:178–90.

Brigandt, I. 2010. The epistemic goal of a concept: Accounting for the rationality of semantic change and variation. *Synthese* 177:19–40.

Brigandt, I. 2020. How are biology concepts used and transformed? In *Philosophy of science for biologists*, eds. K. Kampourakis and T. Uller, 79–101. Cambridge: Cambridge University Press.

Brigandt, I. and Love, A. C. 2012. Conceptualizing evolutionary novelty: Moving beyond definitional debates. *Journal of Experimental Zoology Part B: Molecular and Developmental Evolution* 318:417–27.

Bryant, H. N. and Cantino, P. D. 2002. A review of criticisms of phylogenetic nomenclature: Is taxonomic freedom the fundamental issue? *Biological Reviews* 77:39–55.

Buffon, G. L. Leclerc, comte de. 1749–1789. *Histoire naturelle, générale et particuliére: avec la description du cabinet du Roy*. 36 vols. Paris: Imprimerie royale. (Tome I, 1749, Tome II, 1749; Tome XIV, 1766).

Caesalpinus, A. 1583. *De plantis libri XVI*. Florentiae: G. Marescottus.

Cain, A. J. 1958. Logic and memory in Linnaeus' system of taxonomy. *Proceedings of the Linnean Society of London* 169:144–63.

Cain, A. J. 1994. Rank and sequence in Caspar Bauhin's *Pinax*. *Botanical Journal of the Linnean Society* 114:311–56.

Cain, A. J. 1997. John Locke on species. *Archives of Natural History* 24:337–60.

Cantino, P. D. 2004. Classifying species versus naming clades. *Taxon* 53:795–8.

Caron, J. 1988. 'Biology' in the life sciences: A historiographical contribution. *History of Science* 26:223–68.

Carpenter, J. M. 2003. Critique of pure folly. *Biological Reviews* 69:79–92.

Cockerell, T. D. A. 1920. Biographical memoir of Alpheus Spring Packard 1839–1905. *Biographical Memoirs of the National Academy of Sciences* 9:181–236.

Conix, S. 2018. Integrative taxonomy and the operationalization of evolutionary independence. *European Journal of Philosophy of Science* 154:1–17.

Cubas, P., Vincent, C. and Coen, E. 1999. An epigenetic mutation responsible for natural variation in floral symmetry. *Nature* 401:157–61.

de Queiroz, K. 1997. Misunderstandings about the phylogenetic approach to biological nomenclature: A reply to Lidén and Oxelman. *Zoologica Scripta* 26:67–70.

de Queiroz, K. 1998. The general lineage concept of species, species criteria, and the process of speciation: A conceptual unification and terminological recommendations. In *Endless forms: Species and speciation*, eds. D. J. Howard and S. H. Berlocher, 57–75. Oxford: Oxford University Press.

de Queiroz, K. and Cantino, P. D. 2001. Phylogenetic nomenclature and the PhyloCode. *Bulletin of Zoological Nomenclature* 58:254–71.

de Queiroz, K. and Cantino, P. D. 2020. *International code of phylogenetic nomenclature (PhyloCode)*. Boca Raton, FL: CRC Press. Available from: http://phylonames.org/code/.

Dobzhansky, T. 1937. *Genetics and the origin of species*. New York: Columbia University Press.

Dominguez, E. and Wheeler, Q. D. 1997. Taxonomic stability is ignorance. *Cladistics* 13:367–72.

Enenkel, K. A. E. 2014. The species and beyond: Classification and the place of hybrids in early modern zoology. In *Zoology in early modern culture: Intersections of science, theology, philology, and political and religious education*, eds. K. A. E. Enenkel and P. J. Smith, 57–148. Leiden and Boston, MA: Brill.

Farber, P. L. 1976. The type-concept in zoology during the first half of the nineteenth century. *Journal of the History of Biology* 9:93–119.

Forey, P. L. 2002. PhyloCode – pain, no gain. *Taxon* 51:43–54.

Freudenstein, J. V., Broe, M. B., Folk, R. A. and Sinn, B. T. 2017. Biodiversity and the species concept – lineages are not enough. *Systematic Biology* 66:644–56.

Fuchs, L. 1542. *De historia stirpium commentarii insignes*. Basel: Officina Isingriniana.

Geoffroy Saint-Hilaire, I. 1859. *Histoire naturelle générale des règnes organiques. Tome II*. Paris: Masson.

Gesner, C. 1551. *Historiae animalium liber primus de quadripedibus viviparis*. Tiguri [Zurich]: C. Froschoverus.

Gesner, C. 1560a. *Nomenclator aquatilium animantium. Icones animalium aquatilium in mari & dulcibus aquis degentium*. Tiguri [Zurich]: C. Froschoverus.

Gesner, C. 1560b. *Icones avium omnium*. Editio secunda. Tiguri [Zurich]: C. Froschoverus.

Ghiselin, M. T. 1997. *Metaphysics and the origin of species*. Albany, NY: State University of New York Press.

Gilmour, J. S. L. 1940. Taxonomy and philosophy. In *The new systematics*, ed. J. S. Huxley, 461–74. Oxford: Clarendon Press.

Greuter, W., Hawksworth, D. L., McNeill, J., Mayo, M. A., Minelli, A., Sneath, P. H. A., Tindall, B. J., Trehane, P. and Tubbs, P. 1998. Draft BioCode (1997): The prospective international rules for the scientific names of organisms. *Taxon* 47:127–50.

Griffiths, G. C. D. 1976. The future of Linnaean nomenclature. *Systematic Zoology* 25:168–73.

Gronovius, L. Th. 1760. *Bibliotheca botanica, sive, Catalogus auctorum et librorum qui de re botanica, de medicamentis ex vegetabilibus paratis, de re rustica, & de horticultura tractant, a Joanne Francisco Seguierio Nemausense digestus, accessit Bibliotheca botanica Jo. Ant. Bumaldi, seu potius Ovidii Montalbani Bononiensis, nec non auctuarium in Bibliothecam botanicam Cl. Seguierii*. Lugduni Batavorum: Cornelius Haak.

Gronovius, L. Th. 1760. *Bibliotheca regni animalis atque lapidei, seu Recensio auctorum et librorum qui de regno animali & lapideo methodice, physice, medice, chymice, philologice, vel theologice tractant, in usum naturalis historiae studiosorum*. Lugduni Batavorum: Sumptibus auctoris.

Hjørland, B. 2017. Classification. *Knowledge Organization* 44:97–128. Also in ISKO Encyclopedia of Knowledge Organization, eds. B. Hjørland and C. Gnoli, http://www .isko.org/cyclo/classification.

Hull, D. L. 1998. Taxonomy. In *Routledge encyclopedia of philosophy*, ed. Edward Craig. London: Routledge, vol. 9:272–6.

Hulth, J. M. 1907. *Bibliographia Linnaeana. Matériaux pour servir à une bibliographie Linnéenne*. Uppsala: C. J. Lundström.

International Commission on Zoological Nomenclature. 1999. *International code of zoological nomenclature*. Fourth edition. London: The International Trust for Zoological Nomenclature.

Jenner, R. A. 2006. Unburdening evo-devo: Ancestral attractions, model organisms, and basal baloney. *Development Genes and Evolution* 216:385–94.

Jungius, J. 1662. *Doxoscopiæ physicæ minores, sive Isagoge physica doxoscopica*. Hamburg: Pfeiffer.

Keller, R. A., Boyd, R. N. and Wheeler, Q. D. 2003. The illogical basis of phylogenetic nomenclature. *Botanical Review* 69:93–110.

Kircher, A. 1675. *Arca Noë*. Amstelodami: apud Johannem Janssonium à Waesberge.

Kitcher, P. 1984. Species. *Philosophy of Science* 51:308–33.

Klein, I. Th. 1750. *Historiae avium prodromus: cum praefatione de ordine animalium in genere*. Lubecae: apud Ionam Schmidt.

Klein, I. Th. 1751. *Quadrupedum dispositio brevisque historia naturalis*. Lipsiae: apud Bernhard. Christoph. Breitkopfium.

Lamarck, [J.-B. Monet] chevalier de. 1778. *Flore françoise*. Paris: Imprimerie royale.

Lamarck, [J.-B. Monet] chevalier de. 1815–22. *Histoire naturelle des animaux sans vertèbres*. Paris: Verdière (1 (1815), 2 (1816), 3 (1816); Deterville & Verdière (vols. 4 /1817), 5 (1818); chez l'Auteur (6 (1819, 1822), 7 (1822)).

Lamarck, J.-B.-P.-A. 1809. *Philosophie zoologique, ou exposition des considérations relatives à l'histoire naturelle des animaux*. Paris: Dentu.

LaPorte, J. 2007. In defense of species. *Studies in History and Philosophy of Biological and Biomedical Sciences* 38:255–69.

Laurin, M., de Queiroz, K. and Cantino, P. D. 2006. Sense and stability of taxon names. *Zoologica Scripta* 35:113–4.

Lazenby, E. M. 1995. *The Historia Plantarum Generalis of John Ray: Book I – a translation and commentary*. Thesis submitted for the degree of Doctor of Philosophy in the University of Newcastle upon Tyne:1157–8.

Lidén, M. and Oxelman, B. 1996. Do we need phylogenetic taxonomy? *Zoologica Scripta* 25:183–5.

Linnæus C. 1735. *Systema naturae, sive Regna tria naturae systematice proposita per classes, ordines, genera, & species*. Leiden: de Groot.

Linnæus, C. 1737. *Genera plantarum eorumque characteres naturales secundum numerum, figuram, situm, & proportionem omnium fructificationis partium*. Lugduni Batavorum: apud Conradum Wishoff.

Linnæus, C. 1744a. *Dissertatio botanica de Peloria, quam consensu ampl. Facult. Medicæ in Regia Academia Upsaliensi, præside ... Dn. Doct. Carolo Linnæo ... curiosorum oculis modeste subjicit Daniel Rudberg, Vermelandus*. Upsaliae.

Linnæus, C. 1744b. *Systema Naturae in quo proponuntur naturae regna tria secundum classes, ordines, genera & species. Editio quarta*. Parisiis: sumptibus Michaelis-Antonii David.

Linnæus, C. 1749. *Pan svecicus, quem...præs... Carolo Linnæo...publ. examini...submittit Nicolaus L. Hesselgren, Wermelandus*. Upsaliæ.

Linnæus, C. 1751. *Philosophia botanica, in qua explicantur fundamenta botanica cum definitionibus partium, exemplis terminorum, observationibus rariorum, adjectis figuris aeneis*. Holmiae: G. Kiesewetter.

Linnæus, C. 1753. *Species plantarum, exhibentes plantas rite cognitas, ad genera relatas, cum differentiis specificis, nominibus trivialibus, synonymis selectis, locis natalibus, secundum systema sexuale digestas*. Holmiæ: impensis Laurentii Salvii.

Linnæus, C. 1758. *Systema Naturae per regna tria Naturae secundum classes, ordines, genera, species, cum characteribus, differentiis, synonymis, locis. Tomus I*. Editio decima, reformata. Holmiae: Laurentius Salvius.

Linnæus, C. 1759. Petrificatet *Entomolithus paradoxus*. *Kongliga Vetenskaps Academiens Handlingar* 20:19–24.

Linnaeus, C. 1768. *Systema naturae: per regna tria natura, secundum classes, ordines, genera, species, cum characteribus, differentiis, synonymis, locis*. Ed. 12, reformata. Tomus III. *Regnum Lapideum*. Holmiae: Laurentii Salvii.

Linnaeus, C. 1793. *Systema naturæ per regna tria naturæ, secundum classes, ordines, genera, species; cum characteribus, differentiis, synonymis, locis. Editio decima tertia, aucta, reformata*. / cura Jo. Frid. Gmelin. Tomus III. Lipsiæ: Impensis Georg. Emanuel. Beer.

Marchant, [J.]. 1719. Observations sur la nature des plantes. *Histoire de l'Académie royale des sciences, avec les mémoires de mathématique et de physique*. Paris. Mémoires 1719:59–66.

Mattioli, P. A. 1568. *I discorsi di M. Pietro Andrea Matthioli sanese, medico cesareo [...] nelli sei libri di Pedacio Dioscoride Anazarbeo della materia Medicinale*. Venezia: Valgrisi.

Mayden, R. L. 1997. A hierarchy of species concepts: The denouement in the saga of the species problem. In *Species: The units of diversity*, eds. M. F. Claridge, H. A. Dawah and M. R. Wilson, 381–423. London: Chapman and Hall.

Mayr, E. 1982. *The growth of biological thought*. Cambridge, MA: The Belknap Press of Harvard University Press.

McLaughlin, P. 2002. Naming biology. *Journal of the History of Biology* 35:1–4.

Minelli, A. 2008. Zoological vs. botanical nomenclature: A forgotten 'BioCode' experiment from the times of the Strickland Code. In *Updating the Linnaean heritage: Names as tools for thinking about animals and plants*, eds. A. Minelli, L. Bonato and G. Fusco. *Zootaxa* 1950:21–38.

Minelli, A. 2019a. *Biologia. La scienza di tutti i viventi*. Udine: Forum.

Minelli, A. 2019b. The galaxy of the non-Linnaean nomenclature. *History and Philosophy of the Life Sciences*, 41, 31.

Minelli, A. 2020a. Biology and its disciplinary partitions – intellectual and academic constraints. *Science & Philosophy* 24:105–26.

Minelli, A. 2020b. Disciplinary fields in the life sciences: Evolving divides and anchor concepts, *Philosophies* 5, 34.

Minelli, A. 2020c. Taxonomy needs pluralism, but a controlled and manageable one. *Megataxa* 1:9–18.

Mishler, B. and Donoghue, M. 1982. Species concepts: A case for pluralism. *Systematic Zoology* 31:491–503.

Moffett, T. 1634. *Insectorum sive minimorum animalium theatrum*. London: Cotes.

Neto, C. 2020. When imprecision is a good thing, or how imprecise concepts facilitate integration in biology. *Biology and Philosophy* 35:58.

Nieremberg, J. E. 1635. *Historia naturae, maxime peregrinae*. Antwerp: Balthazar Moretus.

Nixon, K. C., Carpenter, J. M. and Stevenson, D. W. 2003. The PhyloCode is fatally flawed, and the "Linnaean" system can easily be fixed. *Botanical Review* 69:111–20.

Nyhart, L. K. 2019. A historical proposal about prepositions. In *Philosophy of biology before biology*, eds. C. Bognon-Küssand and C. T. Wolfe, 188–97. London and New York: Routledge.

Olivi, G. 1792. *Zoologia Adriatica ossia Catalogo ragionato degli Animali del Golfo e delle Laguna di Venezia; preceduto da una Dissertazione sulla Storia fisica e naturale del Golfo; e accompagnato da Memorie, ed Osservazioni di Fisica Storia naturale ed Economia*. Bassano: [Remondini].

Owen, R. 1848. *On the archetype and homologies of the vertebrate skeleton*. London: vanVoorst.

Pallas, P. S. 1766. *Elenchus zoophytorum, sistens generum adumbrations generaliores et specierum cognitarum succinctas descriptiones, cum selectis auctorum synonymis*. Hagae Comitum et Francofurti ad Moenum: Varrentrapp.

Pallas, P. S. 1778. *Novae species quadrupedum e glirium ordine cum illustrationibus variis complurium ex hoc ordine animalium*. Erlangen: Walther.

Pavlinov, I. Y. 2020. Multiplicity of research programs in the biological systematics: A case for scientific pluralism. *Philosophies* 5:7.

Pavlinov, I. Y. 2022. The species problem from a conceptualist's viewpoint. In *Species and beyond*, eds. J. S. Wilkins, I. Pavlinov and F. Zachos. Boca Raton, FL: CRC Press.

Pleijel, F. and Härlin, M. 2004. Phylogenetic nomenclature is compatible with diverse philosophical perspectives. *Zoologica Scripta* 33:587–91.

Pleijel, F. and Rouse, G. W. 1999. Least-inclusive taxonomic unit: A new taxonomic concept for biology. *Proceedings of the Royal Society of London B* 267:627–30.

Pleijel, F. and Rouse, G. W. 2003. Ceci n'est pas une pipe: Names, clades and phylogenetic nomenclature. *Journal of Zoological Systematics and Evolutionary Research* 41:162–74.

Ray, J. 1660. *Catalogus plantarum circa Cantabrigiam nascentium*. Cambridge: John Field.

Ray, J. 1677. *Catalogus plantarum Angliae, et insularum adjacentium: tum indigenas, tum in agris passim cultas complectens*. Ed. 2. Londini: A. Clark, impensis J. Martyn.

Ray, J. 1686. *Historia plantarum. Tomus Primus*. Londini: Typis Mariæ Clark.

Ray, J. 1693. *Synopsis methodica animalium quadrupedum et serpentini generis*. London: S. Smith and B. Walford.

Ray, J. 1710. *Historia insectorum*. Londini: A. & J. Churchill for the Royal Society.

Reydon, T. A. C. 2020. What is the basis of biological classification? In *Philosophy of science for biologists*, eds. K. Kampourakis and T. Uller, 216–34. Cambridge: Cambridge University Press.

Richards, R. A. 2010. *The species problem – A philosophical analysis*. Cambridge: Cambridge University Press.

Rieppel, O. 2016. *Phylogenetic Systematics*. Boca Raton, FL: CRC Press.

Rondelet, G. 1554. *Libri de piscibus marinis*. Lugduni: apud Matthiam Bonhomme.

Simpson, G. G. 1940. Types in modern taxonomy. *American Journal of Science* 238:413–31.

Sober, E. 1984. Sets, species, and evolution: Comments on Philip Kitcher's "Species". *Philosophy of Science* 51:334–41.

Spring, A. Fr. 1838. *Ueber die naturhistorischen Begriffe von Gattung, Art und Abart, und die Ursachen der Abartungen in den organischen Reichen. Eine Preisschrift*. Leipzig: Friedrich Fleischer.

Stamos, D. N. 2003. *The species problem. Biological species, ontology, and the metaphysics of biology*. Lanham, MD: Lexington Books.

Stengers, I. (Ed.). 1987. *D'une science a l'autre: des concepts nomads*. Paris: Seuil.

Sterelny, K. 1999. Species as ecological mosaics. In *Species, new interdisciplinary essays*, ed. R. A. Wilson, 119–38. Cambridge, MA: MIT Press.

Strickland, H. E. 1845. Report on the recent progress and present state of ornithology. *Report of the Fourteenth Meeting of the British Association for the Advancement of Science held at York in September 1844*: 170–221.

Strickland, H. E., Henslow, J. S., Phillips, J., Shuckard, W. E., Richardson, J. B., Waterhouse, G. R., Owen, R., Yarrell, W., Jenyns, L., Darwin, C., Broderip, W. J. and Westwood, J. O. 1842. *Report of a Committee appointed to consider the rules by which the nomenclature of Zoology may be established on a uniform and permanent basis*. London: John Murray, for the British Association for the Advancement of Science.

Suppe, F. 1989. Classification. In *International encyclopedia of communications*, ed. E. Barnouw, vol. 1:292–6. Oxford: Oxford University Press.

Szalay, F. S. and Bock W. J. 1991. Evolutionary theory and systematics: Relationships between process and patterns. *Zeitschrift für Zoologische Systematik und Evolutionsforschung* 29:1–39.

Tournefort, J. P. de. 1694. *Élémens de botanique, ou méthode pour connoître les plantes*. Tome 1. Paris: Imprimerie Royale.

Turland, N. J., Wiersema, J. H., Barrie, F. R. et al. (eds.) (2018) *International code of nomenclature for algae, fungi, and plants (Shenzhen Code) adopted by the Nineteenth International Botanical Congress Shenzhen, China, July 2017* (Regnum Vegetabile 159). Glashütten: Koeltz Botanical Books.

Turpin, P. J. F. 1837. Esquisse d'organographie végétale, fondée sur le principe d'unité de composition organique et d'évolution rayonnante ou centrifuge, pour servir prouver l'identité des organes appendiculaires des végétaux et la métamorphose des plantes de Goethe. In *Œuvres d'histoire naturelle de Goethe [...] traduits et annotés par Ch. Fr. Martins, Atlas*. Paris – Genève: Cherbuliez.

Wasmann, E. 1910. *Modern biology and the theory of evolution*. St. Louis, MO: Herder.

Waters, C. K. 2014. Shifting attention from theory to practice in philosophy of biology. In *New directions in the philosophy of science*, eds. M. C. Galavotti, D. Dieks, W. J. Gonzalez, S. Hartmann, T. Uebel and M. Weber, vol 5:121–39. Dordrecht: Springer.

Wilkins, J. S. 2009a. *Species. A history of the idea*. Berkeley, CA: University of California Press.

Wilkins, J. S. 2009b. *Defining species. A sourcebook from antiquity to today*. New York: Peter Lang.

Wilkins, J. S. 2011. Philosophically speaking, how many species concepts are there? *Zootaxa* 2765:58–60.

Wilkins, J. S. 2018. *Species: The evolution of the idea*. Boca Raton, FL: CRC Press.

Willughby, F. 1676. *Ornithologiae libri tres*. Londini: impensis Joannis Martyn.

Willughby, F. 1686. *De historia piscium libri quatuor*. Oxonii: e Theatro Sheldoniano.

Witteveen, J. 2016. Suppressing synonymy with a homonym: The emergence of the nomenclatural type concept in nineteenth century natural history. *Journal of the History of Biology* 49:35–189.

Zachos, F. E. 2011. Linnaean ranks, temporal banding, and time-clipping: Why not slaughter the sacred cow? *Biological Journal of the Linnean Society* 103:732–4.

Zachos, F. E. 2016. *Species concepts in biology. Historical development, theoretical foundations and practical relevance*. Basel: Springer.

11 Taxonomic Hierarchies as a Tool for Coping with the Complexity of Biodiversity

Julia D. Sigwart

CONTENTS

11.1 INTRODUCTION

The impacts of human activities on Earth's biodiversity are driving species to extinction at rates comparable to the past mass extinction events that punctuate the fossil record (Barnosky et al. 2011). Stemming this loss presents a new urgency to the grand project to understand the diversity of life on Earth. There are at least ~8 million living eukaryote species, and more than 85% remain undescribed (Mora et al. 2011); these numbers may be vastly larger, on the order of 1 billion living species, realistically accounting for uncertainty in cryptic species of insects and fungi (Larsen et al. 2017; Tedersoo et al. 2018). Taxonomic shortfalls disproportionately impact megadiverse developing countries, which have limited or no infrastructure to support specialist taxonomical work (Paknia et al. 2015). Loss of diversity results in declining ecosystem resilience, which in turn has practical consequences for human livelihoods (Oliver et al. 2015; Molina-Venegas et al. 2021). Describing species ultimately contributes to understanding how ecosystems will respond to new human-induced perturbations in the Earth system. The foundation of how we measure biodiversity is tied to species and thus, depends absolutely on taxonomy and systematics (Wheeler 2018; Winston 2018; Stuessy 2020).

DOI: 10.1201/9780367855604-14

Species and species groups are key units for biodiversity, explicitly and implicitly used as units of assessment in palaeontology, ecology, conservation, developmental biology, and many other areas of geo- and biosciences. In these contexts, species 'groups' may be clades, or functional groups, or other groups of convenience. Yet, all approaches use ranked taxonomy, even if it is in combination with other means of organising species of interest. The established seven nested taxonomic ranks referred to as the Linnean system (species, genus, family, order, class, phylum, kingdom) comprise the best-known as well as the only universal system of species groups.

Biological diversity represents a complex system of species that are interconnected by evolution, ecology, and ancestor–descendent linkages through time. Living species are participants in, not the products of, evolution and are constantly responding to abiotic and biotic factors that may impact selection (Thompson 1999). The variability of lineages, shifting over time, can be visualised as a fuzzy, buzzing, constantly moving tree structure with reticulating branches and hybrids and individual mutations. This is conceptually too complex to use, so we replace it with simplistic wire diagrams of species phylogeny. This is not to undermine the determinability of species: complex and non-linear systems are characterised by identifiable equilibria, the periods of lineage stability (and isolation) that reflect diagnosable species. Articulating the nature of biological diversity depends on identifying patterns within the system and emergent phenomena, and we require a lexicon to enable clear communication about the system and its elements (Vences et al. 2013).

Classification is a cognitive tool to make sense of complex systems, and this function has long been recognised in ethnobiological or traditional non-scientific naming structures that identify species (Guasparri 2019). Taxonomic ranks are important not only because the standard modern hierarchy is taught in school, but actually it is enshrined in international law. For example, the Cartagena Protocol on Biosafety refers to family-level combination as the limit for genetically modified organisms under its purview (United Nations 2005). Other points of reference, such as function, ecological niche, or even phylogenetic position, are not always known, especially for newly discovered species. Standard taxonomic ranks are not the only way to partition species diversity into working units, but it is the only universal classification. That is, they are the only established system that applies to all living and fossil species. So, the question for modern phylogenetic systematics is not whether we should have ranked taxonomy, but rather, how to use this system appropriately to make sense of our modern understanding of global biodiversity.

Conservation, evolutionary biology, phylogenetics, and ecology all use species and species groupings to articulate and to test patterns of life in the natural world. 'Patterns' can include many different things – phylogeny is fundamental, but equally important are patterns of distribution, ecological associations, and more. Any discussion of the classification of species (systematics) must bear this responsibility in mind. Indeed, the necessity of good communication about taxonomy has led to some calls from conservationists for increased governance of species descriptions (Thomson et al. 2018). Well-curated species lists can be extremely important tools for taxonomists and non-taxonomists alike, but it is important that these efforts do not stymie active research that may result in nomenclatural changes (Garnett et al.

2020). The ideal solution for one user group, such as conservationists or phylogeneticists, may not be equally effective for another.

Within the fields of evolutionary biology and phylogenetics, it is clear that Linnean ranks can be misinterpreted when used to describe and classify true patterns of relative relatedness among organisms (Eldredge and Cracraft 1980). A 'class' is not a universal measure of evolutionary scope – Aves has an origin in the late Cretaceous (ca 80 Mya), whereas Polyplacophora is known since the Cambrian (ca 500 Mya) (Hedges et al. 2015). There is no consistent metric of the diversity within a genus or a phylum. This straw man comparison neglects several important points. First, hierarchical groups are a utilitarian tool to make sense of discovery-driven issues in biodiversity (newly discovered species and other groups that have no robustly established phylogenetic context). Second, taxonomic tools are applied to other fields of life sciences. Linnean ranks such as genera and families are routinely applied in biodiversity and ecology. Finally, there is also no universal 'amount' of diversity in a species, because all species are continuously evolving, experience different selective pressures, and have different divergence times separating them from the common ancestor they share with their sister species. This is simply not what taxonomic units are for.

Taxonomic practice is not only relevant for taxonomists; it also has cascading impacts on all areas of sciences that use species. The assignment of species groups to ranked categories is often controversial. Some of this controversy comes from several issues in the way that ranks are considered, which are discussed in detail here: discovery gaps, relativistic thinking, and emergent patterns that help make sense of the complexity. The larger scientific goal, to describe life on Earth, is very complex and requires tools with broad applicability across many disciplines of life sciences. Taxonomic ranks are one such tool to cope with describing the complexity of the diversity of life.

11.2   DISCOVERY GAPS

Taxonomy and systematics are not only the science of discovering and describing new species, but the intimately connected business of arranging the names in some kind of usable order. It is worth reflecting briefly on the origins of this approach. If species can be distinguished, could each stand on its own merits without the need for hierarchical sorting of groups? In a single ecosystem, this is potentially achievable. In a temperate forest there are relatively few species of trees, some smaller plants, fungi, insects, birds, mammals, and soil invertebrates. The kinds are quite different, they occupy different niches, and if there are only a few hundred species, this may not necessarily require further partitioning. This list rapidly reaches a natural limit of what an ordinary person can track without imposing basic organisation (plants, fungi, insects). Species are now traditionally placed into nested groups with standardised ranks referred to as the Linnean system; however, the modern ranked system is not the same as the original arrangement proposed by Linnaeus and has developed substantially in the intervening centuries (Schuh 2003). The Linnean system was initially conceived to help make sense of global biodiversity in the first efforts to make a comparative framework for species worldwide and not only in one bioregion.

Formal taxonomic description of a new species requires it to be compared in a global context. We cannot reasonably compare species x to the other ~2 million species descriptions and then come up with a new one. There has to be a comparative aspect – what is the most similar lineage(s), and how does our new species differ? Thus, we have already imposed two levels of hierarchy: the proximal group of most similar things, and the larger group of all things. It follows logically that it might be useful to have some other groupings so that you can hone in quickly on where the new species ought to fit. This is the origin of systematics and dates not to classical Greek scholarship but to the origins of human language (Simpson 1990).

The natural groupings identified by ethnographers in myriad cultures, with independent vernacular descriptions of species, include five or six widely recognised natural levels (Berlin 1992). This applies to sets of hundreds of names in traditional societies and often includes many species that are recognised for their own sake and not based on cultural or practical utility. These levels are not directly equivalent to Linnean ranks but have some interesting overlap. The typical levels are the folk-kingdom (animals, plants: kingdom), life-form (bird: more or less taxonomic class), intermediate level (water bird: order or family), folk-generic (oak: family or genus), folk-specific, and varietals (species and morphotypes) (Medin and Atran 2004). Ranked groups are a natural, and necessary, device to communicate a basic understanding of biodiversity, especially among non-scientists.

The continuous discovery of new species is a fundamental motivating issue in taxonomy and systematics. Species discovery includes both 'new' discoveries, based on entirely new observations of rare or rarely seen organisms, and also revisions of those that have been seen, such when one widespread highly variable species is reclassified as several separate lineages. But, this distinction is less stark than it appears in most groups of organisms. For example, a new species was recently described from Malaysia, but local records show it had been actively fished for at least 160 years (Sigwart et al. 2021). This is not a rare chance find, but part of the vast discovery gap that most of taxonomy still has to contend with.

The task of taxonomy is to cope not only with the complexity of life but also with the complexity of missing data. As the global taxonomic endeavor expands, the rate of naming of new species is continually increasing (Costello et al. 2013). This does not, however, indicate that there is any end in sight (Larsen et al. 2017). A few familiar groups in which most living species have been described could be considered to be taxonomically 'saturated', such as mammals and birds (Figure 11.1). For these well-studied organisms, scientific practice has moved beyond the initial sorting and inventory and on to understanding more granular issues of ecological and evolutionary mechanisms. The rich available data for these particular organisms dictate baseline assumptions about species lineage dynamics, which may be very misleading when applied to larger, more diverse groups (Sigwart et al. 2018a). It is critical to bear in mind that systematics deals with the discovery, description, and classification of all life on Earth, which means it should be focussed on the less-studied groups and the overwhelming majority of diversity that remains unknown. To be useful, any

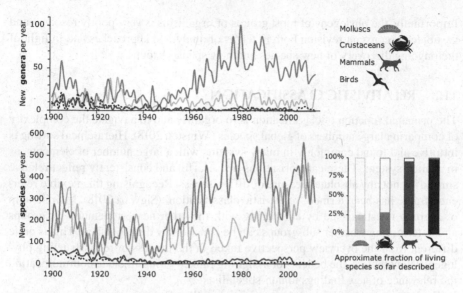

FIGURE 11.1   The number of new genera described each year (top) and new species (bottom) is consistently lower in mammals and birds, which are unusually well studied compared with most organisms. The number of new species (bottom) is much higher and increasing compared with the number of new genera (top). The inset bar chart at lower right illustrates the relative taxonomic saturation of the four groups depicted. (Data from Sigwart, J. D., M. D. Sutton, and K. D. Bennett, *Zoological Journal of the Linnean Society*, 183, 237, 2018b.)

ideas about managing the complexities of systematics must be applicable to all species and not just a few well-studied clades.

Larger or more visible species are better studied; their distributions, phylogeny, and ecology are understood in more detail. Meanwhile, smaller, better-camouflaged, inconspicuous species actually form a larger proportion of local and global species richness. This has been shown in comparing brightly colored tree snails with their small brown counterparts found in leaf litter, where the gaps in basic biology lead to under-estimation of human impacts and hinder the development of effective conservation strategies (Yeung and Hayes 2018).

Phylogeny, or more generally, the evolutionary relationships among organisms, is the fundamental principle that is the basis for all levels of classification. This applies both to the recognition of species as units themselves and to the arrangement of species into groups of genera, families, and higher-ranked groupings in the Linnean system. The aspiration of phylogenetic systematics is to unite knowledge of the tree of life and the formal scientific classification of organisms. This has made the science of systematics and phylogenetics more robust as a discipline, but it has also incidentally created a lot of controversy over almost every aspect of its application. Some of the controversy has been contained within the specialist community, as one would expect for the technical details of discipline-specific debate. But because species names are used by everyone, these issues have a much wider resonance, and the echoes of these arguments are felt in all areas of life science and beyond.

Importantly, the phylogeny of most groups of organisms is very poorly resolved and is subject to constant revision both in terms of analytical approaches and in light of the ongoing discovery of new species and new species groups.

11.3   RELATIVISTIC CLASSIFICATION

The principal function of classification is to organise and to navigate the complexity of comparing large numbers of global species (Winston 2018). Hierarchical sorting is intuitive and found commonly in many systems with a large number of elements, as in a filing system. These categorisations necessarily and consistently reflect relative similarity, not any absolute measure of differentness. Recognising this flexible reference frame has been termed 'relativistic classification' (Sigwart 2018). Membership of a group is established by comparing with the other nearest elements or groups and determining whether sub-groups can be consistently distinguished. This is quite different from the overview perspective taken in the process of reconstructing phylogeny. Nonetheless, the benefits of assigning ranked groups include communicating the relevance of new findings to non-specialists.

Evolutionary relatedness is the most universally informative basis for comparison at a global scale. Furthermore, other sources of data for comparative assessments (morphology, genetics) ultimately arise from evolutionary processes. Evolutionary similarity, and shared origin, is therefore the strongest and most stable foundation for a system of classification. Likewise, the basic utility of hierarchical classification is non-controversial. Thus, the question is whether the existing standard ranks are fit for purpose and whether they are compatible with a more nuanced view of phylogeny and phylogenetic systematics. As the number of species increases, so does the number of nodes in a phylogeny, and the complexity of identifying consistent groups. The ideal of phylogenetic systematics is to name only monophyletic clades, which is laudable but often impractical and even undesirable in groups with many fossil species (Hörandl and Stuessy 2010). For names to be 'phylogenetically meaningful' (Vences et al. 2013), monophyly is not required. The ideal of phylogenetic systematics is to unite phylogeny and classification, but in fact, not all clades need be named, and not all named groups need be clades.

Phylogenetic systematics traditionally rejects paraphyletic taxa. The idea that all named groups should be monophyletic stems from the start of phylogenetic systematics, but it deserves greater scrutiny. The most important reason to avoid non-monophyletic names is that monophyletic groups are supposedly more evolutionarily informative (Ward et al. 2016). Henning (1966) explicitly made the argument that paraphyletic groups would undermine comparisons of higher-ranked groups; however, he provided no evidence for this, and there is no evidence yet from modern modelling approaches that this is true. Phylogenetic systematics is not the only form of taxonomy, and monophyly is entirely optional for named groups (Nordal and Stedje 2005). Yet, phylogenetic systematics could also benefit from further interrogation of these basic assumptions.

In higher-ranked groups, avoiding paraphyly is not controversial. In fact, paraphyletic groups, such as fish (or, invertebrates), disproportionately include groups that were traditionally called 'lower' animals or plants. In fact, these groups contain multiple independent (and deeply divergent) lineages. The collective group name gives an erroneous impression that the group is homogeneous. However, in shallower radiations, splitting a paraphyletic group can over-emphasise differences. Imagine a clade where one species in the 'middle' undergoes a rapid evolution that makes it radically different from the remainder and requires a new genus name. Should the other 'left-behind' species be reclassified into multiple genera or left in a unified paraphyletic genus that excludes their unusual cousin? Species within this paraphyletic genus are more similar than they are different; there are no synapomorphies for the sub-groups. Rather than arguing about paraphyletic taxa, it would be more informative to adopt a convention where they are explicitly flagged, as in 'genus *Leptochiton* (paraphyletic) Gray, 1847' (Sigwart 2018). Paraphyly is not a problem; it is an interesting reflection of real, messy evolution (Hörandl and Stuessy 2010; Seifert et al. 2016).

Although new higher-ranked taxonomic groups are unusual, a new phylum of fungi Entorrhizomycota was erected from re-analysis of the genus *Entorrhiza* Weber, 1884 (Bauer et al. 2015), and ten other clades were elevated to phylum level based on a large-scale phylogenetic analysis (Tedersoo et al. 2018). Other examples of major revisions in the highest ranks include the ongoing debate about the applicability of the kingdom label to groups formerly lumped into 'Protozoa' (Cavalier-Smith 2018). The phylum Annelida has been expanded to incorporate two other clades of worms, the hydrothermal vent-endemic Pogonophora and the spoon worms Echiura (Struck 2011; Weigert et al. 2014). In these cases, the revisions are made either to recognise the unique position of *Entorrhiza* or on the basis of new data showing that non-segmented worms are derived from segmented body plans. Naming these clades as 'phylum' or 'kingdom' functions as a communication tool. To say that *Entorrhiza* is a clade sister to the remaining Dikarya (Bauer et al. 2015) is not actually informative unless you are already familiar with fungal taxonomy. Saying that *Entorrhiza* is its own phylum is meaningful to anyone.

Ranks provide a general sense of the position, importance, or distinctiveness of a named group (Giribet et al. 2016). This does not contain any absolute measure of phylogenetic distance, but neither do any unranked clades. Species themselves represent a broad spectrum of morphological and molecular disparity, depending on how recently a species split from its sister group, genetic bottlenecks, or an unknown future split or 'incipient species'. The species we sample in the living world or in the fossil record are dynamic participants in ongoing evolutionary processes, each with an independent history and trajectory. This is intrinsic to the nature of species evolution. Since species themselves are variable in disparity and divergence time, there is therefore no rational reason to expect that groups of species should be determined by any consistent metrics. Rather, we require tools that can estimate and express the inherent complexity in the underlying system.

Classification can progress via a bottom-up sorting of species into genera, and genera into families. This process is followed when we differentiate whether a species is placed into an established genus or needs a new genus in order to be adequately described. However, classification can also reflect top-down sorting, such as the original establishment of kingdoms, phyla, and classes, which are also reflected in folk taxonomies. Larger groups may be easier to distinguish because of distinct morphological characters and also because they are separated by gaps created by the extinction of intervening stem lineages (Marshall 2017). Interestingly, the middle level of the seven standard Linnean ranks, the order, sits in between these top-down and bottom-up approaches. This matches remarkably with the 'intermediate' level found more or less universally within the six levels of folk taxonomies (Berlin 1992). In practice, the rank of order may be applied differently in different groups of organisms – in larger groups with a long fossil record, there may be debates about whether a certain group should be considered a sub-class or an order (all living polyplacophoran mollusks form a group Neoloricata, which is conventionally the order), or an order or a superfamily (Neoloricata is divided into several clades; some authors consider Neoloricata a sub-class and these subsidiary clades as orders). By contrast, the orders of birds are taxonomically important and used as a major sorting tool to differentiate important groups (Strigiformes, the owls; Falconiformes, the falcons).

This type of apparent inconsistency does not hamper the main function of systematics, which is to communicate information about the relative – not absolute – scale of evolutionary separation between proximate groups. Groups may be paraphyletic and still represent an important evolutionary pattern. And yet, it is clear that some macroevolutionary patterns are robust to variable definitions of rank membership (Sigwart et al. 2018b). It remains important to recognise the inherent variability in clade diversity when designing analyses using higher -level groups as units.

11.4   RANK CONTROVERSY

The discovery and the naming of species are quite separate scientific processes. There are many unnamed species awaiting description and also many more that have not yet been discovered. Natural history collections maintained in major museums contain potentially 0.5 million unnamed species (Costello et al. 2013). The scientific report of a new species can influence the group-assignments of its relatives; or, further research on the phylogeny of known species can prompt a reinterpretation of relevant species groupings. Although living birds are an exceptionally well-described group, some hobbyist bird watchers insist that English vernacular names are 'more stable' than correct scientific names, out of frustration with nomenclatural revision (Thomson et al. 2018).

The assignment of standard rank designations to a group of species frequently causes controversies in phylogenetic systematics because of disagreements over how groups should be conscribed (especially when previous assumptions of monophyly are overturned by further analysis). This is not only about monophyly but about selecting the appropriate rank within the scheme provided by the Linnean system. For example, the bivalve *Crassostrea* (*Magallana*) *gigas* (Thunberg, 1793) is the most widely cultivated oyster in the world, and there has been extensive controversy over whether this species and others should be part of a large globally distributed genus *Crassostrea* Sacco, 1897 (Bayne et al. 2019) or a separate genus *Magallana* Salvi & Mariottini, 2016 (Salvi et al. 2014; Salvi and Mariottini 2020), or if the clade *Magallana* should be classified as a sub-genus (Sigwart et al. 2021). In this case, the

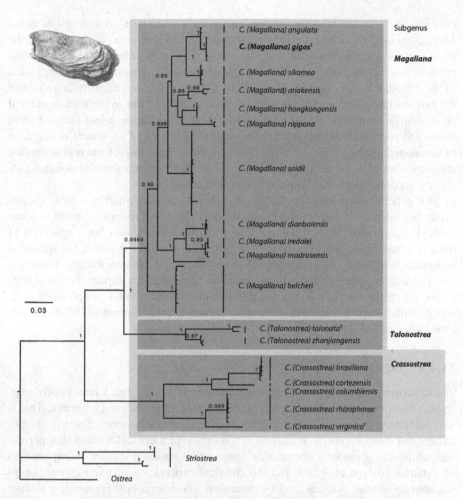

FIGURE 11.2 Phylogeny of the genus *Crassostrea* based on multi-locus Bayesian analysis. The boxes indicate the three sub-genera of *Crassostrea*. The sub-genera had previously been revised as genus-level groups; there are no consistent synapomorphies that diagnose the groups, but the groupings are important for communicating biodiversity of the group and on this basis, were revised to sub-genera. (Redrawn from Sigwart, J. D., Wong, N. L. W. S., and Esa, Y., *Marine Biodiversity*, 51, 83, 2021; inset photo of *Crassostrea gigas* by Christine Morrow, courtesy of iNaturalist, CC-BY-NC, www.inaturalist.org)

phylogeny is not in question, only the practical consideration of how the group should be named (Figure 11.2). The main arguments for maintaining the larger worldwide genus *Crassostrea* are these: there are no morphological characters that diagnose the sub-groups, and the name *Crassostrea gigas* is more familiar to the global oyster industry (Bayne et al. 2019). The main argument for maintaining a separate genus *Magallana* is to reflect the phylogeny (Salvi and Mariottini 2020). There are molecular characters that separate *Magallana* spp., although this has not been tested in all species, since the relevant fragments have not been sequenced in all species. This data gap prevents any new species (and any fossils) from being diagnosed as either *Magallana* or *Crassostrea s.s.* In consequence of this, the most recent study on the topic proposed recognising the clades at sub-generic rank, which is a kind of compromise that is afforded by the flexibility of additional categories already integrated in the Linnean ranks (Sigwart et al. 2021). Global oyster aquaculture is valued in the billions of dollars worldwide (Olivier et al. 2020), but the oyster threatens local ecosystems as a marine invasive species in large areas where it has escaped from cultivated populations (Zwerschke et al. 2018). This type of argument is common in taxonomy, although this one has been more heated because of the real economic consequences of re-labelling global aquaculture products and the vast scientific literature supporting the development of oyster mariculture.

The general framework for sufficient evidence for designating a new species group has been clearly articulated (Vences et al. 2013) but remains subject to some individual interpretation. This is not a problem with the system but a reflection of healthy scientific debate (Thomson et al. 2018). It is naturally easier for specialist systematists to focus solely on the solution that is best for systematics. There are many examples of the tension this creates when name changes impact species of high public interest. Ranks can be used to improve understanding of a complex situation, and systematists must use them to communicate important biological information to broad non-specialist audiences.

11.4.1 Patterns, Complexity, and Taxonomy

The description and naming of a species or species group (e.g., a new family) represents the formal statement of an hypothesis about an evolutionary lineage. This is well understood by taxonomists but not by other areas of science that use species names and species groups. And, this is the origin of a critical tension that permeates all issues of modern systematics: the conflict between species as expressions of dynamic lineage evolution and the classical expectation of fixed essences represented by names. Species (and by extension, species groups) represent a lineage equilibrium: ancestor–descendent population lineages that extend through time but which demonstrably maintain a stable equilibrium as a discrete lineage with a particular evolutionary trajectory over extended periods of time. There is necessarily uncertainty in defining species boundaries or the 'fuzzy edges' of lineages (Sigwart 2018). The goal of modern taxonomy is to take an integrative approach combining natural history, morphology, molecular data, and autecology in a total evidence approach that establishes lineage identities (Winston 2018).

The distance between species, and species groups, is neither fixed nor consistent. The metrics that are used to differentiate specific lineages are not transferable among groups. Even within a large clade, rates of evolution can be very different in species in temperate or tropical ecosystems (Rabosky et al. 2012) and can be influenced by local issues of selection that impact the rate of evolution in one particular lineage. While 'molecular clock' approaches still represent a tool for reconstructing evolutionary timescales by using broad-scale average rates, it is now clear that clocklike rates of change are variable and easily disrupted (Ho 2020). Some authors have suggested that taxonomic ranks should be coordinated with lineage divergence times (x million years for a genus, y million years for a family; e.g., Avise and Mitchell 2007), but this is an over-simplification that has no utility outside of a few well-constrained and recently evolved clades (Rabosky et al. 2012; Lücking 2019). The genetic diversity represented by a species or species group is not straightforward. Genus-level divergence of the DNA barcode marker COI increases linearly with species richness in some vertebrates but not other groups of similar species richness (Sigwart and Garbett 2018). There may be many mechanisms for this, such as differing divergence times, sampling, and taxonomic saturation, but the primary result remains that well-studied groups are simply not an appropriate null model for genus-level diversity in other clades.

The underlying complexity of lineage evolution contrasts with a need to distinguish groups of species; groupings help to articulate the dynamics of the larger system, and a large amount of missing data, since most species remain undiscovered and undescribed. In this complex system, there are nonetheless strong emergent patterns; we can differentiate species, and species groups, and see that these lineages occupy particular ecological roles. Further, stochastic dynamics of speciation and extinction lead to emergent size-frequency distribution of species groups (Sigwart et al. 2018a). The recognition of patterns in complex systems provides a predictive power to help guide the discovery of undescribed and unrecognised species, and also to find an anchor to stabilise the application of ranked groups.

11.4.2 Skew Distributions

Where taxa cannot be identified or quantified at the species level, higher taxonomic levels can be used as a substitute parameter in an approach known as 'taxonomic surrogacy' (Balmford et al. 1996). Substitution of genera and family-level units in calculations about diversity occurs in palaeontology, community ecology, conservation biology, and studies of environmental impacts in scientific literature and industry reports (Bertrand et al. 2006).

The approach of taxonomic surrogacy is based on an assumption that there is a consistent and predictable relationship between species counts and the number of higher taxonomic ranks they represent (Balmford et al. 1996). Where species cannot be counted directly, counts of higher taxa provide an indirect quantitative estimate. At global scales, all of the past mass extinction events, the 'big 5', were initially identified by changes in the fossil record of genera and families of hard-shelled marine invertebrate faunas (Sepkoski 1986). However, non-equivalent species groups have been raised as a potential problem for robust quantitative analysis (Quental and Marshall 2010).

In ecological studies of living species assemblages, the same approach is applied where specimens cannot be identified to the species level because of resource limitations (expertise or time). Thus, taxonomic surrogacy is applied at two separate scales of biodiversity: large-scale global patterns, such as changes over an extinction horizon, and the way that global diversity is subsampled in a local ecosystem in the past or in the present. Most genera contain multiple species, and those species usually do not co-occur. At local scale, however, there is a linear relationship between the number of species that occur at a site and the number of genera or families. This approach has validity because of emergent patterns, not because it is straightforward to compare one genus-level group with another.

Higher taxa are not comparable units and not the same type of things as species; this is the most important epistemological weakness of ranked taxonomy. Higher-ranked groups are proximate clusters of species lineages, and there are emergent macroevolutionary properties of these groups that come out of phylogeny (Sigwart et al. 2018b). So, higher taxa are 'real' in that they are objectively definable units, though a genus or an order cannot really be thought of as a biological entity (Eldredge and Cracraft 1980). A genus, or a family, is composed of species. While the species evolve, it is doubtful that any evolutionary process can be seen as acting on higher taxonomic units.

There is a consistently long-tailed distribution in genus size, and this has been recognised since the 1920s (Yule 1925), although it has somehow not been fully incorporated into general thinking about taxonomy (Williams and Gaston 1994; Strand and Panova 2015). About 30% of genera, globally, are monotypic, meaning they contain only one species in the genus. Thus, among genera of all sizes, the biggest fraction is these monotypic genera, but that still leaves an overall majority of genera with multiple species. This pattern is a predictable, expected outcome of diversification dynamics (Sigwart et al. 2018b). There are many ways to create a small genus: a single lineage that never splits, or a large clade that gets split up by extinction events, or some combination of these effects (Quental and Marshall 2010). There is effectively only one way to evolve a large genus, which is to have a relatively rapid radiation. Therefore, smaller genera are naturally far more frequent than large genera (Sigwart et al. 2018b).

These same patterns are found in purely theoretical models, which use many different ways to define genera (based on phylogenetic topology, characters of simulated species, or chrono-taxonomy with groups determined by divergence time). Regardless of how group membership is defined, the same skew distribution patterns of group size emerge with absolute consistency. This type of skew or hollow-curve distribution is also one of the hallmarks of complex systems (Sornette 2007).

This overall skew distribution is a very strong pattern in all animal and plant groups when sampled at sufficiently large scale. As in any natural system, there is a high level of variance, which in this case is affected both by the underlying phylogeny that generates kinship relationships and secondarily, our imperfect interpretation of those relationships through taxonomic classification. Deviations from a perfect or predicted long-tail distribution may be correlated with the evolutionary histories within that group; it may be because the taxonomy is poorly studied and genera reflect excessive lumping or splitting, or it could be simply the effect of a small sample size. At global scale, across broad taxonomic sampling, these coalesce into a strong central pattern.

These skew distributions are reflected in grouping at all ranks and have been used to model the total diversity of life on Earth (Mora et al. 2011). There are more species than genera, more genera than families, and so on. *Ipso facto*, the discovery gap – knowledge of how many species remain undescribed, undiscovered, or in the case of fossils, unpreserved – is larger at the species level than at higher ranks. At a global scale, distribution patterns of genera or families are probably more informative than data about species. Higher-level groupings are more taxonomically saturated (new phyla are rarely reported, new species are discovered constantly; Figure 11.1), and these nested skew distributions can be used to infer what is still missing at lower ranks. Thus, the patterns of skew distributions present a tool that complements some of the systemic issues presented by discovery gaps. Taxonomic surrogacy, substituting higher-ranked groups for species-level identification, should be very effective at capturing global diversity patterns in large datasets.

Macroevolutionary patterns in group size-frequency distributions are robust, and the same patterns arise under very different models that determine how members are assigned to groups (Sigwart et al. 2018b). As noted earlier, different algorithms to assign species to genera still produce the same emergent patterns. In these *in silico* approaches, a complete phylogeny was partitioned into groups based on a priori criteria; this is very different from the experience of working taxonomists. A perfect, or even an adequately sampled, phylogeny is almost never available, with the exception of the few well-studied and densely sampled groups of organisms, which should not be taken as a guide to normal practice. Poor taxon sampling is well known to confound phylogenetic reconstruction (Heath et al. 2008; Nabhan and Sarkar 2012). Species may be missing from phylogenetic analysis because of missing data and because of undiscovered additional species. The character data available for the available species or groups may be patchy; gene fragments or molecular orthologs may be absent for some taxa, or specific morphological characters or traits may not be documented for all species. The scale of missing data for most species and groups means that this is not a temporary problem but a systemic issue that requires other, robust approaches.

Skewed distribution patterns that emerge from large-scale analyses of taxonomic groups demonstrate how ranked taxonomy is compatible with evolutionary patterns. This underscores the usefulness of ranked groups, especially for confronting the discovery gap that obscures much biodiversity data. Understanding the nature of skewed distributions should also be a useful insight for practicing taxonomists; for example, to support naming new monotypic genera or families where appropriate.

11.5   CONCLUSIONS

The science of discovering, naming, and understanding species and their inter-relationships is intrinsically important, it encapsulates the excitement of discovery, and it produces myriad accessory benefits to humanity. In the study of species, well-meaning scholars have written that ecologists cannot use species names without understanding phylogenetics, or that taxonomists cannot name species without integrating environmental biology. Different branches of life sciences inform and enhance each other; taxonomy and phylogenetics are interlinked but not the same. To confront modern biodiversity loss requires an accelerating taxonomic effort, which is more urgent than resolving phylogenetic questions (Goodwin et al. 2020). It is important not to alienate other sub-disciplines from engaging with taxonomy and systematics, and we are obliged to approach these issues with a very broad potential audience in mind, even for the most obscure-seeming organisms. Ranked taxonomy is an excellent communication tool both within the systematics community, to express hypotheses about evolutionary groupings, and to communicate findings to other fields.

The mission to describe life on Earth is inherently complex and an ambition that is increasingly under pressure from the rapidly mounting threat of global species extinctions. It is therefore important to identify tools that facilitate this goal. Systematics suffers from a legacy of conflict between classical and obsolete ideas of species or species groups, as fixed entities with fixed membership, and our modern understanding of species lineage equilibrium dynamics. It is reasonable to ask how Linnean ranked taxonomy, which was conceived not only in the eighteenth century but as a common product of human language describing the diversity of everyday species, can be usefully applied to communicate the complex patterns shaped by evolution. Systematics needs to escape the question of naming nodes in the tree and think about the original utilitarian role of ranks: taxonomic ranks are, above all, a practical tool to cope with the complexity of diversity of life.

REFERENCES

Avise, J. C., and D. Mitchell. 2007. Time to standardize taxonomies. *Systematic Biology* 56:130–133.

Balmford, A., M. J. B. Green, and M. G. Murray. 1996. Using higher-taxon richness as a surrogate for species richness: I. Regional tests. *Proceedings of the Royal Society of London. Series B: Biological Sciences* 263:1267–1274.

Barnosky, A. D., N. Matzke, S. Tomiya, et al. 2011. Has the Earth's sixth mass extinction already arrived? *Nature* 471:51–57.

Bauer, R., S. Garnica, F. Oberwinkler, et al. 2015. Entorrhizomycota: A new fungal phylum reveals new perspectives on the evolution of fungi. *PLoS ONE* 10:e0128183.

Bayne, B., M. Anglès d'Auriac, T. Backeljau, et al. 2019. A scientific name for Pacific oysters. *Aquaculture* 499:373.

Berlin, B. 1992. *Ethnobiological Classification: Principles of Categorization of Plants and Animals in Traditional Societies.* Princeton, NJ: Princeton University Press.

Bertrand, Y., F. Pleijel, and G. W. Rouse. 2006. Taxonomic surrogacy in biodiversity assessments, and the meaning of Linnaean ranks. *Systematics and Biodiversity* 4:149–159.

Cavalier-Smith, T. 2018. Kingdom Chromista and its eight phyla: A new synthesis emphasising periplastid protein targeting, cytoskeletal and periplastid evolution, and ancient divergences. *Protoplasma* 255:297–357.

Costello, M. J., R. M. May, and N. E. Stork. 2013. Can we name Earth's species before they go extinct? *Science* 339:413–416.

Eldredge, N., and J. Cracraft. 1980. *Phylogenetic Patterns and the Evolutionary Process.* New York: Columbia University Press.

Garnett, S. T., L. Christidis, S. Conix, et al. 2020. Principles for creating a single authoritative list of the world's species. *PLoS Biology* 18:e3000736.

Giribet, G., G. Hormiga, and G. D. Edgecombe. 2016. The meaning of categorical ranks in evolutionary biology. *Organisms Diversity & Evolution* 16:427–430.

Goodwin, Z. A., P. Muñoz-Rodríguez, D. J. Harris, et al. 2020. How long does it take to discover a species? *Systematics and Biodiversity* 18:784–793.

Gould, S. J. 2000. A tree grows in Paris: Lamarck's division of worms and revision of nature. In *The Lying Stones of Marrakech*, ed. S. J. Gould, 115–143. Cambridge, MA: Harvard University Press.

Guasparri, A. 2019. Polysemy revisited. *Altorientalische Forschungen* 46:61–87.

Heath, T., S. Hedtke, and D. Hillis. 2008. Taxon sampling and accuracy of phylogenetic analyses. *Journal of Systematics and Evolution* 46: 239–257.

Hedges, S. B., J. Marin, M. Suleski, et al. 2015. Tree of life reveals clock-like speciation and diversification. *Molecular Biology and Evolution* 32:835–845.

Hennig, W. 1966. *Phylogenetic Systematics.* Reprinted in translation, 1999. Chicago, IL: University of Illinois Press.

Ho, S. Y. W. 2020. The molecular clock and evolutionary rates across the tree of life. In *The Molecular Evolutionary Clock*, ed. S. Y. W. Ho, 3–23. Cham: Springer International Publishing.

Hörandl, E., and T. F. Stuessy. 2010. Paraphyletic groups as natural units of biological classification. *Taxon* 59:1641–1653.

Larsen, B. B., E. C. Miller, M. K. Rhodes, and J. J. Wiens. 2017. Inordinate fondness multiplied and redistributed: The number of species on earth and the new pie of life. *The Quarterly Review of Biology* 92:229–265.

Lücking, R. 2019. Stop the abuse of time! Strict temporal banding is not the future of rank-based classifications in fungi (including lichens) and other organisms. *Critical Reviews in Plant Sciences* 38:199–253.

Marshall, C. R. 2017. Five palaeobiological laws needed to understand the evolution of the living biota. *Nature Ecology & Evolution* 1:1–6.

Medin, D. L., and S. Atran. 2004. The native mind: Biological categorization and reasoning in development and across cultures. *Psychological Review* 111:960–983.

Molina-Venegas, R., M. Á. Rodríguez, M. Pardo-de-Santayana, et al. 2021. Maximum levels of global phylogenetic diversity efficiently capture plant services for humankind. *Nature Ecology & Evolution*. https://doi.org/10.1038/s41559-021-01414-2.

Mora, C., D. P. Tittensor, S. Adl, et al. 2011. How many species are there on earth and in the ocean? *PLoS Biology* 9:e1001127.

Nabhan, A. R., and I. N. Sarkar. 2012. The impact of taxon sampling on phylogenetic inference: A review of two decades of controversy. *Briefings in Bioinformatics* 13:122–134.

Nordal, I., and B. Stedje. 2005. Paraphyletic taxa should be accepted. *Taxon* 54:5–6.

Oliver, T. H., N. J. B. Isaac, T. A. August, et al. 2015. Declining resilience of ecosystem functions under biodiversity loss. *Nature Communications* 6:10122.

Olivier, A. S., L. Jones, L. L. Vay, et al. 2020. A global review of the ecosystem services provided by bivalve aquaculture. *Reviews in Aquaculture* 12:3–25.

Paknia, O., H. Rajaei Sh., and A. Koch. 2015. Lack of well-maintained natural history collections and taxonomists in megadiverse developing countries hampers global biodiversity exploration. *Organisms Diversity & Evolution* 15:619–629.

Quental, T. B., and C. R. Marshall. 2010. Diversity dynamics: Molecular phylogenies need the fossil record. *Trends in Ecology & Evolution* 25:434–441.

Rabosky, D. L., G. J. Slater, and M. E. Alfaro. 2012. Clade age and species richness are decoupled across the eukaryotic tree of life. *PLoS Biology* 10:e1001381.

Salvi, D., A. Macali, and P. Mariottini. 2014. Molecular phylogenetics and systematics of the bivalve family Ostreidae based on rRNA sequence-structure models and multilocus species tree. *PLoS ONE* 9:e108696.

Salvi, D., and P. Mariottini. 2020. Revision shock in Pacific oysters taxonomy: The genus *Magallana* (formerly *Crassostrea* in part) is well-founded and necessary. *Zoological Journal of the Linnean Society.* https://doi.org/10.1093/zoolinnean/zlaa112.

Schuh, R. T. 2003. The Linnaean system and its 250-year persistence. *The Botanical Review* 69:59.

Seifert, B., A. Buschinger, A. Aldawood, et al. 2016. Banning paraphylies and executing Linnaean taxonomy is discordant and reduces the evolutionary and semantic information content of biological nomenclature. *Insectes Sociaux* 63:237–242.

Sepkoski, J. J. 1986. Phanerozoic overview of mass extinction. In *Patterns and Processes in the History of Life*, eds. D. M. Raup and D. Jablonski, 277–295. Berlin: Springer Verlag.

Sigwart, J. D. 2018. *What Species Mean: A User's Guide to the Units of Biodiversity*. Boca Raton, FL: CRC Press.

Sigwart, J. D., K. D. Bennett, S. M. Edie, et al. 2018a. Measuring biodiversity and extinction—present and past. *Integrative and Comparative Biology* 58:1111–1117.

Sigwart, J. D., and A. Garbett. 2018. Biodiversity assessment, DNA barcoding, and the minority majority. *Integrative and Comparative Biology* 58:1146–1156.

Sigwart, J. D., M. D. Sutton, and K. D. Bennett. 2018b. How big is a genus? Towards a nomothetic systematics. *Zoological Journal of the Linnean Society* 183:237–252.

Sigwart, J. D., N. L. W. S. Wong, and Y. Esa. 2021. A newly described species of oyster from SE Asia and a global controversy in oyster systematics (Bivalvia: Ostreidae: Crassostreinae). *Marine Biodiversity* 51:83.

Simpson, G. G. 1990. *Principles of Animal Taxonomy*. New York: Columbia University Press.

Sornette, D. 2007. *Probability Distributions in Complex Systems*. arXiv:0707.2194.

Strand, M., and M. Panova. 2015. Size of genera – biology or taxonomy? *Zoologica Scripta* 44:106–116.

Struck, T. H. 2011. Direction of evolution within Annelida and the definition of Pleistoannelida. *Journal of Zoological Systematics and Evolutionary Research* 49:340–345.

Stuessy, T. F. 2020. Challenges facing systematic biology. *Taxon* 69:655–667.

Tedersoo, L., S. Sánchez-Ramírez, U. Koljalg, et al. 2018. High-level classification of the Fungi and a tool for evolutionary ecological analyses. *Fungal Diversity* 90:135–159.

Thompson, J. N. 1999. The evolution of species interactions. *Science* 284:2116–2118.

Thomson, S. A., R. L. Pyle, S. T. Ahyong, et al. 2018. Taxonomy based on science is necessary for global conservation. *PLoS Biology* 16:e2005075.

United Nations. 2005. *Cartagena Protocol on Biosafety to the Convention on Biological Diversity*. Montreal, 29 January 2005. *UN Treaty Series* 2226:208–372.

Vences, M., J. M. Guayasamin, A. Miralles, and I. D. L. Riva. 2013. To name or not to name: Criteria to promote economy of change in Linnaean classification schemes. *Zootaxa* 3636:201–244.

Ward, P. S., S. G. Brady, B. L. Fisher, and T. R. Schultz. 2016. Phylogenetic classifications are informative, stable, and pragmatic: The case for monophyletic taxa. *Insectes Sociaux* 63:489–492.

Weigert, A., C. Helm, M. Meyer, et al. 2014. Illuminating the base of the annelid tree using transcriptomics. *Molecular Biology and Evolution* 31:1391–1401.

Wheeler, Q. 2018. Blank canvas: The case for descriptive taxonomy. *Integrative and Comparative Biology* 58:1118–1121.

Williams, P. H., and K. J. Gaston. 1994. Measuring more of biodiversity: Can higher-taxon richness predict wholesale species richness? *Biological Conservation* 67:211–217.

Winston, J. E. 2018. Twenty-first century biological nomenclature—the enduring power of names. *Integrative and Comparative Biology* 58:1122–1131.

Yeung, N. W., and K. A. Hayes. 2018. Biodiversity and extinction of hawaiian land snails: How many are left now and what must we do to conserve them—a reply to Solem (1990). *Integrative and Comparative Biology* 58:1157–1169.

Yule, G. U. 1925. II.—A mathematical theory of evolution, based on the conclusions of Dr. J. C. Willis, F. R. S. *Philosophical Transactions of the Royal Society of London. Series B, Containing Papers of a Biological Character* 213:21–87.

Zwerschke, N., J. Kochmann, E. C. Ashton, et al. 2018. Co-occurrence of native *Ostrea edulis* and non-native *Crassostrea gigas* revealed by monitoring of intertidal oyster populations. *Journal of the Marine Biological Association of the United Kingdom* 98:2029–2038.

Section 4

Metaphysics and Epistemologies

12 The Species Problem from a Conceptualist's Viewpoint

Igor Ya. Pavlinov

CONTENTS

> It seems that the human reason should first independently devise the structures before they can be detected in the things.[*]
>
> **A. Einstein**

12.1 INTRODUCTION

In the basic thesaurus of biology, the *species concept*[†] in its most general sense takes a special position along with those of organism, gene, cell, ecosystem, evolution, etc. It is considered fundamental in most schools of systematics, because the entire

---

[*] The original quotation in German appeared in Einstein's 'Mein Weltbild' (My World-view) (1931). I thank Frank Zachos for this seemingly adequate translation. With this, one of the rather 'liberal'" English translations of this quotation looks very attractive from a conceptualist standpoint: 'Whether you can observe a thing or not depends on the theory which you use. It is the theory which decides what can be observed' (https://en.wikiquote.org/wiki/Albert_Einstein#Mein_Weltbild_(My_World-view)_(1931)).

[†] In distinguishing the terms 'concept' and 'conception', I follow Wilkins (2018).

DOI: 10.1201/9780367855604-16

'Linnaean' classification hierarchy is built up with respect to just the species. In the evolutionary concepts concentrating around synthetic theory of evolution, species is understood as a nodal point of the evolutionary process. In recent years, species is usually considered as a basic unit of the structure of biological diversity (*aka* biodiversity). Moreover, species performs a universal reference function: actually, biological knowledge is usually considered scientifically meaningful if it is linked properly to the particular species of living beings (Volkova and Filyukov 1966). Such fundamental significance of species is evidenced by a proposal to single out a biological discipline dealing with various species issues: it is designated as *hexonomy* (Skvortsov 1967), or *eidology* (Zavadsky 1968) (not in the sense of Husserl), or *eidonomy* (Dubois 2011). Establishing The International Institute for Species Exploration (IISE, www.esf.edu/species/) especially signifies this general trend.

A specific 'enclosure' of the general species concept is formed by an equally fundamental and general *species problem*. Its main content and historical origins can be thought of in different ways (Ereshefsky 1992, 2009, 2010; Ruse 1995; de Pinna 1999; de Queiroz 2005a; Reydon 2005; Pavlinov 2009, 2013a; Wilkins 2009, 2018; Ellis 2011; Kunz 2012; Bartlett 2015; Zachos 2016; Nathan 2017; Maxwell et al. 2020; Mishler 2021), but in general, it is caused by a variety of different particular species conceptions not reducible to any single universal or particular one. This *theoretical species uncertainty* (Hey et al. 2003) can be traced back to Late Antique times; its understanding as an essential part of the species problem was recognised and denoted explicitly as early as in the first half of the twentieth century (Robson 1928; Hawkins 1935). Since then, it has been actively discussed, especially during several of the last decades, with several monographs and collections published on that matter (Volkova and Filyukov 1966; Roger and Fischer 1987; Ghiselin 1997; Wilson 1999; Wheeler and Meier 2000; Hey 2001a; Morgun 2002; Stamos 2003, 2007; Wilkins 2009, 2018; Richards 2010; Pavlinov 2013b; Zachos 2016; Sigwart 2018). These hot debates, with no clear 'light at the end of the tunnel', motivated Scott Atran to point out that 'ongoing inquiry into the "species problem" surely counts as one of the basic set of scientific puzzles by which to measure success' of theoretical biology (Atran 1987: 270).

Both the general species concept and the respective species problem can be considered from various standpoints – empirical and theoretical, biological and philosophical, ontological and epistemological, etc. However, according to a standpoint elaborated by contemporary conceptualism (Swoyer 2006), no empirical issues have scientific significance without the theoretical considerations, and the latter have no justification without the philosophical context, so it is this context that is of prime importance in the theoretical considerations of the species problem (e.g., Kendig in this volume). This is because, as Ernst Mayr said, scientific and philosophical concepts 'run into each other, [so] every basic biological concept is also a philosophical concept' (cited after Greene 1992: 259). The fundamental questions about the reality or nominality of species, about possible ways of defining it, about the scientific status of both knowledge about species and methods of species exploration, etc., all are philosophical ones.

This chapter considers a possible format of philosophical and theoretical consideration of both the general species concept and the species problem within the context of just-mentioned contemporary conceptualism, as I see it as a 'philosophically minded' biologist. I shall first consider briefly the historical roots of both the species concept and the problem to show why and how they became fundamental for biology. Next will be a brief, sketchy consideration of the structure of cognitive situation, within which both the concept and the problem may be considered from a conceptualist's standpoint. This will allow me to restore the reduction cascade of the species conceptions of descending levels of generality, which will lead to the issue of how to define the general notion of 'natural species' as one of the structural units of evolving, functioning, and structuring biota. I will then show that one of the ways to comprehend the 'species in general' as a natural phenomenon is to consider it as endowed with an evolving multi-faceted conjectural essence, *specieshood*, as an integrated part of the whole natural history of organisms, which evolves together with development of biota and therefore, whose manifestations are specific to the particular branches of the Tree of Life. The References section happens to be rather extensive, though I tried to limit myself to the publications most essential to my consideration; it may be useful as a source for the important literature on the species problem (including its definitive 'philosophical framing').

12.2 WHY SPECIES?

According to one of the bright aphorisms born from non-equilibrium thermodynamics, any developing non-equilibrium system is in a sense a 'victim of its history' (Brooks and Wiley 1986). This follows from the initial conditions model, according to which previous stages of the development of such a system affect to a degree its subsequent stages. From the standpoint of evolutionary epistemology, scientific knowledge in general, and its conceptual constructions in particular, can be thought of as a developing non-equilibrium information system (Toulmin 1972; Hull 1988; Bradie and Harms 2016), so this aphorism seems to be quite appropriate here. Therefore, in order to understand why an idea of species is both fundamental and problematic for the whole of contemporary biology, causing theoreticians and philosophers to pay considerable attention to it, we should recall the principal stages in the conceptual history of the general notion of species.

I do not permit myself to go too deeply or in too much detail into this history, which was thoroughly discussed in the recent review by Wilkins (2009, 2018a; see also Minelli's chapter in this volume). I begin with considering the most important events in this history with the ideas of Late Antique neo-Platonists, who in their interpretations of the natural philosophical and cognitive systems of Plato and Aristotle, paid particular attention to the relationship between concepts of genus and species as the key categories of the logical genus–species argumentation scheme. Neo-Platonists' concern for the need to recognise the fundamental status of species as one of these two categories had been expressed by Boëthius' aphorism: 'if we do not realise what species is, nothing would secure us from delusion'.[*] Thus,

---

[*] *'Si enim quid sit species intellegimus, nihil impediti errore turbamur'* (Boëthius 1906). Thanks to John Wilkins for providing me with the source of this original quotation.

those biologists and philosophers who would have subsequently been animatedly discussing the species problem for centuries were and are, to a degree, 'Boëthians'. According to this scheme, genus and species are logical judgments (predicables) within the framework of the classical logic of definitions: any object of the material and/or ideal world can be properly defined through the 'generic universal and species particular'. In this scheme, such 'genera' and 'species' do not have any fixed natural meaning; they just denote the sequential steps of logical divisions of respective notions from more to less inclusive.

Neo-Platonists were followed by Medieval scholastics, who affirmed the fundamental role of the genus–species scheme for the entire cognitive system based on the Aristotelean categories. They developed it as a universal scheme of describing a hierarchical structure of the qualitative diversity of nature by respective hierarchy of notions. It is clear that the same features – both fundamentality and universality – were ascribed to the notion of species as an important part of this scheme.

The genus–species scheme began to play an important role in describing diversity of living beings in the Herbal epoch of the fifteenth and sixteenth centuries. Indeed, the complete title of one of the major works of that epoch, ΠΙΝΑΞ Theatri botanici by Caspar Bauhin, clearly emphasised its priorities: 'A methodical description [of plants] according to their genera and species' (Bauhin 1623). However, herbalists had no clearly fixed biological connotation of those 'genera' and 'species'; therefore, Arthur Cain, in his study of Bauhin's conceptual system, preferred to designate his concepts as 'generoid' (as if genus) and 'speciate' (as if species) (Cain 1994).

Aristotelians of the sixteenth and seventeenth centuries dealing with the systematization of wildlife (Cesalpino, Jung, Rivinus, etc.) employed the scholastic logical genus–species scheme as a universal tool for elaborating consistent classifications of plants and by this, laid the foundation of biological systematics. For example, Joachim Jung began his opus De plantis speciebus with the assertion of a necessity of ordering plants into certain genera and terminal species based on a solid reason (Jungius 1662).[*] From then until the end of the eighteenth century, systematics took a leading position in the emerging life science; and it was engaged mainly in the elaboration of classifications of organisms by means of that scheme. This attitude led to the fixation of both fundamental and universal status of genera and species as the classification units, though still without an evident biological meaning.

The next important step was marked by a shift from the nominal (logical) to real (natural) treatment of species, which happened in the seventeenth and eighteenth centuries within the framework of the then prevailing Biblical worldview. John Ray, in his Historia plantarum species, explicitly substantiates the real existence of species by reference to their Divine creation (Ray 1686). Thereafter, Joseph Python de Tournefort, in his Élémens de botanique (Tournefort 1694), seriously modified the scholastic classification scheme by replacing its rankless hierarchy with a ranked one, in which species was assigned the status of a certain (seemingly biological) entity with fixed position in that hierarchy. Both of those ideas were consolidated in the

---

[*] Minelli (this volume) considers the fragment of Jung's text in question as evidence of his having recognised the ranks of genus and species.

mid-eighteenth century by the authoritative Carl Linné, who supported the real status of species by stating in his *Philosophia botanica* that 'the Creator has entrusted the generation of species to Nature, and not to men' (Linnaeus 1751: Section 259). It was noteworthy that Linnaeus' opponent Georges-Louis Leclerc de Buffon agreed with him on this point by beginning his small opus *De la Nature* with the statement that 'Species are the only creatures of Nature, eternal and unchanging like herself' (Buffon 1843: 52). Due to all these considerations, reality and fundamentality were ascribed both to the species unit and to the species rank.

In the nineteenth century, transformist natural philosophy, according to which the material world appeared not due to its Divine creation but as a result of its self-development, began to play a key role in maturation of the species concept. Indeed, the title of Charles Darwin's book *On the Origin of Species* clearly indicated that one of his main tasks was to explain how and why some species of living beings had supposedly turned into others in the course of biological evolution (Darwin 1859). This evolutionary interpretation gave a new natural philosophical support to an idea of the real (natural instead of logical) status of species as a natural unit. With this, however, Darwin deprived 'Linnaean' species of its flagged fundamental position in the hierarchy of life by equalising ontological status of species and geographic races as the sequential stages within the continuous process of microevolution (Beatty 1992; Stamos 1996, 2007; Ereshefsky 2009).

In the mid-twentieth century, adherents of the then most popular 'synthetic theory of evolution' regained the fundamental status of species by recognising speciation as a nodal point in the evolutionary process (Mayr 1963). Thus, in the framework of this evolutionary theory claiming to be both fundamental and universal, the same features appeared to be assigned to species: it regained its privileged position in the hierarchy of life. More recently, this view of species was supported by its recognition as a fundamental unit of biodiversity deserving special consideration within the framework of the current strategy of nature conservation (Claridge et al. 1997a; Wheeler et al. 2012; Sigwart 2018; Costello 2020; Padial and De la Riva 2020; Raposo et al. 2020).

Contrary to this, also in the mid-twentieth century, adherents of positivist philosophy of science proposed to exclude the species concept from the biological thesaurus as a metaphysically surplus and obsolete notion (Burma 1954; Michener 1962; Sokal and Sneath 1963). They were joined more recently by those biologists (conservationists, ecologists, cladists, etc.) who on their own grounds, rejected species category as having no sound biological meaning (Vrana and Wheeler 1992; Riddle and Hafner 1999; Hendry et al. 2000; Zachos 2016; Mishler and Wilkins 2018; Mishler 2021). This 'defundamentalisation' of the species category seems to revive the old scholastic rankless genus–species scheme in a new format and adds its own accent to the species problem.

12.3   WHY THE SPECIES PROBLEM?

The main reason for the ascent and persistence of the species problem in its traditional sense was indicated earlier: it is due to the existence of different species conceptions. However, it seems to be only a part of the problem in question. The entire cognitive situation construed around the general species concept (see Section 12.4) is problematic not just because of the diversity of particular conceptions as such but rather, due to the contradiction between the aspiration and the inability of biologists and natural science philosophers to reduce this diversity to a single species conception, be it the most general ('omnispective') or a privileged particular one (Pavlinov 2013a, 2017, 2018).

From historical and natural philosophical perspectives, the reasons for focusing on searching for a unified concept can be seen in the latent influence of Antique natural philosophy based on idea of the fundamental 'first principle' of nature. It was strengthened subsequently by the Biblical faith in that the Antique 'first principle' was actually the Divine plan. In the classical natural sciences, this has led to an aspiration to describe and explain everything existing in the world by a unified and thus, universal theory with its similarly universal notions and concepts. In classical biology, the universal Natural System of organisms appeared to be such a theory, one of the basic universals of which was the notion of species. Therefore, the classical paradigm was and is aimed at searching for such a species concept, embodied by a kind of 'absolute notion of species' (in the sense of de Pinna 1999) that would be universal for all groups of organisms; this is the so-called *species monism*. In this respect, there is not much difference between earlier adherents of an idea of species as a unit of the universal Natural System and contemporary advocates of treating species as a lineage fragment of the universal phylogenetic Tree of Life. Each of these standpoints presumes a certain 'final theory' with its universal conceptual framework (in the sense of Weinberg 1992), beyond which there should be no serious conceptual development but just some clarifications of its basic concepts and notions.

In non-classical philosophy of science, as it is defined by Ilyin (2003) and Stepin (2005), any search for such a 'final decision' is considered a pseudo-task. It is declared instead that any natural phenomenon (object or process) is too complex to be exhaustively represented by a single more or less finalised concept. This is because the latter is but a simplified representation of the phenomenon proper that is unable to represent it in an exhaustive form (see Section 12.5). So in our case, each particular species conception captures but a particular manifestation of species as a natural phenomenon (not in the sense of Wilkins 2018b; also in this volume) that can be aptly termed *natural species* (Kunz 2012). The latter is an inherent attribute of nature, so it is endowed with an ontological meaning; the *phenomenological species*, as it is defined by Sterelny (1999), seems to be its epistemological equivalent. At any rate, the most adequate 'omnispective' representation of such generally understood species can be gained by a combination of all particular conceptions. This presumes conceptual *species pluralism* as a normal state of (not yet existing) 'species theory', which is acknowledged by many authors, although on various grounds (Mishler and Donoghue 1982; Kitcher 1984; Ereshefsky 1992, 1998; Stanford 1995; Dupré 1999; Hull 1999; Pavlinov 2009, 2013a; Zachos 2016; Nathan 2017; Minelli 2020a).

Considering the conceptual species pluralism from a more general perspective, it is to be emphasised that this standpoint seems to be supported by two lines of argumentation. One of them is provided by the above-mentioned evolutionary epistemology, which presumes diversification of concepts as a normal state of any normally developing natural science (Hull 1988; Fadda 2020). It seems to be reasonable to suppose that biological systematics, to which scope the species problem mostly belongs, is no exception (Pavlinov 2018, 2021). Another argument is drawn by supposition that there are different 'kinds of species' peculiar to particular groups of organisms with their peculiar natural histories that are generalised by different species conceptions (Pavlinov 2009, 2013a, 2021; Mishler 2021) (see Sections 12.7 and 12.9).

Trying to put species monism and pluralism into some balance, it seems to be reasonable to distinguish between two major levels of conceptual generalisations about species. One of these is an idea of 'species in general', which is a legacy of that stage in the development of natural science when species was thought of as a part (element) of the natural-philosophically understood universal System of Nature. The latter became outdated subsequently, at least in its classical sense of a universal law of nature, but an integrating effect of the general *idea of species* remained. This idea can be designated informally as a *general concept of species*: it is not the same an integrated species concept of Giray (1976), nor is it the general lineage concept of de Queiroz (1999, 2005b), though they are identified by Ereshefsky (2010). A lower level of the generalisations in question corresponds to more or less formalised particular species conceptions representing various specifications of the general idea/concept, which are quite numerous. From this standpoint, the general species concept can be thought of as an 'umbrella' concept integrating ('sheltering') all its particular interpretations (Reydon 2005; Amitani 2017). So, one of the key issues in the species problem, as it is seen from a conceptualist pluralistic standpoint, is to understand interrelations between general and particular species conceptions within such a hierarchy (Mayden 1997; Naomi 2011; Pavlinov 2013a).

12.4 FRAMING THE SPECIES PROBLEM: COGNITIVE SITUATION

Modern conceptualism, coupled with cognitive science (Swoyer 2006; Velichkovsky 2006), acknowledges a fundamental role of three-partitioned *cognitive* (or *knowledge*) *situation* for the analysis of initial conditions, constituents, and processes of any cognitive activity (Wood 1940; Yudin 1997). In our case, we are talking about the cognitive situation developed with regard to both the species concept and the species problem in biology (Pavlinov 2013a, 2017, 2018, 2021). As this problem (according to the viewpoint of the author) does not exist outside the respective cognitive situation, any serious consideration of the former should begin with clarifying how the latter can be imagined.

Speaking formally, a three-partitioned cognitive situation is composed of ontological, epistemological, and subjective basic components. They interact with each other in a complex manner, and it is their interaction that features the structure of the cognitive situation. Aphoristically, the latter can be represented as a kind of *cognitive triangle*: its vertices correspond to these three components, while its edges signify their mutual interrelations (Pavlinov 2017, 2018, 2021).*

---

* There are several versions of the triadic representation of the things related to cognitive activity; some of them ascend to Pierce's semiotic triad, whereas some (for instance, in cognitive psychology, behaviorism, etc.) are more original (e.g., Surov 2002; Mechkovskaya 2007; Atkin 2013; Innis 2020). The concept outlined here does not pretend to be completely original; it is introduced just to emphasis a mutual interrelation between three basic components of a cognitive situation.

The *ontological component* is about *what* specifically is represented and studied in this particular cognitive situation; this 'what' refers to an object being cognised, species in our case. It is important to emphasise that the component in question, strictly speaking, is shaped not by natural species as such, i.e., as a natural phenomenon outside and besides a cognitive situation, but by its representation in the latter in a form of a certain concept designated by a certain notion. It is this general concept of species that frames the ontological substantive *conceptual space* constituting the basis for both theoretical and eventually, empirical research (Gärdenfors 2000). This general concept, according to one of its possible interpretations, can be considered a *cognitive model* (in the sense of Wartofsky 1979) of what natural species is or may be. Being a model, it functions as a kind of heuristic designed to arrange the diversity of organisms into an ordered set of certain units traditionally called species (Pavlinov 2013a, 2017; Wells et al. 2021).

The main task of developing the ontological component is to make the species concept an adequate representation of natural species, which simultaneously makes the model in question most reliable as an heuristic. Since this concept is obviously not a natural phenomenon itself (i.e., the 'species in nature') but its representation (cognitive model), an ontological gap arises between them (Williamson 2000), which is often not taken into account when both the general species concept and its particular interpretations are discussed. The competence of the ontological component includes consideration of the metaphysics of species: what does species make species, what is its ontological status (natural kind, quasi-individual, historical unit, or other), how does it relate to other manifestations of the structure of biodiversity, i.e., how 'species' differs from various 'non-species' (such as life form), etc.

One important ontological issue, which is among those most relevant to both the general species concept and the species problem, concerns the relation between the notions of species unit and species rank. They refer to different manifestations of the natural species phenomenon and therefore, are usually considered separately (Mayr 1963, 1970; Wilkins 2003, 2009, 2018; Bock 2004; Pavlinov 2013a, 2017; Zachos 2016). This standpoint resulted from adoptions of Woodger's (1937) suggestion to distinguish between classification units and classification categories. However, the concepts of species unit and species rank (category) are intercorrelated: in fact, species rank is attributed to a certain unit (i.e., species unit), while the latter is presumed to be of a certain fixed level of generality (i.e., species rank). Therefore, when either of them is considered, the other is always present in mind.

The *epistemological component* is about *how* – and why in such a way and no other – the natural species phenomenon can and should be represented and explored in the cognitive situation. Its main task is to develop some episteme as a means of comprehending and dealing with the phenomenon being conceptualised in one way or another. This episteme, in the most general case, includes (a) criteria, according to which this phenomenon can be acknowledged or not as really existing and knowable, (b) principles of cognition of species by means of respective definitions of its attributes, and eventually, (c) methods of distinguishing the particular species as certain units of biodiversity.

An irremovable occurrence of the epistemological component in the cognitive situation inevitably leads to an epistemic pitfall, which is caused by ignoring or downplaying the above-mentioned ontological gap (Williamson 2000). Due to this, according to one of the Zen Buddhism aphorisms, 'a person mistakes his finger pointing at the moon for the moon itself'. In our case, when biologists analyse the species diversity, they scarcely take into account that it is a particular species conception (a 'finger') that makes them see in nature a certain phenomenon they call 'species' (a 'moon'). Due to this, 'biologists often behave as if the species problem is solved, while they should be fully aware that it is not' (Amitani 2015: 289).

The significance of the *subjective component* of the cognitive situation is determined by the very fact that such a situation cannot exist without a knowing subject (Toulmin 1972; Haack 1979; Maturana 1988). It is the latter that builds and configures the entire situation by fixing, in one way or another, the two basic components and their interrelations. In our case, it is the subject (scientific community, research school, person) that is responsible for both elaborating and implementing particular species conceptions, according to which it is decided what species is and what it is not, whether it is real or not, etc. Thus, as the subject's influence is inevitably present in any scientific knowledge, one should not ignore its formative input into the species problem.

Therefore, taking into account the framing effect of the subjective component, one can say that the entire species problem is a by-product of a certain cognitive activity and is psychological to a certain extent (Hey 2001a, 2001b; Ellis 2011). Actually, this problem emerged as a fascinating 'mind game' of theoreticians, who 'have established a minor industry devoted to the production of new definitions for the term species' (de Queiroz 1998: 57). However, practitioners also contribute to this 'game' by requiring them to consider the species problem basically from a pragmatic perspective (Seifert 2014; Stijn 2018); from a conceptualist standpoint, this is but another side of an effect of the same subjective component.

The above-mentioned interrelation of these three components of the cognitive situation means that due to specific features of the knowing subject's activity, ontological premises depend to a degree on epistemic ones, and vice versa. This is formalised by the *principle of onto-epistemic correspondence*, according to which the basic (for the given cognitive situation) statements, relating to both ontology and epistemology, should be meaningfully compatible with each other and thus, interrelated (Pavlinov 2011, 2018, 2021). For instance, an epistemic requirement to deal with the observable entities only leads to a strongly reduced ontological worldview, in which there is no species in nature at all. On the other hand, if a basic ontological concept (premise) refers to the phylogeny as a primary cause of species diversity, then certain epistemic instruments are to be developed to allow just the phylogenetically defined species, and not any other kind, to be recognised. Considering a species concept as a cognitive model with certain heuristic functions endows it with both ontological and epistemological interrelated meanings.

To clarify how the cognitive situation is generally shaped by the interrelations stressed earlier, three basic modes of ontologising the natural species phenomenon should be flagged. According to *nominalism*, species does not exist at all; it is only a cognitive artifact. According to *realism*, the species phenomenon certainly exists, and, moreover, its existence does not require any particular justification. According to *conceptualism*, an initial assumption asserts that there exists a multifold structure of the diversity of living matter, organised in a complex way, one of whose elements is species. To recognise the latter and to distinguish between 'species' and 'non-species' phenomena, a certain theoretical construction – namely, a general concept of species – is needed (Reig 1982; Pavlinov 2009, 2017; Richards 2010). Particular conceptions result from subjects focusing attention on different aspects of the general concept, so that they recognise different species units within that multifold structure. Something between these basic modes is *bionominalism*, borrowing a little from each: it presumes that there are various levels of integrity and reality from highest (organism) to lowest (biota), with species taking some intermediate position (Mahner and Bunge 1997). Considered conceptually, the ontology of the species phenomenon might be thought of as a 'cross-section' of the properties of two classes of objects, in one of the terminological systems termed 'fiat' (artifactual) and 'bona fide' (natural) ones (De Florio and Frigerio 2019).

The above-considered distinction between two basic levels of species concepts, general and particular, makes it reasonable to distinguish between similar levels in the structure of overall cognitive situations. One of these corresponds to particular rather simply construed individual situations, each dealing with a particular species conception. So, the latter can be thought of as a focal center of the respective cognitive triangle. Another kind corresponds to a common meta-situation that combines particular ones, just as particular species conceptions are combined some way by the general one. The respective 'meta-triangle' has in its focal center not only the general concept of species but also the species problem. The latter does not evidently exist in any one of the particular situations but encompasses all the uncertainties generated by attempting to combine them.

12.5   REDUCTION CASCADE OF SPECIES CONCEPTIONS

According to one of the principal ideas put forward in the previous section, any multi-faceted natural phenomenon is represented in a cognitive situation not by itself but by its cognitive model in the form of a certain more or less formalised concept. Another important idea is that such a model (concept) is simpler as compared with the respective natural phenomenon proper being modeled (conceptualised). Thus, construing such a cognitive situation involves a certain ontological reduction, which means both 'cutting' a phenomenon to be investigated (species in our case) from its natural 'environment' and simplifying it. This reduction is an obligatory part of the epistemic component of any cognitive situation.

Epistemological justification of ontological reduction is of two kinds. On the one hand, it is due to the limited ability of human cognitive means emphasised earlier. On the other hand, its objective is to elaborate an operational conception as a tool for the analysis of species diversity in nature. From this perspective, 'a species concept that unambiguously reflects one aspect of variation may be preferable to one that ambiguously reflects several things' (Mishler and Donoghue 1982: 495). The general lineage concept suggested by Kevin de Queiroz seems to be the one that meets this criterion (de Queiroz 1998, 1999, 2005b).

The entire reduction operation begins not with species itself but at a higher level of generality and can be represented in the form of a deductive (descending) step-wise reduction cascade. It begins with singling out living matter from the Universe; it becomes biota as a certain structured, functioning, and developing unit. Then, some process or structure is identified within the biota, a participant or part/element in which is supposed to be species; for example, it may be either the evolutionary process or the functional structure of ecological communities or the stationary structure of biodiversity. In the next step, 'species in general' is singled out as a supposed natural phenomenon, with any other natural 'non-species' phenomena (biomorphs, syntaxa, guilds, etc.) possibly involved in those processes and/or structures being discarded. The next is reduction of the general species concept to certain particular conceptions, each defined with reference to the particular causes shaping species diversity; these may be evolutionary or phylogenetic or ecological conceptions. The whole cascade proceeds by the subsequent reduction of these conceptions to far more particular ones; for instance, the lineage conception can be singled out within the phylogenctic one.

At each step of this reduction cascade, a cognitive model (in the form of a particular conception) is elaborated for a particular phenomenon (biota or inert matter, evolution or biodiversity, species or biomorph, etc.) by indication of (a) its significance for the process or structure fixed at the respective step and (b) its own fundamental properties. All this is evidently justified by certain theoretical considerations. For example, first comes speculation about the System of Nature or about evolutionary process or about ecosystem interactions or about something else that might be responsible for functioning and structuring of biota. Within each theoretical context thus defined, it is then justified why it is species that should be considered as a fundamental unit of the System of Nature or a phyletic lineage or an element of the ecological communities, etc. This makes the entire reduction cascade evidently conceptualised from the very beginning to the end.

Nature is a hierarchical system organized in a complex way, in which constituting elements and processes of different levels of generality interact with each other both 'horizontally' (within each level) and 'vertically' (between levels). It is evident that the more such interactions are 'cut off' by succeeding reduction steps, the less a conception thus construed retains reference to the properties of real nature. Consequently, the more significant and noticeable become inputs of the epistemic and subjective components of the cognitive situation to the entire ontological reduction. Therefore, the more steps are involved in the reduction cascade leading to a particular species conception, the less the latter is 'natural' (referring to something actually existing in nature) and the more it is 'artificial' (referring to something that does not exist outside a particular cognitive situation). This means that the longer the cascade, evidently, the wider is the above-indicated ontological gap between an idea of the natural species phenomenon and the particular species conceptions.

Looking at this reduction cascade from an ontological perspective, the complex evolutionary species conception of Simpson can be thought of (from a certain theoretical standpoint, of course) as corresponding to something really existing in nature. With regard to the genealogical species conception of cladists, such an assumption is more than doubtful; actually, scarcely anything like genealogical lineage as such exists in nature, simply because genealogy as such, i.e., grasped from a more broadly understood evolutionary process of self-development of biota, does not exist. Of course, each of these two conceptions (taken only as examples) can serve effectively as a heuristic in the respective cognitive situation; however, an intriguing metaphysical question remains as to whether there anything exists in nature that a particular reductionist conception is designed to signify.

However, when considering the same cascade from an epistemological perspective with its pragmatically defined tasks cognitive activity, we face a serious *problem of NP-completeness*, as explained by Garey and Johnson (1979). In the case under consideration, it means the following: the more content-rich and, therefore, complex a particular species conception is, the less attainable appears a classification that might adequately represent the manifestation of the species diversity presumed by this conception. Therefore, again, a more reductionist and, therefore, more operational species conception is preferable as compared with a more content-rich one.

12.6   MULTIPLICITY OF SPECIES CONCEPTIONS: CONCEPTUAL PYRAMID

As far as the above-discussed reduction means simplification, there is a potential for any complex natural phenomenon to be represented by several cognitive models that are adequate from certain theoretical standpoints (Wartofsky 1979). Thus, 'species in general' can be reduced to either ecological or phylogenetic species conceptions, and the latter can be decomposed into narrower conceptions depending on the treatment of species as various fragments of the phylogenetic tree (Mayden 1997; Baum and Shaw 1995; de Queiroz 1999; Mishler and Theriot 2000; Kornet and McAllister 2005; Zachos 2016; Mishler 2021). Therefore, the reduction cascade leads inevitably to a consequential multiplication of particular species conceptions: the more general and inclusive are fewer in number than the more particular.

Accordingly, such a hierarchy of species conceptions generated by a reduction cascade can be shaped in the form of a hierarchical *conceptual pyramid* as another representation of the cognitive situation dealing with the species problem. It formalises an idea of Richard Mayden about hierarchical relationships between species conceptions of various levels of generality (Mayden 1997, 1999; Naomi 2011; Pavlinov 2013a; Zachos 2016). This hierarchy presumes that higher-order conceptions serve as a kind of conceptual framework for respective lower-order ones, thus making them *framework-relative* (Bartlett 2015). The highest level of this pyramid corresponds to a concept of the most general order providing an understanding of what natural 'species in general' is or may be. Its lower levels include more particular theoretical conceptions, for example, generative or ecological ones. Each of them is then specified by adding certain details to its definition – for example, different interpretations (phylogenetic, genealogical, etc.) of the generative species conception (Wilkins 2009, 2018). The lowermost levels include operational conceptions that elaborate certain criteria to delineate particular species units for given particular empirical data. This kind of conceptualist interpretation of a 'pyramidal' hierarchy of species conceptions leads to an inevitable recognition of their plurality and shows that the monistic reduction of the species problem (Lam 2020) is untenable.

It is of prime importance to emphasise that at each level of generality (save for the highest one) of the conceptual pyramid thus construed, the lower-level species conceptions arise not by themselves but as particular interpretations of the higher-level conceptions. These are such interpretations that make respective particular conceptions meaningful according to a certain biologically sound conception or theory operating at higher levels of generality. Correspondingly, without reference to the latter, any particular species conception seems to be introduced *ad hoc* without serious justification. This leads to an important conclusion concerning operational species conceptions: it is the theoretical ones that make them biologically sound. Therefore, empirical conceptions of the least recognisable or least inclusive unit (Cracraft 1989a; Pleijel and Rouse 2000), though preferable from a pure pragmatic standpoint (Brasier 1997; Claridge et al. 1997b; Seifert 2014, 2020), seem to be 'biologically empty' in so far as they do not refer to any biologically sound metaphysics. The conception of the smallest named and registered clade (SNaRC) (Mishler and Wilkins 2018) seems to be an example of the validly interpreted operational unit as far as it refers to phylogeny. So, figuratively speaking, it is the particular theoretical conception that 'dictates' to a researcher how empirical data should be analysed and which units of biodiversity should be recognised as species. This conclusion, followed from the above-mentioned framework-relative epistemology (Bartlett 2015), agrees with the general conceptualist idea of the theory-ladenness of any empirical observations in natural sciences (Quine 1969; Carrier 1994).

Indeed, when a certain group of organisms is recognised by a certain classification method, on what basis should it be treated as a species? Undoubtedly, on the basis of a biologically meaningful theoretical species conception underlying that method; otherwise, it remains unclear why one should identify such a group as a species. On the other hand, when different groups are recognised by different specialists based on different (or occasionally, the same) classification methods, what makes these groups comparable as just species? Again, a particular species conception implied by the respective specialists and justifying both the methods employed by them and the interpretations of the obtained results. Thus, each species conception functions in empirical studies as a kind of integrating factor bringing a common 'denominator' into a set of results of different empirical investigations. With this, it should not be obligatory for such a conception to be stated in an explicit form; in fact, it may belong to an implicit 'intuitive ontology' shaped by education, belonging to a certain scientific tradition, etc. That is why, as Charles Darwin said, 'every naturalist knows vaguely what he means when he speaks of a species' (Darwin 1859: 37). Such an empirical position is justified philosophically by the conception of 'species as phenomenon' (Mishler and Wilkins 2018; Wilkins 2018b, 2021, in this volume). However, without such a conception, in whatever guise, any pure empirical findings, be they results of personal perceptions or numerical calculations, may provide classificatory units with drastically different biological meaning, such as species or biomorphs or developmental stages, or others. To distinguish between them, a certain conceptual background is needed, without which the 'practicalities of doing natural science' (Wilkins, in this volume) have scarcely any scientific meaning.*

This conclusion becomes more evident if we consider exploration of species diversity as advancing and testing particular scientific hypotheses about the structure of this diversity. From a conceptualist perspective, no one such hypothesis can be meaningfully elaborated if it has no conceptual framework in its background (Popper 1959). In our case, any species classification treated as a taxonomic hypothesis (in the sense of Cartmill 1981; Pavlinov 1995), to be scientifically sound and substantively meaningful, should be developed within the framework of a certain theoretical species conception referring to a certain natural biological phenomenon. Whatever this particular conception might be – whether 'biospecies' or 'ecospecies' or 'phylospecies' or else – it should at any rate be implied. This makes any biologically sound particular species classification context-dependent and eventually, theory-laden.

12.7   DEFINING SPECIES

Accordingly to the conceptualist standpoint in its extreme expression, 'A species [...] is exactly what a given definition claims it is' (Bartlett 2015: 33). With this, one of the profound meanings of the above-discussed conceptual pyramid is that it implies certain limitations to the fundamental issue of defining species as a natural phenomenon. These limitations seem to be in effect regardless of the particular standpoints from which we consider the latter, be they either epistemic or ontic (Barker 2019); they face the following epistemic pitfalls.

One of them is imposed by the above-considered hierarchical genus–species scheme, according to which any particular conception – in our case, that of species as a biological entity – is to be consistently defined logically as a 'species particular' in the context of the respective 'generic universal'. From this perspective, the most popular definition of species as a 'group of organisms' (either just similar or cross-breeding or other) cannot be considered logically consistent, as it refers not to a certain higher-level 'generic universal' (evolutionary process, structure of biota, etc.) but to a lower-level 'species particular' (organisms, life cycles, etc.). Unlike this, say, definition of species as a phylogenetic lineage looks more consistent, as it presumes reference to a higher-level phenomenon, i.e., phylogenetic pattern. Thus, following this argumentation scheme, it seems that no species conception of a certain level of generality can be rationally defined outside a framework context provided at a higher, more inclusive level within the respective conceptual pyramid.

In this regard, a question inevitably arises: how to elaborate a certain general species concept that would be more 'constructive' than the general idea of species and

---

* When supposing that 'Linnaeus and his predecessors named species, in the absence of much scientific theory', John Wilkins seems not to take into consideration the rich metaphysical context of Linnaeus' conception of the Natural System and genera and species as its natural units, which served for him as a background theory (Pavlinov 2021b).

provide at least a provisional and rough definition of species as a natural phenomenon. This question seems to be similar in a way to the one imposed by David Hull about an 'ideal' species concept (Hull 1997). If the earlier argumentation referring to the genus–species scheme is correct, then a valid answer to this question can hardly be obtainable within a conceptual pyramid with the general species concept taking its top position. It is evident (at least to me) that in order to answer this question correctly, it is needed to raise the pyramid's highest level of generality to consider natural 'species in general' within the context of some more general biological concept/theory. Otherwise, we will still have to operate with a non-defined general notion as a kind of undefinable 'unique beginner' (in the sense of Berlin et al. 1973).

Such an argumentation scheme presumes a necessity to denote, first of all, such a 'logical genus', by which division it would become possible to get 'logical species' containing a general definition of the natural species that biology deals with (Pavlinov 2013a, 2018, 2021). Keeping in mind that species are not uniquely real biological entities (Mishler 2010) and in order to avoid a logical 'genus–species' tautology, it seems necessary to define, in the same more general context ('logical genus'), some other biological units (other 'logical species') of the same level of generality as natural species proper, but which are certainly not species. Such an approach is aimed at understanding what the species phenomenon is and how it differs from any 'non-species' phenomena. Otherwise, it appears impossible to decide positively why we think of a particular phenomenon as being species and not anything else. For instance, it is important to distinguish between species and life form (biomorph), which is another significant unit of biodiversity (Chernov 1991; Pavlinov 2007).

If this consideration is correct, then the following question becomes of prime importance: which natural phenomenon should be considered the 'logical genus', with reference to which structural/functional units of the same level of generality, including species, life forms, syntaxa, guilds, etc., could be defined as its various 'logical species'. Taking into consideration that both 'species' and 'non-species' units are thought of as the elements of evolving and structured biota, it seems reasonable to fix a framework concept of the latter at the top of the respective conceptual pyramid. In this regard, treating biota as a developing non-equilibrium system seems very attractive: it allows one to place the main emphasis on those natural causes (factors) that operate at the level of biota and structure it as it develops and functions, thus generating and individualising its various structural supra-organismal units (Pavlinov 2013a). The joint co-action of these causes yields *dynamic stability* of both biota and its various structural-functional units, including species, as one of their most fundamental properties to be comprehended (Brooks and Wiley 1986; Kendig in this volume).

So, to get a metaphysically consistent understanding (and eventually definition) of natural 'species in general', the entire conceptual pyramid should be construed as a descending cascade of the causes that structure biota and provide dynamic stability of species. Such a standpoint would yield a causal and by this, explanatory understanding/definition of the species phenomenon aimed, first of all, at comprehension of the very existence of this phenomenon. This means that any substantive species conception, to be biologically (ontologically) sound, is to be '*cause-focused*' in the sense of Matt Barker (2019, in this volume). There are several general causes (external and internal, initial and proximal, etc.), and they interact in a complex manner with each other, so no one of them, taken separately, can explain the natural species exhaustively. Particular cause-focused species conceptions accentuate particular causes (ecological, historical, etc.) considered the most important for individuation of species (Ereshefsky 1992; Pavlinov 2013a; Barker 2019). However, if a multi-causal nature of 'species in general' is acknowledged, then its sought definition should be not reductive, by indicating only one particular cause, but as exhaustive (omnispective) as possible to incorporate all causes ensuring its existence (Sluys 1991; Wilkins 2009, 2018a; Pavlinov 2013a; Nathan 2017). With this, it seems that a conjoint consideration of these causes responsible for the dynamic stability of species units in different groups of organisms may become one of the main issues in (non-existing) eidology. On the one hand, this would provide a comprehension of what makes 'species in general' just the species and not some other unit of the structure of developing and functioning biota. On the other hand, this would allow us to recognise more clearly particular causes responsible for particular manifestations of the natural 'species in general' reflected in respective particular conceptions.

An attempt to transform a general understanding of species, satisfactory enough from a metaphysical standpoint, into a more or less rigorous species definition faces another epistemic pitfall. In this case, the latter has the form of the principle of *inverse relation between the rigor and meaningfulness of concept definition* (Voyshvillo 1989): the more strictly a concept is defined, the less likely there is something in nature to which it may correspond. A good example is provided by the definitions of mathematical objects: as Albert Einstein asserted it, 'as far as they are certain, they do not refer to reality' (cited after Hempel 1956: 1646). In the case of species, a 'negative' effect of this principle is evident when different rigorous interpretations of species as a historical unity are involved (Hull 1997; Hey 2001b; Pavlinov 2009, 2013a; Kunz 2012): all attempts to make its definition more rigid lead to the introduction of more narrow conceptions (cladospecies, apospecies, paraspecies, etc.; see Wilkins 2018a; Mishler 2021 for review) departing from the initial general idea and thus becoming less 'natural'.

A more profound vision of this epistemic pitfall is provided by the *principle of incomprehensibility* of knowledge of particular natural phenomena, according to which a complete and exhaustive formalised knowledge, in the form of an array of rigid definitions, is impossible in natural sciences (Minati 2019). In our case, it means that any species conception is and will remain more or less imprecise and vague, because it seems to be impossible to give it a proper rigid definition due to an irremovable ontological gap mentioned earlier (see Section 12.4). Indeed, if it appeared unfeasible to elaborate unambiguous and precise definitions for homology, gene, organism, ontogenesis (Hall 1994; Rheinberger 2000; Folse and Roughgarden 2010; Brigandt 2012; MacLeod 2012; Pavlinov 2012; Minelli 2014; Pradeu et al. 2016), why should we chase such definition for species, which is certainly no less complex a natural phenomenon than those just indicated?

As far as 'species in general' is concerned, a search for its metaphysically satisfactory non-reductionist definition run into one more pitfall. As was presumed earlier, different manifestations of the natural species phenomenon, as they are captured by different particular conceptions, result from different, not very correlated causal factors, of which most the evident are historical and ecological. Therefore, they appear to be complementary and linked by a kind of uncertainty relation understood in a broad epistemic sense (Armand 2008). In our case, it means that fixing unequivocally any one of those aspects of natural species entails the impossibility of fixing its other aspects with an equal unequivocality (Pavlinov 2009, 2013a). Due to this, incorporating all of them into one 'omnispective' species definition makes the latter a rather long list of the properties supposed to be essential for species (Zavadsky 1968). This is one of the evident causes of the above-mentioned (see Section 12.5) preference for simple and therefore, more operationally definable conceptions over more sophisticated ones.

An apparently reasonable means to circumvent these epistemic pitfalls is provided by realising that any definition of species as a rather complex natural phenomenon, claiming to be meaningful biologically, is deemed to be non-rigid semantically. Because of this, the epistemic component of the cognitive situation dealing with the species problem should include some elements of fuzzy logic. One of the latter's important features is that it formalises the possibility of operating with non-rigid context-dependent definitions (Kosko 1993). In a conceptual space 'fuzzily' understood, a conception is represented by a probabilistic distribution of its possible particular meanings, with the latter being fixed contextually in particular cognitive situations (Nalimov 1979). Employing this logic presumes that when defining species by means of a particular conception, it makes sense to fix somehow the latter's 'core' and not to try to formalise its 'periphery'. Consequently, different species conceptions of the same level of generality (e.g., different interpretations of generative conception) will inevitably overlap by their semantic 'peripheries'.

A fuzzy (imprecise) character of species definitions entails the unfeasibility of strict and unambiguous applications of respective conceptions in the studies of biodiversity. First, such definitions cannot be invariably applied to all groups of organisms; their explications seem to be dependent on the contexts set by biological specifics of these groups (Pavlinov 2013a, 2017, 2021; Maxwell et al. 2020; Mishler 2021). This conclusion follows from a metaphysical supposition that the essential properties of species as a natural phenomenon, whatever they might be, do not exist by themselves but are constituents of the whole natural history of organisms (see Section 12.9). Besides, definition fuzziness leads to an impossibility, in some groups, of unambiguous decision as to how interpret respective units of the structure of their diversity. For example, in fungi, it is usually difficult to distinguish unambiguously between the species, be they defined either reproductively or genealogically, and the life forms (biomorphs) defined mainly morpho-ecologically (Parmasto 1986; Brasier 1997). For the same reason, various species units distinguished in different groups on the basis of apparently the same conception may not strictly coincide in a biologically meaningful way.

It is to be noted in the margin that a kind of fuzziness should be attributed not only to the conceptions but also to the concrete species units recognised in practical research. Regardless of any presumed conceptions, such species are 'fuzzy' just because of variation of the traits used for their delineation, which results in the *empirical species uncertainty* (Hey et al: 2003; Maxwell et al. 2020). Such fuzzy treatment of species units and their boundaries seems to be universal in its application to all groups of organisms (Knowlton and Weigt 1997; Sites and Marshall 2004; Pavlinov 2013a, 2017, 2018; Zachos 2016, 2018), but especially to viruses and prokaryotes (Van Regenmortel 1997; Hanage et al. 2005; Hanage 2013). It is opposed to the so-called 'xenotaxonomy' that treats species as absolutely discrete classification units (McCabe 2008).

12.8   A NOTE ON FUZZY SPECIES RANK

The fuzzy vision of both species conceptions and species units makes the entire species problem 'fuzzy' with all its particular ingredients, including species ranking. This implies that the latter should not be treated as rigorously and discretely ordered, as in the classical 'Linnaean' hierarchy, but rather, as more or less blurred. Indeed, it seems illogical to anticipate that any species rank definition might be less fuzzy than any species unit definition. Such an attitude would make surplus most of the debates concerning precise fixation of the 'neighbor' ranks, be they species, mesospecies, semispecies, allospecies, superspecies, etc., as far as they cannot be both unambiguously defined in theory and distinguished in practice. So, it might be more reasonable to speak about some 'around-species' ranks in a fuzzy hierarchy rather than about discretely distinguished particular ranks to which particular units should be definitely allocated following the 'either–or' rule begotten by a 'discrete thinking' (Mikhailov 2003; Pavlinov 2017; Pfander 2018).

But, it seems to me that such fuzziness of both the whole hierarchy and particular ranks in it by no means implies something like rankless hierarchy of such 'around-species' units of different levels of generality. Indeed, if one cannot distinguish unequivocally between, say, low mountains and high hills and other things rising above the plain, this does not mean that these descriptive landscape categories are of neither conceptual sense nor practical use in physical geography (Wilkins 2018b). Another substantiation of a system of (fuzzily) fixed ranks employed by taxonomists for description of the hierarchical structure of biodiversity may be its analogy to the system of Cartesian coordinates used to represent and analyse some aspects of diversity patterns in many branches of natural science (Lyubarsky 2018). Hardly anyone would insist that the latter system is a natural phenomenon, but hardly anyone would reject its great value as a kind of reference system; so it seems to hold for a ranked hierarchy of the 'around-species' units.

12.9   AN ELUSIVE AND EVOLVING SPECIESHOOD

It seems to be clear from these considerations that attempts to find a metaphysically sound general understanding and eventually, definition of natural species seems to plunge the entire species problem into the context of 'new essentialism' quite pertinent to modern biology (Ellis 2001; Devitt 2008; Wilson et al. 2009; Rieppel 2010; LaPorte 2017; Maxwell et al. 2020). This means acknowledging that if species is not reducible to just a sum of its constituent organisms, it must be endowed with something like a certain emergent intrinsic property (Sober 1980; Devitt 2008), even though of a fuzzy character. This intrinsic property, considered as a whole, may be informally designated as *specieshood* of whatever content, distinguishing the species from any 'non-species' with their own essences or '-hoods' (Pavlinov 1992, 2009, 2013a, 2018, 2021; Griffiths 1999; Wilkins 2007; LaPorte 2017; Barker 2019).

From a metaphysical perspective, one of the main issues of the current species problem becomes searching for such specieshood, i.e., a kind of unknown and elusive essence of natural species. So, Boëthius' aphorism (see Section 12.2) seems to be no less relevant today than in his time; though one cannot forget, as well, that Charles Darwin had warned of 'the vain search for the undiscovered and undiscoverable essence of the term species' in his *On the Origin of Species* (Darwin 1859: 485).

The first noticeable step toward a biologically substantive understanding of specieshood seemed to be defining species as a syngameon at the very beginning of the twentieth century (Poulton 1904). It was subsequently transformed into a biological (actually, interbreeding) species conception by Dobzhansky–Mayr based on an amphimictic breeding system. With this, it became clear that for different organisms possessing different systems of reproduction, there should be different species conceptions referring to different 'kinds of species' (Romanes 1895; Cain 1954; Mayr 1963; Richards 2010). Their multiplicity was emphasised in botanical biosystematics with its very detailed classifications of such 'around-species' units (Sylvester-Bradley 1952; Dean 1980; Stace 1989).

Linking hypothetical specieshood to the species reproductive system, viewed from a more general metaphysical perspective, turns out to be quite important for understanding its possible biological meaning. Such perspective presumes consideration of the specieshood, however conceptualised, as an *integrated part of the overall natural history of organisms* (Pavlinov 2013a, 2018, 2021). It incorporates, in a certain unobvious way, particular mechanisms responsible for the dynamic stability of particular species – their self-reproduction and mutual isolation, their place in the niche structure of ecosystems, their persistence as genealogical lineages, etc.

Considering species as an element of the functional structure of evolving biota, and specieshood as a part of the overall natural history of organisms, seems to mitigate a contradiction between the interpretations of species as entity and process, or between intrinsic and relational treatments of the species essence (see on this, e.g., Okasha 2002; Devitt 2008; Barker 2010; Dupré 2020). The particular species exist in the whole of the networks of interactions, making biota in general and local ecosystems in particular the holons (in the sense of Allen and Starr 1982). So, they are integrated and individuated both intrinsically by species-specific interactions among their constituent organisms and their own properties, and extrinsically by species-specific interactions with other species and their abiotic environment.

This understanding evidently yields an assumption that specieshood depends, to a greater or lesser degree, on other aspects of the integrated natural history of organisms. This inevitably makes it group-specific: even if we suppose that some 'species in general' may be endowed with certain emergent properties common to all (or to the vast majority of) living beings, it may have different manifestations depending on the particular biological properties of the particular groups of organisms. So first, the existence of different 'kinds of species' associated with different breeding systems is but one of the outcomes of such group-specific specieshood manifestations. And second, it is reasonable to suppose that the above-stressed 'fuzziness' of particular species definitions (see Section 12.7) may in part be a consequence of this dependence, as it is hardly possible to demarcate, with needed certainty, different group-specific modes of natural history.

Taking this dependence for granted, it might be assumed next that the presumed specieshood manifestations change together with other biological properties of organisms in the course of biological evolution along with the development of functional and structural organisation of biota in general and its various elements in particular (Pavlinov 2013a, 2018, 2021). So, we have here something like an 'evolving essence', which might be pertinent to the so-called 'historical essentialism' (Griffiths 1999; Pedroso 2012; Maxwell et al. 2020) and responsible for a steady dynamics of the 'species as process' (in the sense of Dupré 2020, in this volume) and making the latter a kind of phylocreod (in the sense of Waddington 1962) as a 'historical entity' (as the latter is understood, say, by Wiley 1978; Kluge 1990; Velasco 2008). Among other things, this means a supposition that the biological mechanisms responsible for the dynamic stability of the species and constituting their respective specieshoods were quite loose at the beginning of the biological evolution. At its end, more perfect intrinsic mechanisms shaping the specieshood became developed in more advanced organisms, thus making respective species units more cohesive and discrete (Dobzhansky 1970; Eldredge 1985; Brooks and Wiley 1986; Ghiselin 1989). So, the loosely and fuzzily integrated 'around-species' units of viruses, prokaryotes, and lower eukaryotes may properly be termed *quasispecies* or *pseudospecies* (Dobzhansky 1970; Eigen 1983; Nowak 1992; Van Regenmortel 1997; Domingo 2002; Stamos 2003; Wilkins 2006; Ereshefsky 2010; Andino and Domingo 2015; Dupré in this volume), though a 'full-species' status may be ascribed to them occasionally with some reservations (Bobay and Ochman 2017). In contrast to these, the more integrated units that emerged at the final phase of the evolution of the specieshood may be termed *euspecies*.* Quasi-individuality of the former seems to be provided mostly extrinsically by specific ecological niche occupancy, while that of the latter is supported greatly by certain intrinsic mechanisms of within-species interactions (Brooks and Wiley 1986; Van Regenmortel 1997; Richards 2010; Pavlinov 2013a). One may suppose that the existence of 'bad' and 'good' species recognised by specialists (Amitani 2015, in this volume; Wilkins in this volume) is in part explained by these differences between particular manifestations of the evolving specieshood. Such consideration is relevant to the 'species-as-status' conception of Reydon (2005, in this volume; Reydon and Kunz 2019): in fact, all species units reaching a certain level of evolutionary independence may be ascribed such status, but their particular specieshoods would be different.

---

* This term has been coined to designate a particular step of the speciation process (Dillon 1966), but it seems to be quite apt to use it in this more general sense.

It is to be noted that it might be an important task of comparative eidology/eidonomy to study the distribution of various kinds of species, with their specific specieshoods, among the groups of organisms with different modes of natural histories. Resolving this task would be aimed at revealing whether there are indeed any intercorrelations between these '-hoods' and 'histories', which would provide a more reasonable ground to discuss possible causes of the evolution of both these intercorrelations and specieshood itself.

How might the attempts to uncover still quite enigmatic and elusive specieshood bear on general understanding of what the natural species phenomenon is or may be and accordingly, what might the entire species problem be at the next stage of the latter's conceptual development?

Were theoreticians lucky enough to reveal the sought specieshood, they would be within their rights to declare that 'species in general', as a universal natural phenomenon common to all (or at least to the vast majority of) living organisms, does actually exist, albeit in different manifestations. It seems to be reasonable to suppose that such substantiation of fuzzy species unit would simultaneously mean substantiation of fuzzy species rank in the fuzzily structured biota that cannot be reduced to either lower or higher ranks.

On the other hand, the failure to reveal any (quasi-)universal specieshood would mean that there is no species as a (quasi-)universal natural phenomenon inherent in biota. Instead, there are essentially different units of the functional-structural organisation of biota intrinsic to the different groups of organisms with different natural histories. Such units should be formalised not as 'species' but rather, as different 'species-like' units with different particular '-hoods'. In this case, it might appear reasonable, at least in the biological research, to reject the traditional general notion of species and to invent new terms to denote individually every such 'species-like' biological entity (Dubois 2011; Nathan 2017). Such failure would most probably lead also to rejection of the fixed ('Linnaean') species rank traditionally understood, so an idea of the rankless 'around-species' hierarchy may gain more support. With this, the general notion of species could be preserved for its original (inherited from scholasticism) classification usage (Ereshefsky 1992, 1998; Mahner 1993; Baum 2009), which could probably make such 'taxonomic species' a kind of the above-mentioned 'umbrella concept' for all others with their particular biological '-hoods'.

The contemporary usage of the general notion of species in classifications usually presumes that it is endowed with a certain biological meaning. However, some such usages in certain fields of explorations of biological diversity seem to be more properly called 'logical species'. The first to come to mind is the practice of 'parataxonomy' with its 'parataxa' or 'form-taxa' that cannot be directly associated with 'eutaxa' thought of as biologically sound units (Krell 2004). The diversity of such (mostly fossil) entities as spores in botany or traces, eggs, and burrows in zoology is routinely described based on delineation of 'species' and 'genera' as certain parataxonomic units without evident biological connotation (Donovan 1994; Hughes 2006).[*]

---

[*] In this regard, they are probably analogous to 'species' recognized in chemistry and mineralogy (e.g., Holleman 1953).

Another field, where such 'species-like' parataxonomic units as a rather formal classification tool are usually recognised, is 'turbo-taxonomy', designed for accelerated description of the diversity of extant organisms (Krell 2004; Riedel et al. 2013). Recognition of the so-called 'metagenomic species' (MGS, MOTU) based on the environmental DNA data (Ficetola et al. 2008; Sunagawa et al. 2013; Dadi et al. 2017; Milanese et al. 2019; Marcelino et al. 2020) seems to be also nearly formal, as these units cannot be associated with certain biological entities traditionally called species.

12.10   HOW TO DEAL WITH SPECIES PLURALISM?

Thus, the conceptualist standpoint outlined earlier leads to acknowledgement of the irreducible plurality of species conceptions and adoption of species pluralism (Pavlinov 2009, 2013a; Bartlett 2015), even if the latter term is considered ambiguous by itself (Nathan 2017). This conclusion obliges us to touch briefly on two basic outcomes, 'speculative' philosophical and 'constructive' operational.

The most important issue in philosophical considerations of species pluralism concerns the problem of reality versus nominality of species as a natural phenomenon (Stamos 2003; Wilkins 2009, 2018; Claridge 2010; Richards 2010; Pavlinov 2013a). Adherents of the strong dichotomy of these two categories take inability to elaborate a unified rigid and with this, deliberately operational definition of species for an convincing argument for its nominality, and this yields one radical solution of the species problem (Stanford 1995). Consideration of species units as natural kinds that can be delineated differently by researchers using different characteristic features to resolve different tasks also serves as another indirect support of the nominalistic viewpoint (Kitcher 1984). However, as considered earlier, the conceptualist standpoint presumes a more flexible look at this particular problem: indeed, our inability to define a multifold something may point just at our inability and not at the inexistence of that 'something'.

A 'constructive' issue is aimed at attempts to turn the species problem into a resolvable task. Its main content, from this perspective, could be set properly by a question as to how to combine together different conceptions defining essentially different 'around-species' or 'species-like' units without eliminating any one of them from the respective conceptual space. One of the possible solutions of the species problem thus reformulated may appear to be the development of something like a *faceted classification* (see Kwasnik 2000 and Broughton 2006 for the latter). The latter provides a possibility to combine different particular species conceptions, based on different causalities, into a single pool by means of an appropriate meta-language with an exhaustive substantive ontology for a general cognitive meta-situation fragmented by the particular conceptions that make it a kind of 'a collection of interacting patchwork structures' (Novick and Doolittle 2021: 72). One of the main objectives of developing such a meta-language is to make various particular conceptions (conceptual 'patches') mutually interpretable within the framework of common conceptual space. This seems to allow us to overcome in some way the problem of substantive, or 'taxonomic' (in the sense of Sankey 1998), *incommensurability* (or ontological disunity; see Reydon in this volume) of particular species conceptions addressing different biological properties of 'species in general'.

For such faceted classification of species conceptions to be theoretically well-founded, instead of being just a 'basket' to gather them for a certain pragmatic aim, it would be important to reconsider the whole of the diversity of these conceptions and to find a broad, meaningful basis for making them mutually compatible and comparable. The above-considered evolutionary interpretation of the diversity of supposed specieshoods might be of use for outlining this basis and arranging the species conceptions into an appropriate conceptual pyramid. With this, a typological approach might provide a significant input into the elaboration of such a hierarchical faceted classification by figuring a prototypical 'good species' concept, as explained by Amitani (2015) and Wilkins (in this volume), for each cluster of particular conceptions for species units sharing the same specieshood.

This pluralistic approach will hopefully legalise the possibility to apply different species conceptions as heuristics (research instruments), in different groups of organisms most adequate to their peculiar biological properties, including those responsible for the structure of their diversity (Richards 2010; Pavlinov 2013a, 2017; Bzovy 2016; Maxwell et al. 2020; Wells et al. 2021; see also Sections 12.7 and 12.9). The above-mentioned phenomenological species conception (Sterelny 1999) may serve as a support for this viewpoint: species diversity in different groups of organisms can be effectively explored on (roughly) the same phenomenological basis, albeit conceptual treatments of species in these groups might be different. This approach seems to be of great importance for making different methods of delimiting species mutually compatible and comparable in a biologically meaningful way within the same conceptual pyramid. This would provide a constructive basis for incorporating diverse operational methods and criteria for delimiting 'phenomenological' species into a conceptually integrated toolkit (Sites and Marshall 2004) and counting total number of species on Earth without a uniform understanding of what the 'specieshood' is (Claridge et al. 1997b; Casiraghi et al. 2016; Reydon 2019; Garnett et al. 2020; Amitani in this volume). This might appear to be more productive than the endless and fruitless debates about the appropriateness of one or another particular species conception to all groups of living beings.

12.11   INSTEAD OF CONCLUSION: IF NOT SPECIES, THEN WHAT?

One of the most radical solutions of the species problem, suggested repeatedly, appeared to be exclusion of the notion of species altogether from the contemporary biological thesaurus; indeed, if there is no species, there is no species problem at all, right? Thus, Benjamin Burma was sure that the very notion of species had become outdated and that biology must discard it, just as physics had once discarded the notion of ether (Burma 1954). This suggestion was supported by several authors from various positions (Michener 1962; Sokal and Sneath 1963; Riddle and Hafner 1999; Hendry et al. 2000) and summarised by agitating for 'biology without species' (Kober 2008). Something similar is seemingly presumed by a rejection of the species rank together with a definite species unit suggested by some adherents of the rankless taxonomic hierarchy (Michener 1962; Sokal and Sneath 1963; Mishler 2010, 2021, in this volume; Zachos 2016; Mishler and Wilkins 2018).

However, when considering such a far-reaching suggestion, one should take into serious consideration that both the general concept and the notion of species are sturdily embedded in the thesaurus of many fundamental and applied biological disciplines (Cracraft 1989b; Ruse 1995; de Queiroz 1999; LaPorte 2007; Pavlinov 2013a, 2021; Minelli in this volume). So, realisation of this suggestion may lead to a substantial reorganisation of a large part of the conceptual space of the whole of biology. The reason is quite obvious: such rejection would entail serious correction of other concepts and notions associated, one or another way, with that of species.

For example, replacing species by phylogroup should obviously entail replacing speciation by something else, such as phyliation (Skarlato and Starobogatov 1974), which means emergence of a new phylogroup (in the sense of Haeckel, clade in the sense of Huxley). Generally speaking, there is nothing critically wrong with such a replacement in the case of a strictly phylogenetic interpretation of biological evolution. However, it is not evident that other biological disciplines taking the latter in a broader sense would enjoy abandoning the concept of speciation in their descriptions of the historical changes studied by them.

In ecology, the notion of species is not essential for the structural descriptions of the local ecosystems, as far as such descriptions are based on biomorphs (see Section 12.7). However, in comparative analyses of different ecosystems, there is an evident need for a certain unified basis of comparison (the above-mentioned 'denominator') that would allow linking to each other the biomorphs (or guilds or other) recognised within each of the local ecosystems. It happens to be species that currently fulfills this function; as a matter of fact, for the evolutionary ecologists, biomorphs exist in the local ecosystems not by themselves but as manifestations of the local populations of the widespread species (Schwarz 1980).

It seems to me that this particular aspect of the species problem is not just a consequence of the conceptual conservatism of biologists but reflects one of the universally valid epistemological principles. According to this, in order to explore any differences between organisms, one must have some more or less solid unified basis for their comparison, making them components of a single commonality (elements of the same class, tokens of the same natural kind, etc.) by possessing some fundamental feature(s) in common. For many research tasks in biology, the currently acknowledged basis means *conspecificity*, i.e., belonging of the organisms, differing from each other by whatever features, to the same species as a really existing natural unit. This reference function of species seems not to come up against the above-mentioned problem of the substantive incommensurability of the species units recognised in different groups of organisms according to different conceptions. In fact, in each group with its specific specieshood, conspecificity presumes the same naturally defined and clearly understood meaning, i.e., the belonging of certain organisms to the same particular species.

From this perspective, it is clear that in order to get rid of the species concept in biology, it is necessary to introduce some other, no less general concept that might provide another, no less solid unified natural basis of biologically sound comparisons.

It is evident that this replacement would require serious justification by reference to a certain fundamental, sufficiently comprehensive biological theory embracing various branches of biology, and not only contemporary phylogenetics.

Maybe, this will occur in the future; maybe not. After all, the content of the general notion of species has changed drastically over the long history of biology, making it a kind of 'nomadic concept' (Minelli 2020b); therefore, another conceptual shift would hardly look too radical from either the historical or the theoretical perspective. With this, it is worth keeping in mind that any new general concept destined to replace that of species in an imaginary 'beyond-species biology' will certainly face the same particular ontological and epistemological problems outlined earlier, i.e., ontic and epistemic reductions, incomprehensibility, non-rigidity, etc. So, the species problem will not be actually 'closed' or abandoned but will simply turn into another one, hardly monistic and therefore no less problematic. Accordingly, such a shift from the general species concept to another no less general one, presuming to resolve the current species problem most radically, will most probably require a thorough analysis from the conceptualist standpoint outlined in the present chapter.

ACKNOWLEDGEMENTS

This contribution was supported by the Governmental Theme no. 121032300105-0 implemented by the Research Zoological Museum, Lomonosov Moscow State University.

I am indebted to John Wilkins and Frank Zachos for their comments on early drafts of this contribution, as well as for their help with translating some quotations from Latin and German sources.

REFERENCES

Allen, T. F. H., and T. B. Starr. 1982. *Hierarchy: Perspectives for ecological complexity.* Chicago, IL: The University of Chicago Press.

Amitani, Y. 2015. Prototypical reasoning about species and the species problem. *Biological Theory* 10:289–300.

Amitani, Y. 2017. The general concept of species. *The Journal of Philosophical Ideas, Special Issue* 89–120.

Andino, R., and E. Domingo. 2015. Viral quasispecies. *Virology* 479–480:46–51.

Armand, A. D. 2008. [*Two in one: A law of complementarity.*] Moscow: LKI Press. (in Russian).

Atkin, A. 2013. Peirce's theory of signs. In *The Stanford encyclopedia of philosophy*, ed. E. N. Zalta. https://plato.stanford.edu/archives/sum2013/entries/peirce-semiotics/.

Atran, S. 1987. Origin of the species and genus concepts: An anthropological perspective. *Journal of the History of Biology* 20:195–279.

Barker, M. J. 2010. Specious intrinsicalism. *Philosophy of Science* 77:73–91.

Barker, M. J. 2019. Eliminative pluralism and integrative alternatives: The case of species. *The British Journal for the Philosophy of Science* 70:657–81.

Bartlett, S. J. 2015. *The species problem and its logic.* https://arxiv.org/pdf/1510.01589.pdf

Bauhin, C. 1623. *ΠΙΝΑΞ Theatri botanici Caspari Bauhini [...] sive Index in Theophrasti, Dioscoridis, Plinii et botanicorum qui à seculo scripserunt opera: Plantarum circiter sex millium ab ipsis exhibitarum nomina cum earundem synonymiis & differentiis methodicè secundùm earum & genera & species proponens*. Basileae Helvet.: Typis Ludovici Regii.

Baum, D. A. 2009. Species as ranked taxa. *Systematic Biology* 58:74–86.

Baum, D. A., and K. L. Shaw. 1995. Genealogical perspectives on the species problem. In *Experimental and molecular approaches to plant biosystematics*, ed. Hoch, P. C., and A. G. Stephenson, 289–303. St. Louis, MO: Missouri Botanical Garden.

Beatty, J. 1992. Speaking of species: Darwin's strategy. In *The units of evolution*, ed. M. Ereshefsky, 227–46. Cambridge, MA: MIT Press.

Bobay, L.-M., and H. Ochman. 2017. Biological species are universal across Life's domains. *Genome Biology and Evolution* 9:491–501.

Bock, W. J. 2004. Species: The concept, category and taxon. *Journal of Zoological Systematics and Evolutionary Research* 421:178–90.

Boëthius. 1906. Anicii Manlii Severini Boethii in Isagogen Porphyrii commentorum editionis secundae. In *Corpus Scriptorum Ecclesiasticorum Latinorum* (38), ed. S. Brandt. Vienna/Leipzig: Tempsky/Freitag. http://www.logicmuseum.com/wiki/Authors/Boethius/isagoge/CPorII.

Bradie, M., and Harms, W. 2016. Evolutionary epistemology. In *Stanford encyclopedia of philosophy*, ed. E. N. Zalta. https://plato.stanford.edu/archives/spr2016/entries/epistemology-evolutionary/.

Brasier, C. M. 1997. Fungal species in practice: Identifying species units in fungi. In *Species: The units of biodiversity*, ed. M. F. Claridge, H. A. Dawah, and M. R. Wilson, 135–70. London: Chapman & Hall.

Brigandt, I. 2012. The dynamics of scientific concepts: The relevance of epistemic aims and values. In *Scientific concepts and investigative practice*, ed. U. Feest, and F. Steinle, 75–104. Baltimore, MD: John Hopkins University Press.

Brooks, D. R., and E. O. Wiley. 1986. *Evolution as entropy*. Chicago, IL: University of Chicago Press.

Broughton, V. 2006. The need for a faceted classification as the basis of all methods of information retrieval. *New Information Perspectives* 58:49–72.

Buffon. 1843. *De la Nature. Second vue. In Oeuvres choisies de Buffon, contenant les discours académiques, des extraits de la théorie de la terre, les epoques de la nature, la génésie des minéraux, l'histoire naturelle de l'home ed des animaux, T.1:52–64*. Paris: Pourrat Fréres.

Burma, B. H. 1954. Reality, existence, and classification: A discussion of the species problem. *Madroño* 12:193–209.

Bzovy, J. 2016. *Species Pluralism: Conceptual, ontological, and practical dimensions*. PhD Thesis, University of Western Ontario. https://ir.lib.uwo.ca/cgi/viewcontent.cgi?article=5976&context=etd.

Cain, A. J. 1954. *Animal species and their evolution*. London: Hutchinson.

Cain, A. J. 1994. Rank and sequence in Caspar Bauhin's Pinax. *Botanical Journal of the Linnean Society* 114:311–56.

Carrier, M. 1994. *The completeness of scientific theories*. Dordrecht: Kluwer Academic Publishers.

Cartmill, M. 1981. Hypothesis testing and phylogenetic reconstruction. *Zeitschrift für Zoologische Systematik und Evolutions-forschung* 19:73–96.

Casiraghi, M., A. Galimberti, A. Sandionigi, et al. 2016. Life with or without names. *Evolutionary Biology* 43:582–95.

Chernov, Yu. 1991. [Biological diversity, its essence and related problems.] *Uspehi sovremennoj biologii* 111:499–507. (in Russian).

Claridge, M. F. 2010. Species are real biological entities. In *Contemporary debates in philosophy of biology*, ed. F. J. Ayala and R. Arp, 91–109. The Atrium Southern Gate: John Wiley & Sons.

Claridge, M. F., H. A. Dawah, and M. R. Wilson (eds.). 1997a. *Species: The units of biodiversity*. London: Chapman & Hall.

Claridge, M. F., H. A. Dawah, and M. R. Wilson. 1997b. Practical approaches to species concepts for living organisms. In *Species: The units of biodiversity*, ed. M. F. Claridge, H. A. Dawah, and M. R. Wilson, 1–15. London: Chapman & Hall.

Costello, M. J. 2020. Taxonomy as the key to life. *Megataxa* 1:105–13.

Cracraft, J. 1989a. Speciation and its ontology: The empirical consequences of alternative species concepts for understanding patterns and processes of differentiation. In *Speciation and its consequences*, ed. D. Otte, and J. Endler, 28–59. Sunderland: Sinauer Association.

Cracraft, J. 1989b. Species as entities of biological theory. In *What the philosophy of biology is: Essays dedicated to David Hull*, ed. M. Ruse, 31–52. Dordrecht: Kluwer Academic. Publishers.

Dadi, T. H., B. Y. Renard, L. H. Wieler, et al. 2017. SLIMM: Species level identification of microorganisms from metagenomes. *PeerJ*. 5:e3138.

Darwin, C. 1859. *On the origin of species by means of natural selection, or the preservation of favoured races in the struggle for life*. London: John Murray.

De Florio, C., and A. Frigerio. 2019. A conceptualist view in the metaphysics of species. In *Natural and artifactual objects in contemporary metaphysics*, ed. R. Davies, 131–8. London: Bloomsbury Academic.

de Pinna, M. C. C. 1999. Species concepts and phylogenetics. *Reviews in Fish Biology and Fisheries* 9:353–73.

de Queiroz, K. 1998. The general lineage concept of species, species criteria, and the process of speciation. In *Endless forms: Species and speciation*, ed. D. J. Howard and S. H. Berlocher, 57–75. Oxford: Oxford University Press.

de Queiroz, K. 1999. The general lineage concept of species and the defining properties of species category. In *Species: New interdisciplinary essays*, ed. R. A. Wilson, 49–89. Cambridge, MA: The MIT Press.

de Queiroz, K. 2005a. Different species problems and their resolution. *BioEssays* 27:1263–9.

de Queiroz, K. 2005b. A unified concept of species and its consequences for the future of taxonomy. *Proceedings of the California Academy of Sciences*, ser. 4, 56 (Suppl. I):196–215.

Devitt, M. 2008. Resurrecting biological essentialism. *Philosophy of Science* 75:344–82.

Dillon, L. S. 1966. The life cycle of the species: An extension of current concepts. *Systematic Zoology* 15:112–26.

Dobzhansky, T. 1970. *Genetics of evolutionary process*. New York: Columbia University Press.

Domingo, E. 2002. Quasispecies theory in virology. *Journal of Virology* 76:463–5.

Donovan, S. K. (ed.). 1994. *The palaeobiology of trace fossils*. Chichester: Wiley & Sons.

Dubois, A. 2011. Species and 'strange species' in zoology: Do we need a 'unified concept of species'? *Comptes rendus de l'Académie des Sciences, Series IIA, Earth and Planetary Science* 10:77–94.

Dupré, J. 1999. On the impossibility of a monistic account of species. In *Species: New interdisciplinary essays*, ed. R. A. Wilson, 3–21. Cambridge, MA: The MIT Press.

Dupré, J. 2020. Life as process. *Epistemology & Philosophy of Science* 57:96–113.

Eigen, M. 1983. Viral quasispecies. *Scientific American*, 269:42–9.

Eldredge, N. 1985. *Unfinished synthesis: Biological hierarchies and modern evolutionary thought*. New York: Oxford University Press.

Ellis, B. 2001. *Scientific essentialism*. Cambridge: Cambridge University Press.

Ellis, M. 2011. The problem with the species problem. *History & Philosophy of the Life Sciences* 33:343–63.

Ereshefsky, M. 1992. Eliminative pluralism. *Philosophy of Science* 59:671–90.

Ereshefsky, M. 1998. Species pluralism and anti-realism. *Philosophy of Science* 65:103–20.

Ereshefsky, M. 2009. Darwin's solution to the species problem. *Synthese* 175:405–25.

Ereshefsky, M, 2010. Microbiology and the species problem. *Biology and Philosophy* 25:553–68.

Fadda, A. 2020. Population thinking in epistemic evolution: Bridging cultural evolution and the philosophy of science. *Journal for General Philosophy of Science* 52:351–69.

Ficetola, G. F., C. Miaud, F. Pompanon, et al. 2008. Species detection using environmental DNA from water samples. *Biology Letters* 4:423–5.

Folse, H. J., and J. Roughgarden. 2010. What is an individual organism? A multilevel selection perspective. *The Quarterly Review of Biology* 85:447–72.

Gärdenfors, P. 2000. *Conceptual spaces*. Cambridge, MA: MIT Press.

Garey, M. R., and D. S. Johnson. 1979. *Computer and intractability: A guide to the theory of NP-completeness*. San Francisco, CA: W.H. Freeman.

Garnett, S. T., Christidis, L., Conix, S., et al. 2020. Principles for creating a single authoritative list of the world's species. *PLoS Biology* 18(7):e3000736.

Ghiselin, M. 1989. Sex and the individuality of species: A reply to Mishler and Brandon. *Biology and Philosophy* 4:73–6.

Ghiselin, M. T. 1997. *Metaphysics and the origin of species*. New York: State University of New York Press.

Giray, E. F. 1976. An integrated biological approach to the species problem. *The British Journal for the Philosophy of Science* 27:317–28.

Greene, J. C. 1992. From Aristotle to Darwin: Reflections on Ernst Mayr's interpretation in the growth of biological thought. *Journal of the History of Biology* 25:257–84.

Griffiths, P. E. 1999. Squaring the circle: Natural kinds with historical essences. In *Species: New interdisciplinary essays*, ed. R. A. Wilson, 209–28. Cambridge, MA: The MIT Press.

Haack, S. 1979. Epistemology with a knowing subject. *The Review of Metaphysics* 33:309–35.

Hall, B. K., ed. 1994. *Homology, the hierarchical basis of comparative biology*. San Diego, CA: Academic Press.

Hanage, W. P. 2013. Fuzzy species revisited. *BMC Biology* 11:41.

Hanage, W. P., C. Fraser, and B. G. Spratt. 2005. Fuzzy species among recombinogenic bacteria. *BMC Biology* 3:6. http://www.biomedcentral.com/1741-7007/3/6.

Hawkins, H. L. 1935. The species problem. *Nature* 136:574–5.

Hempel, C. G. 1956. Geometry as empirical science. In *The world of mathematics*, V. 3., ed. J. R. Newman, 1635–46. New York: Simon and Shuster.

Hendry, A. P., M. S. Vamosi, S. J. Latham, et al. 2000. Questioning species realities. *Conservation Genetics* 1:67–76.

Hey, J. 2001a. *Genes, categories, and species. The evolutionary and cognitive cause of the species problem*. New York: Oxford University Press.

Hey, J. 2001b. The mind of the species problem. *Trends in Ecology and Evolution* 16:326–9.

Hey, J., R. Waples, M. Arnold et al. 2003. Understanding and confronting species uncertainty in biology and conservation. *Trends in Ecology and Evolution* 18:597–603.

Hull, D. L. 1988. *Science as a process*. Chicago, IL: University of Chicago Press.

Hull, D. L. 1997. The ideal species concept – and why we can't get it. In *Species. The units of biodiversity*, ed. M. F. Claridge, A. H. Dawah, and M. R. Wilson, 357–80. London: Chapman & Hall.

Hull, D. L. 1999. On the plurality of species: Questioning the party line. In *Species: New interdisciplinary essays*, ed. R. A. Wilson, 23–48. Cambridge, MA: MIT Press.

Hughes, N. F. 2006. *Fossils as information: New recording and stratal correlation techniques*. Cambridge: Cambridge University Press.

Ilyin, V. V. 2003. [*Philosophy of science.*] Moscow: Moscow University Press. (in Russian).

Innis, R. E. 2020. Semiotic framing of thresholds of sense. In: *Between philosophy and cultural psychology*, ed. R. E. Innis, 31–48. Cham, Switzerland: Springer International Publishing.

Jungius, J. 1662. *Joachimi Jungii doxoscopiae physicae minores, sive isagoge physica doxoscopica. In qua praecipuae opiniones in physica … examinantur.* Hamburgi: Johannis Naumanni.

Kitcher, P. 1984. Species. *Philosophy of Science* 51:308–33.

Kluge, A. G. 1990. Species as historical individuals. *Biology and Philosophy* 5:417–31.

Knowlton, N., and L. A. Weigt. 1997. Species of marine invertebrates: A comparison of the biological and phylogenetic species concepts. In *Species: The units of biodiversity*, ed. M. F. Claridge, H. A. Dawah, and M. R. Wilson, 199–220. London: Chapman & Hall.

Kober, G. 2008. *Biology without species: A solution to the species problem*. PhD thesis, Boston University. https://philpapers.org/rec/KOBBWS.

Kornet, D. J., and J. W. McAllister. 2005. The composite species concept: A rigorous basis for cladistic perspective. In *Current themes in theoretical biology. A Dutch Perspective*, ed. T. A. C. Reydon, and L. Hemerik, 95–127. Amsterdam: Springer.

Kosko, B. 1993. *Fuzzy thinking: The new science of fuzzy logic*. New York: Hyperion.

Krell, F.-T. 2004. Parataxonomy vs. taxonomy in biodiversity studies – pitfalls and applicability of 'morphospecies' sorting. *Biodiversity and Conservation* 13:795–812.

Kunz, W. 2012. *Do species exist? Principles of taxonomic classification*. Weinheim, Germany: Wiley-VCH Verlag & Co.

Kwasnik, B. H. 2000. The role of classification in knowledge representation and discovery. *The School of Information Studies*: 147. http://surface.syr.edu/istpub/147.

Lam, K. H. 2020. The realism of taxonomic pluralism. *Metaphysics* 3:1–16.

LaPorte, J. 2007. In defense of species. *Studies in History and Philosophy of Biological and Biomedical Sciences* 38:255–69.

LaPorte, J. 2017. Modern essentialism for species and its animadversions. In *The Routledge handbook of evolution and philosophy*, ed. R. Joyce, 182–93. New York: Routledge.

Linnaeus, C. 1751. *Philosophia botanica in qua explicantur fundamenta botanica cum definitionibus partium, exemplis terminorum, observationibus rariorum, adiectis figuris aeneis*. Stockholmiae: Godofr. Kiesewetter.

Lyubarsky, G. Yu. 2018. [*Origins of hierarchy: The history of taxonomic rank.*] Moscow: KMK Sci. Press. (in Russian).

MacLeod, M. 2012. Rethinking scientific concepts for research contexts: The case of the classical gene. In *Scientific concepts and investigative practice*, ed. U. Feest, and F. Steinle, 47–74. Baltimore, MD: John Hopkins University Press.

Mahner, M. 1993. What is a species? A contribution to the never ending species debate in biology. *Journal for General Philosophy of Science* 24:103–26.

Mahner, M., and M. Bunge. 1997. *Foundations of biophilosophy*. Frankfurt: Springer Verlag.

Marcelino, V. R., E. C. Holmes, and T. C. Sorrell. 2020. The use of taxon-specific reference databases compromises metagenomic classification. *BMC Genomics* 21:184.

Maturana, H. R. 1988. Ontology of observing: The biological foundation of self consciousness and the physical domain of existence. *Irish Journal of Psychology* 9:25–82.

Maxwell, S. J., B. C. Congdon, and T. L. Rymer. 2020. Essentialistic pluralism: The theory of spatio-temporal positioning of species using integrated taxonomy. *Proceedings of the Royal Society of Queensland* 124:1–17.

Mayden, R. L. 1997. A hierarchy of species concepts: The denouement in the saga of the species problem. In *Species. The units of biodiversity*, ed. M. F. Claridge, A. H. Dawah,
and M. R. Wilson, 381–424. London: Chapman & Hall.

Mayden, R. L. 1999. Consilience and a hierarchy of species concepts: Advances toward closure on the species puzzle. *Journal of Nematology* 31:95–116.

Mayr, E. 1963. *Animal species and evolution*. Cambridge, MA: Harvard University Press.

Mayr, E. 1970. *Populations, species, and evolution*. Cambridge, MA: Harvard University
Press.

Mccabe, T. 2008. Studying species definitions for mutual nonexclusiveness. *Zootaxa*
1939:1–9.

Mechkovskaya, B. N. 2007. [*Semiotics: Language. Nature. Culture.*] Moscow: Akademia.
(in Russian).

Michener, C. D. 1962. Some future developments in taxonomy. *Systematic Zoology* 12:151–72.

Mikhailov, K. E. 2003. [Typological comprehension of 'biological species', and the way of
stabilization of near-species taxonomy in birds.] *Ornitologia* 30:9–24. (in Russian).

Milanese, A., D. R. Mende, L. Paoli, et al. 2019. Microbial abundance, activity and population
genomic profiling with mOTUs2. *Nature Communications* 10:1014.

Minati, G. 2019. On theoretical incomprehensibility. *Philosophies*, 4:49.

Minelli, A. 2014. Developmental disparity. In *Towards a theory of development*, ed. A.
Minelli and T. Pradeu, 227–45. Oxford: Oxford University Press.

Minelli, A. 2020a. Taxonomy needs pluralism, but a controlled and manageable one.
Megataxa 1:9–18.

Minelli, A. 2020b. Disciplinary fields in the life sciences: Evolving divides and anchor concepts. *Philosophies* 5:34.

Mishler, B. D. 2010. Species are not uniquely real biological entities. In *Contemporary
debates in philosophy of biology*, ed. F. J. Ayala and R. Arp, 110–22. The Atrium
Southern Gate: John Wiley & Sons.

Mishler, B. D. 2021. *What, if anything, are species?* Boça Raton, FL: CRC Press.

Mishler, B. D., and M. J. Donoghue. 1982. Species concepts: A case for pluralism. *Systematic
Zoology* 31:491–503.

Mishler, B. D., and E. C. Theriot. 2000. The phylogenetic species concept: Monophyly, apomorphy, and phylogenetic species concept. In *Species concepts and phylogenetic theory: A debate*, ed. Q. D. Wheeler, and R. Meier, 44–54. New York: Columbia University
Press.

Mishler, B. D., and J. S. Wilkins. 2018. The hunting of the SNaRC: A snarky solution to the
species problem. *Philosophy, Theory, and Practice in Biology* 10:1–18.

Morgun, D. V. 2002. [*Epistemological foundations of the species problem in biology.*]
Moscow: Moscow State University Publ. (in Russian).

Nalimov, V. V. 1979. [*A probabilistic model of language. On relation between natural and
artificial languages.*] Moscow: Nauka. (in Russian).

Naomi, S.-I. 2011. On the integrated frameworks of species concepts: Mayden's hierarchy of
species concepts and de Queiroz's unified concept of species. *Journal of Zoological
Systematics and Evolutionary Research* 49:177–84.

Nathan, M. J. 2017. Pluralism is the answer! What is the question? *Philosophy, Theory, and
Practice in Biology* 11:1–14. https://www.du.edu/ahss/philosophy/media/documents/
mnathan/nathanpluralismisthequestion.pdf.

Novick, A., and W. F. Doolittle. 2021. 'Species' without species. *Studies in History and
Philosophy of Science* 87:72–80.

Nowak, M. A. 1992. What is a quasispecies? *Trends in Ecology and Evolution* 7:118–21.

Okasha, S. 2002. Darwinian metaphysics: Species and the question of essentialism. *Synthese*
131:191–213.

Padial, J. M., and I. De la Riva. 2020. A paradigm shift in our view of species drives current trends in biological classification. *Biological Reviews* (online version). https://doi.org/10.1111/brv.12676.

Parmasto, E. 1986. [Species problem in fungi.] In [*Species and genus problems in fungi*], ed. E. Parmasto, 9–28. Tallinn: Inst. of Zoology. (in Russian).

Pavlinov, I. Ya. 1992. [If there is the biological species, or what is the 'harm' of taxonomy.] *Journal of General Biology* 53:757–67. (in Russian).

Pavlinov, I. Ya. 1995. [Classification as hypothesis: Entering the problem.] *Journal of General Biology* 56:411–24. (in Russian).

Pavlinov, I. Ya. 2007. On the structure of biodiversity: Some metaphysical essays. In *Focus on biodiversity research*, ed. J. Schwartz, 101–14. New York: Nova Science Publishers.

Pavlinov, I. Ya. 2009. [The species problem: Another look.] In [*Species and speciation: An analysis of new views and trends*], ed. A. F. Alimov, and S. D. Stepanyanz, 250–71. St. Petersburg: Zoological Institute. (in Russian).

Pavlinov, I. Ya. 2011. [How it is possible to elaborate a taxonomic theory.] *Zoologicheskie Issledovania* 10:45–100. (in Russian).

Pavlinov, I. Ya. 2012. The contemporary concepts of homology in biology: A theoretical review. *Biology Bulletin Reviews* 2:36–54.

Pavlinov, I. Ya. 2013a. The species problem: Why again? In *The species problem: Ongoing issues*, ed. I. Ya. Pavlinov, 3–37. Rijeka: InTech Open Access Publ.

Pavlinov, I. Ya. (ed.) 2013b. *The species problem: Ongoing issues*. Rijeka: InTech Open Access Publ.

Pavlinov, I. Ya. 2017. [The species problem in biology, its roots and present.] In [*Species concept in fungi: A new look at old problem*], *Proc. 8th Russian Mycological Conference*, 5–19. Moscow: Moscow State University Press. (in Russian, English translation available at https://www.researchgate.net/publication/315815813_The_species_problem_in_biology_its_roots_and_present).

Pavlinov, I. Ya. 2018. [*Foundations of biological systematics: Theory and history.*] Moscow: KMK Sci. Press. (in Russian).

Pavlinov, I. Ya. 2021. *Biological systematics: History and theory*. Boca Raton, FL: CRC Press.

Pavlinov, I. Ya., and G. Yu. Lyubarsky. 2011. [*Biological systematics: Evolution of ideas.*] Moscow: KMK Sci. Press. (in Russian)

Pedroso, M. 2012. Essentialism, history, and biological taxa. *Studies in History and Philosophy of Science, Part C, Studies in History and Philosophy of Biological and Biomedical Sciences* 43:182–90.

Pfander, P. V. 2018. [Tragedy of near-species taxonomy.] *The Russian Journal of Ornithology* 27, Express Iss. 1558:301–35. (in Russian).

Pleijel, F., and G. W. Rouse. 2000. Least-inclusive taxonomic unit: A new taxonomic concept for biology. *Proceedings of the Royal Society B (Biological Sciences)* 267:627–30.

Popper, K. 1959. *The logic of scientific discovery*. London: Hutchinson & Co.

Poulton, E. B. 1904. What is a species? *Proceedings of the Entomological Society of London* (for 1903):lxxvii–cxvi.

Pradeu, T., L. Laplane, K. Prevot, et al. 2016. Defining 'development'. *Current Topics in Developmental Biology*, 117:171–83.

Quine, W. V. 1969. *Ontological relativity & other essays*. New York: Columbia University Press.

Raposo, M. A., G. M. Kirwan, A. C. C. Lourenço, et al. 2020. On the notions of taxonomic 'impediment', 'gap', 'inflation' and 'anarchy', and their effects on the field of conservation. *Systematics and Biodiversity*. https://www.tandfonline.com/doi/abs/10.1080/14772000.2020.1829157?journalCode=tsab20.

Ray, J. 1686. *Historia plantarum: Species hactenus editas aliasque insuper multas noviter inventas & descriptas complectens. In qua agitur primò de plantis in genere, earúmque partibus, accidentibus & differentiis; deinde genera omnia tum summa tum subalterna ad species usque infirmas, notis suis certis & characteristicis definita, methodo naturæ vestigiis insistente disponuntur Londoni: typus Mariae Clark.*

Reig, O. A. 1982. The reality of biological species: A conceptualistic and a systemic approach. *Studies in Logic and the Foundations of Mathematics* 104:479–99.

Reydon, T. A. C. 2005. On the nature of the species problem and the four meanings of species. *Studies in History and Philosophy of Biological and Biomedical Sciences* 36:135–58.

Reydon, T. A. C. 2019. Are species good units for biodiversity studies and conservation efforts? In *From assessing to conserving biodiversity: Conceptual and practical challenges*, ed. E. Casetta, J. Marquez da Silva, and D. Vecchi, 167–93. Cham: Springer.

Reydon, T. A. C., and W. Kunz. 2019. Species as natural entities, instrumental units and ranked taxa: New perspectives on the grouping and ranking problems. *Biological Journal of the Linnean Society* 126:623–36.

Rheinberger, H.-J. 2000. Gene concepts: Fragments from the perspective of molecular biology. In *The concept of the gene in development and evolution*, ed. P. Beurton, R. Falk, and H.-J. Rheinberger, 219–39. Cambridge: Cambridge University Press.

Richards, R. A. 2010. *The species problem: A philosophical analysis*. Cambridge: Cambridge University Press.

Riddle, B. R., and Hafner, D. J. 1999. Species as units of analysis in ecology and biogeography: Time to take the blinders off. *Global Ecology and Biogeography* 8:433–41.

Riedel, A., K. Sagata, Y. R. Suhardjono, R. Tänzler, et al. 2013. Integrative taxonomy on the fast track – towards more sustainability in biodiversity research. *Frontiers in Zoology* 10:15. http://www.frontiersinzoology.com/content/10/1/15.

Rieppel, O. 2010. New essentialism in biology. *Philosophy of Science* 77:662–73.

Robson, G. C. 1928. *The species problem: An introduction to the study of evolutionary divergence in natural populations*. Edinburgh: Oliver & Boyd.

Roger, J., and J. L. Fischer (ed.). 1987. *Histoire du concept d'espèce dans les Sciences de la Vie*. Paris: Fondation Singer-Polignac.

Romanes, G. J. 1895. *Darwin, and after Darwin: An exposition of the Darwinian theory and a discussion of the post-Darwinian questions. Vol. 2. Post-Darwinian questions: Heredity and utility*. London: Longmans & Green.

Ruse, M. 1995. The species problem. In *Concepts, theories, and rationality in the biological sciences*, ed. G. Wolters, and J. G. Lennox. Pittsburgh, PA: University of Pittsburgh Press.

Sankey, H. 1998. Taxonomic incommensurability. *International Studies in the Philosophy of Science* 12:7–16.

Schwarz, S. S. 1980. [*Ecological patterns of evolution.*] Moscow: Nauka. (in Russian).

Seifert, B. 2014. A pragmatic species concept applicable to all eukaryotic organisms independent from their mode of reproduction or evolutionary history. *Soil Organisms* 86:85–93.

Seifert, B. 2020. The gene and gene expression (GAGE) species concept: An universal approach for all eukaryotic organisms. *Systematic Biology* 69:1033–8.

Sigwart, J. D. 2018. *What species mean: A user's guide to the units of biodiversity*. Boca Raton, FL: CRC Press.

Sites, J. W., and J. C. Marshall. 2004. Operational criteria for delimiting species. *Annual Review of Ecology, Evolution, and Systematic* 35:199–227.

Skarlato, O. A., and Ya. I. Starobogatov. 1974. [Phylogenetics and principles of elaboration of the natural system.] *Proceedings of Zoological Institute AS USSR* 53:30–46. (in Russian).

Skvortsov, A. K. 1967. [The main stages in the development of the concept of species.] *Bulleten Moskovskogo obshestva ispytatelej prirody, Biol.* 72:11–27. (in Russian).

Sluys, R. 1991. Species concepts, process analysis, and the hierarchy of nature. *Experientia* 47:1162–70.

Sober, E. 1980. Evolution, population thinking, and essentialism. *Philosophy of Science*, 47:350–83.

Sokal, R. R., Sneath, R. H. A. 1963. *Principles of numerical taxonomy.* San Francisco, CA: W.H. Freeman & Co.

Stace, C. A. 1989. *Plant taxonomy and biosystematics*, 2nd ed. London: Hodder Arnold.

Stamos, D. N. 1996. Was Darwin really a species nominalist? *Journal of the History of Biology* 29:127–44.

Stamos, D. N. 2003. *The species problem. Biological species, ontology, and the metaphysics of biology.* Oxford: Lexington Books.

Stamos, D. N. 2007. *Darwin and the nature of species.* Albany, NY: State University of New York.

Stanford, P. K. 1995. For pluralism and against realism about species. *Philosophy of Science* 62:70–91.

Stepin, V. S. 2005. *Theoretical knowledge.* Heidelberg: Springer.

Sterelny, K. 1999. Species as ecological mosaics. In *Species, new interdisciplinary essays*, ed. R. A. Wilson, 119–38. Cambridge, MA: MIT Press.

Stijn, C. 2018. Values, regulation, and species delimitation. *Zootaxa* 4415:390–2.

Sunagawa, S., D. Mende, G. Zeller et al. 2013. Metagenomic species profiling using universal phylogenetic marker genes. *Nature Methods* 10:1196–9.

Surov, I. A. 2002. Quantum cognitive triad. Semantic geometry of context representation. *Neurons and Cognition* (q-bio.NC): arXiv:2002.11195. https://arxiv.org/abs/2002.11195.

Swoyer, C. 2006. Conceptualism. In *Universals, concepts, and qualities: New essays on the meaning of predicates*, ed. E. S Trawson, and A. Chakrabarti, 127–54. Boca Raton, FL: CRC Press.

Sylvester-Bradley, P. C. 1952. *The classification and coordination of infraspecific categories.* London: Syst. Assoc.

Toulmin, S. 1972. *Human understanding: The collective use and evolution of concepts.* Princeton, NJ: Princeton University Press.

Tournefort, Pitton de, J. 1694. *Élémens de botanique, ou Methode pour connoître les plantes*, T. 1. Paris: De l'Imprimerie Royale.

Van Regenmortel, M. H. V. 1997. Viral species. In *Species: The units of biodiversity*, ed. M. F. Claridge, H. A. Dawah, and M. R. Wilson, 17–24. London: Chapman & Hall.

Velasco, J. D. 2008. Species concepts should not conflict with evolutionary history, but often do. *Studies in History and Philosophy of Science* 39:407–14.

Velichkovsky, B. M. 2006. [*Cognitive science. Foundations of psychology of knowing.*] Moscow: Academia Publishing. (in Russian).

Volkova, E. V., and Filyukov, A. I. 1966. [*Philosophical issues of the species theory.*] Minsk: Nauka & Tekhnika Publishing. (in Russian).

Voyshvillo, E. K. 1989. [*Concept as a form of thinking: Logical and epistemological analysis.*] Moscow: Moscow State University Press. (in Russian).

Vrana, P., and W. Wheeler. 1992. Individual organisms as terminal entities: Laying the species problem to rest. *Cladistics* 8:67–72.

Waddington, C. H. 1962. *New pattern in genetics and development.* New York: Columbia University Press.

Wartofsky, M. W. 1979. *Models: Representation and scientific understanding.* Boston, MA: Springer Science & Business Media.

Weinberg, S. 1992. *Dreams of a final theory: The scientist's search for the ultimate laws of nature*. New York: Pantheon.

Wells, T., T. Carruthers, P. Muñoz-Rodríguez, et al. 2021. Species as a heuristic: Reconciling theory and practice. *Systematic Biology* 10:1–11.

Wheeler, Q. D., and R. Meier (ed.). 2000. *Species concepts and phylogenetic theory: A debate*. New York: Columbia University Press. 230 p.

Wheeler, Q. D., S. Knapp, D. W. Stevenson, et al. 2012. Mapping the biosphere: Exploring species to understand the origin, organization and sustainability of biodiversity. *Systematics and Biodiversity* 10:1–20.

Wiley, E. O. 1978. The evolutionary species concept reconsidered. *Systematic Biology* 27:17–26.

Wilkins, J. S. 2003. *The origins of species concepts. History, characters, modes, and synapomorphies.* https://www.semanticscholar.org/paper/The-Origins-of-species-concepts-%3A-and-Wilkins/2d5bc18e51cb8bd1f49845eeef4bb13fb2b8349c.

Wilkins, J. S. 2006. The concept and causes of microbial species. *History and Philosophy of Life Science* 28:389–408.

Wilkins, J. S. 2007. The dimensions, modes and definitions of species and speciation. *Biology and Philosophy* 22:247–66.

Wilkins, J. S. 2009. *Species: A history of the idea*. Berkeley, CA: University of California Press.

Wilkins, J. S. 2018a. *Species: A history of the idea*, 2nd ed. Boca Raton, FL: CRC Press.

Wilkins, J. S. 2018b. The reality of species: Real phenomena not theoretical objects. In *Routledge Handbook of Evolution and Philosophy*, ed. R. Joyce, 167–81. Abingdon, UK: Routledge.

Williamson, T. 2000. *Knowledge and its limits*. Oxford: Oxford University Press.

Wilson, R. A. (ed.). 1999. *Species: New interdisciplinary essays*. Cambridge, MA: The MIT Press.

Wilson, R. A., M. J. Barker, and I. Brigandt. 2009. When traditional essentialism fails: Biological natural kinds. *Philosophical Topics* 35:189–215.

Wood, L. 1940. *The analysis of knowledge*. London: Routledge.

Woodger, J. H. 1937. *The axiomatic method in biology*. Cambridge, UK: Cambridge University Press.

Yudin, E. G. 1997. [*Methodology of science. Systemness. Activity.*] Moscow: Editorial URSS. (in Russian).

Zachos, F. E. 2016. *Species concepts in biology. Historical development, theoretical foundations and practical relevance*. Basel: Springer.

Zachos, F. E. 2018. (New) species concepts, species delimitation and the inherent limitations of taxonomy. *Journal of Genetics* 97:811–15.

Zavadsky, K. M. 1968. [*Species and speciation.*] Leningrad: Nauka. (in Russian).

13 (Some) Species Are Processes

John Dupré

CONTENTS

13.1 INTRODUCTION

This chapter will primarily concern the ontology of species, what a species is. For some, taking seriously the origin of the term in Aristotelian logic, this may seem bizarre. There have certainly been questions about whether species are (real) natural kinds, whether the divisions we mark between species correspond to objective divisions in nature, or even, once, whether there must be a universal, some abstract entity that all members of a species have in common. For those, like myself, who are skeptical about any but a very thin sense of naturalness for biological kinds, it has seemed unlikely that there is any uniquely correct or useful way of classifying organisms into kinds. But these are ontological questions about the species category, as philosophers sometimes like to say, rather than the species taxon. My concern here is with the species taxon.

Before getting to the main task of this chapter, it is worth dwelling a moment on this notorious and rather strange distinction between species taxon and category. The species *Sus domesticus*, the domestic pig, is a taxon. What does this mean? A taxon is a unit in a taxonomy, or classificatory scheme. In a taxonomy of furniture, *armchair* might be a taxon, a species of the genus chair. Species (the category) in this taxonomy would be the smallest sets of items of furniture with a distinctive function judged worth distinguishing from other functions. How could anyone confuse the taxa, armchair, desk chair, kitchen chair, etc., with the category, species of furniture? The difference between this case and the case of biological species is that whereas no one supposes that the set of all armchairs constitutes some kind of entity, at least in any sense beyond that in which any arbitrary set of objects constitutes an entity, the

DOI: 10.1201/9780367855604-17

set of all pigs, past, present, and to come, is often supposed precisely to constitute an entity, the species. The term 'species', in the biological context, is, if not straightforwardly homonymous, at least polysemic (see Reydon, this volume). And it is polysemic in a rather remarkable way, in that it can be used to refer to ontologically quite distinct kinds. '*Sus domesticus* is a species' can mean either that *Sus domesticus* is a classification at the lowest level of the taxonomic hierarchy, or that *Sus domesticus* is a certain kind of individual entity. Questions about the 'species category' are questions about what makes a set of individuals a species taxon. It may be that there is no more to say than that it is a particular rank in the taxonomic hierarchy. 'Species taxon' refers to a specific instance of that category. One question about species taxa is whether their existence is an objective fact in the world or something that exists only in relation to a classificatory term. One way for the first possibility to be true is if species taxa are concrete, if diffuse, material individuals; though even then, calling them *species* taxa presupposes an answer to the species category question. There is, no doubt, a possibility of providing an answer to the species category question that implies species taxa that do not coincide with the objective entities or kinds in the world. But this seems like a biological mistake rather than a conceptual confusion. I still struggle to see where such confusion might come from.*

   The idea that species taxa could be understood as a kind of individual is famously associated with Michael Ghiselin (1966, 1974) and David Hull (1976). Generally, this has been interpreted as claiming that a species was something like a thing; and this has raised worries about the disconnection of its parts or more generally, its great difference from more familiar things. My thesis in this chapter is that the source of many such problems has been in the hasty assumption of thingness. The problems largely disappear when we appreciate that a species is an individual process.† This

---

* Michael Devitt (2008) seems to think that confusion between species categories and taxa is endemic – I am one of many who allegedly perpetrated it – and largely explains the widespread lack of enthusiasm for biological essentialism. He writes:

   'What is it to be a poodle not a bulldog?' is an instance of the taxon problem 1, 'What is it for poodles to be a subspecies not a species?' is an instance of the category problem 2. The distinction between the two problems may seem obvious and yet it is easily conflated by certain forms of words. In particular, consider the question, 'What is a species?' or 'What is the nature/ definition of a species?' These questions are ambiguous. They could be asking what sort of a nature any group has that happens to be a species, an instance of the taxon problem 1. But they are more likely to be asking what is it for any group to be a species, an instance of the category problem 2.

Indeed they are! At least, that is what the words appear to mean (taking the question about poodles to disambiguate the quantifier 'any' in the penultimate sentence). The explanation does nothing that I can see to show how anyone could be the victim of such an obvious and gross confusion; I certainly deny having been one myself.

† I first proposed this idea in Dupré (2017). I failed to note then that a very similar thesis had been defended by Olivier Rieppel (2009), for which I must now apologise. His characterisation of species as open-ended processual systems is very close to my own, as is his view that we need an alternative to the dichotomy of presentist endurantism and eternalist perdurantism, one provided exactly by the idea of an open-ended process. Rieppel calls this 'futuralism'; I have occasionally preferred to describe the view as three and a half dimensionalism. The closest concept in mainstream metaphysics is probably the growing block theory of time.

will also, for better or worse, show that the ontological species taxon is even further from the classificatory species category than is generally supposed.

Here, I shall suggest that the key to better understanding of these issues is to start at the beginning rather than, as is more often the preferred option, the end. By that, I mean we should start with microbes* rather than with vertebrate animals such as ourselves. The latter are, of course, a very late addition to the biosphere and a dangerous model for life generally. In the microbial realm, we find that species-like entities can be identified, but they are by no means universal (not every microbe belongs to such an entity), and they differ considerably in the degree to which they are species-like. Only at the other end of evolutionary history does species formation start to look like a general feature of life.

This chapter, then has two main theses. The first is that species as individuals should be understood as processes, not things. The second is that species formation is a diverse process, is not universal, and hence, that there are different degrees to which individual species realise the important characteristics of species as individuals. A consequence of these two theses is that the connections between the species category and the concrete species taxon are more complex and diffuse than is often assumed. And finally, this reinforces my earlier thesis (Dupré 2001) that species as a classificatory category must be sharply distinguished from species as a theoretical topic in evolutionary biology.

I should say a bit more here about this last distinction. It is widely assumed that all classification is theory-laden, and no doubt this is typically the case. Classifications, it is widely hoped, will reflect real divisions in nature and will therefore be apt for stating significant truths about nature. A paradigm is the periodic table of the elements. In the case of species, as just discussed, it is often supposed that species names will refer to real entities in nature, species processes. However, unlike the periodic table, there is no reason to suppose that every organism is part of such a species process. There is, on the other hand, an excellent reason for assigning every organism to a classification (Dupré 2001). This provides the system in which knowledge about the living world is stored and communicated. An opposing paradigm is the classification of places with latitude and longitude. No one imagines that this reflects any facts in the world, but it is also essential for many practical purposes. It is hoped that the species level in the taxonomic hierarchy will contain terms that refer only to real entities of the kind referred to in biological (generally evolutionary) theory, but there is no guarantee that such referents will be available, in which case the classificatory category *species* will be broader than the referent of the theoretical term. Given that this is at least a possibility, the terms are strictly different in meaning.

One final complication. As I shall discuss, the familiar examples of species as individuals are drawn from complex eukaryotes, but more or less similar phenomena are found in the microbial world. I say 'more or less'. Some, who believe they are less similar and think it important to distinguish the two, prefer to refer to the microbial

---

* Given the controversial nature of the term 'prokaryote' (Pace 2006), I shall use microbe here to refer to bacteria and archaea. I shall not discuss viruses.

analogues as paraspecies or pseudospecies.* In the present chapter, I will refer to any real process that (re)produces similar organisms through time and provides an at least partly isolated channel for evolutionary change as a species. The important contrast is between these and species names as merely classificatory terms that may have no more ontological significance than a degree of longitude.

13.2   SPECIES AS INDIVIDUAL PROCESSES

The Ghiselin/Hull thesis that species are individuals has often struck people as odd. The first reason for this, of course, is just the default assumption that a species name is a classificatory term rather than the proper name of a concrete individual. But more seriously, a species doesn't seem at all like a paradigm individual. Not only are its parts physically dispersed – and we generally expect an individual to have some degree of spatial coherence – but most of them no longer exist. While species are often thought of as consisting of organisms connected by reproductive relations, it is also understood that species may consist of populations that are not presently in contact with one another, so their only reproductive links are mediated by such no longer existing, perhaps long dead, individuals.

The solution to such worries, I believe, is to recognise that a species is not an individual thing, or substance, but an individual process (Dupré 2017; Rieppel 2009). By this I mean, most fundamentally, an entity that persists only through change or activity. Whereas it is generally assumed of a thing that what requires explanation is any changes it undergoes, stability being the default, for a process, its stability is every bit as much in need of explanation as its changes and generally is the more fundamental question. Physiology is the science that aims to explain the stability of organisms, and medicine is its application to maintaining the stability of a particular species of organism.

There is a tangle of issues in metaphysics here that I can only briefly mention. Species in the sense that interested Ghiselin and Hull are clearly entities that persist through time and undergo change (and so, not Mayr's 'non-dimensional' species). But a traditional philosophical view contrasts continuants (substance, things) with occurrents (events, processes) and holds that only the former can undergo change. Occurrents just are changes. They have temporal parts that differ, but these parts do not change. So, the idea of a continuant process, a process that persists through time undergoing change, is contradictory. I shall say here only that these ideas have become increasingly debated in recent years. Rowland Stout (2016) believes that processes are both occurrents and continuants. Helen Steward (2015) rejects this possibility but argues in a partially parallel way that processes persist in a way that is distinct from both perdurance and endurance. Endurance is what substances are most widely supposed to do and is associated with the idea that a substance, say a human, is 'wholly present' at any time that it is present at all. It does not, therefore, have temporal parts, as these would be present only at particular times. Perdurance is what processes and events do; they have temporal parts (timelessly; i.e., whenever

---

* I might also mention viral quasispecies, though these are a very different phenomenon.

they exist, they have other temporal parts not currently present) and therefore do not change. Thoroughgoing four-dimensionalists (e.g., Lewis 2002) hold that everything is in this sense an occurrent.

I won't attempt to engage in any detail with any of these arguments but will just state as minimal as possible a position that makes sense of the idea that species are processes. It will not convince metaphysicians wedded to a different view. But it may be encouraging to philosophers of biology to know that there is space within the current debate that is consistent with the position that I advocate. First, I of course believe that there are continuant, persisting processes. I believe I am one. I am generally sympathetic to perdurantism.* Although one can see the attraction of claiming to be 'wholly present' when present, it is far from clear what it means. The natural non-technical interpretation is that I am all there – either in the colloquial sense of being fully concentrated on the matters at hand or, more literally, that I haven't left a limb (perhaps prosthetic) in another room. Of course, there is not a child in the room at the same time, who existed half a century ago and gradually became me, not least because that child has not existed for some time. If denying that I am wholly present is just denying this, it is hard to see what is wrong with the denial. At any rate, I don't mind too much if someone believes that there are substances that endure in the sense just noted. What matters is that there are also continuant, perdurant processes that undergo change. Certainly, no one supposes a species is wholly present to anyone, either temporally or in almost all cases, physically. A further reason why I hold that species are individual processes is that it fits into a general philosophy of biology that I have defended for some time, according to which all biological individuals are processes. Most importantly, I argue that organisms are processes (Dupré and Nicholson 2018; Dupré 2020; Nicholson 2018).

There is a more biologically interesting reason for attending to the processual nature of species. Individual things or substances, the supposed furniture of the world, are generally thought of, like furniture, as entities that stay pretty much the same unless something happens to them. Processes, on the other hand, typically require activity to persist, and as noted earlier, I take this to be their defining feature. Species clearly take activity to persist. In the first place, they are composed of organisms, which themselves, I argue, are processes; certainly, these must act to persist. Moreover, organisms must specifically engage in reproductive activity for the species to persist beyond the lives of a particular set of members. But there is more, at the level of the species itself. A persisting species doesn't merely involve its members producing more members. The new members either replicate the adaptive qualities of their ancestors or have new properties adapted to different circumstances. To some extent, this may be explained by the flexible development of individual members of the species. But also crucially important is, of course, natural selection. Natural selection is often seen as an inventor or a designer (Dawkins 1996; Dennett 1996). But as selection can only select among the options presented, this is

---

* As noted in footnote 2, I also believe that the future, but not the past, is open. So, unlike most perdurantists, I am not a four-dimensionalist. But I don't believe anything I shall say in this chapter depends strictly on this, so I shall not pursue this point here.

a problematic understanding. Dawkins responds to the problem by giving random mutation almost limitless power to move in the direction of greater adaptation. The idea is that provided there is some sequence of small and adaptation-enhancing steps that lead to the 'designed' innovation, mutations will arise that make these moves, and a sequence of mutations and selection will lead to the complex outcome. If this seems implausible as a general account of the origin of all biological complexity – as I think it should – it is fortunate that more interesting and powerful sources of novelty have been explored by recent evolutionists.* But from a processual perspective, the particular importance of natural selection lies elsewhere, in its ability to stabilise what would otherwise be a rapidly diverging process. The over-production of offspring so strongly emphasised by Darwin, and their subsequent pruning (natural selection), is what keeps the lineage stable at a well-integrated adaptive peak but also allows it to track adaptively small changes in the environment.†

13.3   LINEAGES AND CLADES

Species, in so far as they are concrete entities rather than merely classificatory terms, are parts of lineages. They are the termini of some lineages. It is useful at this point to distinguish sharply between a lineage and a clade. First, and most crucially, as I am using the term, a lineage is a real process, and a clade is not. Or at least, in so far as the classificatory tree maps a real evolutionary tree, the lineages it marks are real processes; when this is not the case, the lines on a classificatory tree don't represent anything in the real world. A lineage is a process within which evolutionary change happens and down which it flows; a lineage is, generally at least, exactly one species thick. A clade, on the other hand, may be as wide as the whole of life, and only historical processes connect its parts. Second, lineages and clades fit quite differently on to the tree of life. Assume, for the moment, that there is a tree of life always diverging permanently at forks and never converging. Then, a clade is best seen as a forward-looking entity. Start with a cross-section of any branch of the tree, considered as representing all the organisms in that branch that were alive at a particular moment. Those organisms and all of their descendants form a clade. Another name for this is a monophyletic group.‡ Defining a lineage, on the other hand, looks backward. Start with a currently existing terminal branch of the phylogenetic tree

---

* For discussion of a wide range of current developments and controversies in evolutionary theory, see the essays in Bateson et al. (2017).

† This does suggest a solution to some common theoretical concerns about how sex came to be as common as it is; namely, that it is an adaptation that stabilises the lineage rather than benefiting the individual organism. Of course, it will be contentious whether lineages are the kinds of things that can have adaptations; but seeing them as processes in much the same sense as organisms should do something to dispel these concerns.

‡ Both 'monophyletic group' and 'clade' are sometimes used atemporally to refer to the currently extant members of the clade. This usage will not be relevant to the present discussion.

and trace all the populations that are ancestral to this population or metapopulation.*
A lineage is typically not a clade. If we work backwards from the current population of humans, we eventually come to a species that is ancestral to both humans
and chimpanzees. The clade stemming from this species includes chimpanzees, but
chimpanzees are not part of the lineage leading to humans.

One might now offer the following criterion for being a species: a species is that
part of a clade that is also a lineage. There is a clade of great apes, but humans and
gorillas are not part of the same lineage process, so that they, but not hominids,
are species. Similarly, at a smaller scale, there is a clade to which I belong deriving
from the family of one my great-great-great grandfathers and his partner. But the
members of this clade are dispersed across the globe and do not form an intelligible
(meta-)population or any kind of connected process. To the extent that lineages are
real processes, the criterion puts at least an upper bound on the size of the species.
The existence of allopatric populations within a species-level metapopulation shows
that we cannot easily define the lower bound.† However, this should not be too surprising, as it is characteristic of processes that they can be individuated in different
ways (how many battles are there in a war or skirmishes in a battle, for instance?).
The point is that we are aiming to pick out a real phenomenon, the stabilisation of
lineages by reproduction and selection, even if this does not provide us with unambiguous individuation of these processes.

So, assume that we have a criterion that identifies humans as an atemporal‡ species. We include in that species every member of the lineage leading to humans that
is also part of the same clade. Suppose the last branch in the tree is the one that leads
to chimpanzees. Then, we exclude the chimpanzees, because they are not part of
our lineage. We exclude parts of the lineage prior to the fork, because these belong
to a clade that includes organisms that are not in our lineage. In principle, we can
continue to trace this species-wide lineage as far back as it consists of a well-defined
species, perhaps to a single-celled ancestor.

I have left open the question of how we identify the atemporal species. Let us suppose for the sake of argument that we can do this in terms of reproductive exchange
and isolation. This works fairly well for vertebrates.§ Hybridisation and gene introgression are commoner than once supposed but may typically make the boundaries
between species vague rather than ambiguous. At any rate, this is not our problem

---

* A species is often taken to consist of several (somehow) connected populations and hence, as typically
being a 'meta-population'. The meaning of this claim should become clearer in due course. I must also
note that 'lineage' is sometimes used to refer to the set of ancestors of much larger taxonomic groups,
e.g., the mammalian lineage. This strikes me as wasteful, as the mammalian lineage, being no longer
an actual process, can just as well be referred to as a clade. Anyone who prefers to maintain this usage,
however, is welcome to offer another term for my concept of a lineage.

† Species are often said to consist of a metapopulation, meaning that they consist of a number of populations too well isolated from one another, generally by allopatry, to be counted as a single population.
This complicates the discussion in ways that I will occasionally note. Thanks to Frank Zachos (personal communication) for reminding me of the importance of this problem.

‡ This is related to Mayr's non-dimensional species concept, but the latter also excludes conspecific
allopatric populations.

§ *Fairly* well, though note the difficulty mentioned earlier about the status of allopatric populations.

here, as I want, rather, to go to the opposite end of the *scala naturae* and consider microbes.

But before turning to the microbial, I must make one further point about reproductive connection and isolation. This is a dynamic concept. Reproductive relations are maintained by reproductive activity, and isolation, where it is not simply provided by physical separation (allopatry), is typically maintained by activities of recognition at the organism level or selection against hybrids at the lineage level. The lineage (though not, of course, the clade) is, once again, a *process*, maintained by the activities that instantiate reproductive connections and preserve reproductive isolation. While it is famously difficult to reconcile the lineage with traditional conceptions of an individual as substance, it is no problem at all to understand it as a process, stabilised and maintained by the activities just mentioned (Dupré 2017).

13.4 MICROBES

So, let me turn to microbial species. Asexuality presents a familiar problem, and it has often been assumed that the Mayrian biological species concept has no relevance here. If microbes formed a wholly divergent tree, then every microbe would potentially be the origin of a clade, and lineages would never be more than one cell thick. Every cell division, put differently, constitutes the beginning of two lineages. Applying the idea that a species is both a lineage and a clade, it appears that every cell is its own species. This is what we should expect: every cell is reproductively isolated from every other. Given this, the restriction of taxa to clades that are lineages makes no sense, and we must allow that any clade might be selected as a taxon. Put in other terms, there is an obvious grouping criterion for bacterial taxonomy, monophyly: any of the countless clades might be identified as a species. But this says nothing about a ranking criterion or which of these countless clades should be identified as any taxonomic unit at all, let alone a species. Perhaps, we could identify sufficiently significant mutations as the initiating events of taxa, but in practice this seems unlikely, given the generally high mutation rates, and in practice, a great deal of microbial taxonomy is indeed based on various morphological criteria, ranging from the traditional grounding in medical pathology to more modern criteria in terms of degrees of genome sequence matching.

But we also now know that microbes are far from forming perfectly divergent trees because of lateral gene transfer. Lateral gene transfer implies that a cell may have ancestors, providers of parts of its genome, that are themselves distantly related. Such a cell is therefore a member of distinct intersecting lineages. It is also, and more significantly, a member of two distinct clades. This means that not only is the ranking criterion for microbial taxonomic units grossly underdetermined, but so is the grouping criterion. We may often, or even typically, have to choose which of several clades a cell should be assigned to. In a pedantic sense, we might note that sexual species do not form a *perfect* tree, as sexually mediated gene exchange happens within the species. Within the species, and unlike the perfect tree imagined for microbes, the tree becomes a net. Such a tree might nonetheless count as perfect if there were perfect reproductive barriers between species; it is simply a matter of properly defining the edges of the branches and hence also the breadth of the lineages. It is now believed that there are some microbial (meta-)populations that approximate this condition by virtue of frequent genetic exchange via homologous recombination within a well-defined group of microbial cells (see, e.g., Achtman and Wagner 2008; for related discussion of microbial species, see Wilkins 2007). If genetic material is circulating within such a group but not entering it from outside, and if this maintains a reasonably high degree of similarity between the members of the group, then the motivation for calling this a species is much the same as that for a reproductively isolated sexual species.* One may suppose that all members share a common ancestor – and are therefore a clade – and also that the network of genetic relations between members, if sufficiently dense, could also be said to constitute them as a lineage.

But the conditions just noted are certainly not universal among microbes.† Indeed, it may be that they are never fully satisfied. One problem is that the distinction assumed between homologous recombination and (other kinds of) lateral gene transfer is not sharp. Homologous recombination is generally understood as the exchange or replacement of genetic material between similar stretches of nucleic acid sequence. But they must be somewhat different, or no detectable event would have occurred. How different? There is no clear answer to that. One might imagine that as long as all the recombination occurs within a clade, this is all that is needed. But a moment's thought should dispel this idea. For suppose I am trying to decide whether a gene exchange between two cells should count as homologous recombination. Are they members of the same clade? The answer is always yes: there is some clade of which all microbes, indeed all life on earth, are members. Certainly, if including an additional cell in a taxon requires a massive increase in the size of the clade, that is a reason not to count their genes as homologous. But then, we have just shifted the question, from how similar must the sequence be for the exchange to count as homologous recombination, to how much would the clade need to be expanded to include this event as an exchange within a taxonomic unit. Neither question admits of an objective answer.

---

* It should be noted, however, that the genetic diversity within a putative bacterial species may be orders of magnitude larger than could occur within a eukaryote species. The *Escherichia coli* pangenome is thought to contain about 13,000, genes of which about 2200 are found in all strains (Rasko et al. 2008). This sounds more like the animal kingdom than an animal species.

† An intriguing recent paper by Bobay and Ochman (2017) argues, on the basis of analysis of a large number of bacterial and archaeal genomes, that discontinuities in homologous gene exchange distinguish the large majority of microbes into well-defined biological species. There are various questions that might be raised about this study, such as whether a relatively small number of named and well-studied species is a good proxy for the whole microbial realm. Moreover, validating that *E. coli*, for instance, is a good biological species is allowing that biological species can encompass massive genomic diversity. If this result holds up, however, it will radically change thinking about microbial species. Since Bobay and Ochman do not suppose that all microbes can be assigned to species (they propose that <15% are unassignable), it does not affect the main conclusion of the present chapter, that discrete lineage processes are not coextensive with taxonomical categories.

In practice, there will no doubt be cases, as is in reality typically the case with species in complex eukaryotes, in which homologous recombination defines a species of microbes quite clearly enough for practical purposes. Perhaps, this is true for most microbes, as Bobay and Ochman (2017) argue. But what is equally clear is that there is no reason to expect this to be the case generally. This could come about through too much lateral gene transfer, lateral gene transfer from distantly related cells that would imply far too much expansion of the clade, or too little. As we have noted, with no gene transfer, there is no genetically based grouping criterion and, *a fortiori*, no ranking criterion.

There may be other ways in which unmistakable discontinuities in the distribution of properties across cells mandate the recognition of a taxonomic unit. The most plausible such case is the sculpting of taxa by ecological forces. It might often be the case that a very precise set of features is optimal for a microbe with a well-defined ecological role, and the rapid and powerful impact of natural selection on fast-reproducing cells will guarantee that intermediate phenotypes between the optimal type and other, ecologically distinguished, types will remain uncommon. Even if the cells sharing this phenotype do not form a clade, as is perfectly possible, they will form an objectively appropriate taxonomic unit. It might even be seen as a species as individual process, but just a process the integrity of which is maintained by ecological pressures rather than by genetic exchange. The crucial point, however, is that consideration of the kinds of species (as individual processes) that might exist in the microbial world does not reveal phenomena that should be expected to engage all the cells in the microbial world. Indeed, they may be rare and exceptional occurrences (though see footnote 14). And this leads me to an important conclusion. Whereas it is an overarching goal of taxonomy to provide an assignment of every organism to a taxon, it is neither a scientific goal nor even a possible one to assign every organism to a species process. If – as I believe – some named species are individuals, this is because the assignment of organisms to species has, as an obvious general principle, that where there is an individual species-like process, we should give the organisms involved in that process a common species name. But since some organisms are not involved in such processes, species as an ontological category must be sharply distinguished from species as a taxonomic category, even though the two have a considerable and non-accidental overlap.

13.5 OTHER SPECIES

What does this tell us about the taxonomy of eukaryotes? The obvious first conclusion is that we cannot assume that all organisms belong to a species individual. This should, perhaps, be an immediate consequence of any species as individuals thesis. Since the species/member relationship is really a part–whole relation, universal species membership would require some kind of *de re* necessity that entities of one kind were parts of entities of another kind. This has sometimes been said to be true of quarks (an issue I will leave to those more expert), but I am not aware of any cases of larger objects, and it is very hard to imagine what the grounds for such a claim could be.

Occasional failure to note the preceding simple point may be due to the failure to acknowledge a point noted at the beginning of this chapter, that the word 'species' is polysemic. Polysemy itself is a diverse phenomenon. Generally, it is distinguished from homonymy, where there are two different words that look and/or sound the same, as in 'bank', a financial institution, and 'bank', the side of a river. The meanings of a polysemous word are somewhat related to one another. So at one relatively unproblematic level, 'species' may, in the context of the widely endorsed species pluralism, be taken to refer to biological species (sensu Mayr), ecological species, phylogenetic species, and so on. Even a monist who thinks that only one of these kinds of species is the legitimate referent for the term as used by biologists might admit that the word could be used in these different ways, though in most of these ways non-referentially.

Interestingly, this last possibility exemplifies a kind of polysemy that is not always recognised, ontological difference. If I were to believe that there were only monophyletic species, say, then the person who refers to morphological species is referring to an empty kind. And an empty kind, surely, is ontologically different from an exemplified kind. Morphological species, for such a theorist, are not like dodos, once exemplified but just not at this point in time, but like unicorns, never exemplified and perhaps even necessarily so (Kripke 1980, p. 24). Species as kinds, at any rate, are clearly ontologically distinct from species as individuals, and the range of the former, we have seen, is greater than that of the latter. In the following discussion, it will be helpful to mark this distinction, and I shall refer to species as kinds as species$_k$ and species as individuals as species$_i$.

For most of the history of life, it would have been possible for an intelligent being who had been present, and is to some extent retrospectively possible for us today, to provide a system of species$_k$ in which to fit every existing organism. Even acknowledging that the division of biological stuff into organisms may be underdetermined by nature (Dupré and Nicholson 2018; Dupré 2020), it is always possible for a classification system to settle by fiat what counts as an individual for its purposes. At some points, species$_k$ began to emerge. Sets of cells (or protocells), probably clades of cells, began to form stabilised and persistent features of the living world. Think of these as partially stable eddies in the flow of life through space and time. Suppose that the stabilisation is maintained by sufficiently faithful reproduction of individuals and some early form of homologous recombination. The boundaries of such a process need not be at all sharp. Cells not part of the original clade (probably part of a slightly larger clade including the founder clade) may engage in the recombination process. Cells generated within the process may mutate in ways that isolate them from the possibility of recombination. Such species$_i$ may be rare, and they may have vague boundaries. As the microbial world persisted over billions of years, such species$_i$ may have become more common but were surely far from universality.

The appearance of obligatory sexual reproduction in eukaryotes was, among other things, the appearance of a device for stabilising species$_i$. The degree of infertility between members of different species, the specialised symbioses between plants and pollinating insects, and the wonderful devices for mate recognition and selection all testify to the evolutionary importance of building barriers around species and thereby maintaining the stability of a well-adapted phenotype. The more widespread sex became, the more theoretically attractive it would become to attempt to align species$_k$ fully with species$_i$. But of course, this was never fully possible. To begin with, even eukaryotes are sometimes wholly asexual. More interestingly, the boundaries between species$_i$ processes are often not sharply defined. Classic and endlessly discussed cases such as the oaks, in which fairly well-differentiated morphotypes persist despite continuous gene flow, are hardly surprising from the point of view of species$_i$ as more or less sharp-edged processes. Whether one prefers to see a group of sympatric, hybridising oaks as one process with well-differentiated subprocesses, or a set of processes interacting with one another, is surely a choice with at most pragmatic significance and no definitive answer to be gleaned from nature.

13.6 CONCLUDING REFLECTIONS

Once it is clear that species as individual processes and species as basal taxa have quite different extensions, a good response might be to try to undermine the deep polysemy in the term 'species' by introducing new terminology for one or the other. Given the origin of the term in Aristotelian logic as a classificatory term marking differentiation within the genus, it is clear that the meaning of species as taxa can claim priority. But it is unlikely that such a linguistic move will catch on, and recalling my discussion earlier of the confusion, real or otherwise, between the species category and the species taxon, I doubt whether the present situation is likely to cause a lot of confusion. Errors are far more likely to be factual than conceptual.

How widespread and well-defined are distinct species processes among eukaryotes? Starting again at what we are inclined to think of as the top, namely, vertebrate animals, while asexual reproduction does occur in some groups, such as reptiles and fish, obligatorily asexual reproduction is very rare. In plants, while there may be many debatable cases, such as that mentioned for groups of oaks, the general contours of the divisions between lineages are generally fairly clear.

It is quite generally the case that in a process ontology sharp boundaries are the exception rather than the rule. When does a wide part of a river become a lake, or which is the river and which the tributary? Just as symbiotic organisms may be engaged in activities of mutual stabilisation, interactions that in some cases blur the boundaries between the individual organisms (Dupré and O'Malley 2009), so it may be that gene flow or ecological interaction between related species is mutualistic and adaptive for the lineages involved.

The extent to which even the general contours of process divisions are clear for plants, fungi, or protists is an empirical question well beyond the scope of this chapter. What I think is fairly uncontroversial is that the progression from animals to plants to fungi to protists involves an increasing distance from the near-universal sexuality defining distinct gaps between most species. Some kinds within all these groups of organisms exhibit various processes for the exchange of genetic material and a tendency for genetic material to be selectively exchanged with relatively similar organisms. In many cases, perhaps most cases of sexually reproducing plants, this will generate well-defined species processes. But at least in fungi and protists, there are major differences between these processes and familiar animal sex, and isolated species-like processes are further from universal. There may, certainly, be other processes that sculpt isolated processes from the flux of reproductive and developmental activity, notably ecological niches driving precise selection. But again, there is little or no reason to suppose this will apply universally.

Species as individual processes, then, are a characteristic feature of vertebrate animals and a common, even typical, occurrence in many eukaryote groups. They are by no means a universal feature of the living world, however. Where such processes exist, it is natural to identify them with the extension of a species term. The nature of the process is such as to create similarity among its parts (members of the species) and distinction from non-parts,* and these are precisely the desirable characteristics for a taxonomic group. But since such identification is not universally available, it would be impossible and misguided to attempt to apply such a strategy universally. This shows that at the very least, the theory supposedly underlying classification must be diverse. Where there is no objective phenomenon such as a lineage process on which to map the species category, it seems inevitable that classification should be given a pragmatic rather than a theoretical basis, and there is no reason to suppose that this pragmatic basis will be the same across very different groups of organisms. Classification, in short, should be allowed a life of its own, and a life that is unashamedly pluralistic and pragmatic. And that, finally, should help to put a final stop to reactionary attempts to extract some life from the ashes of biological essentialism.

13.7   ACKNOWLEDGEMENTS

I am grateful to Frank Zachos and John Wilkins, whose comments on an earlier version have led to many improvements to the text.

BIBLIOGRAPHY

Achtman, M., & Wagner, M. (2008). Microbial diversity and the genetic nature of microbial species. *Nature Reviews Microbiology*, 6(6), 431–440.

Bateson, P., Cartwright, N., Dupré, J., Laland, K., & Noble, D. (2017). Special issue of *Interface Focus*, on New trends in evolutionary biology: Biological, philosophical and social science perspectives. Published online, 18 August 2017. https://doi.org/10.1098/rsfs.2017.0051.

Bobay, L. M., & Ochman, H. (2017). Biological species are universal across life's domains. *Genome Biology and Evolution*, 9(3), 491–501.

Dawkins, R. (1996). *The blind watchmaker: Why the evidence of evolution reveals a universe without design*. New York: WW Norton and Company.

Dennett, D. C. (1966). *Darwin's dangerous idea: Evolution and the meaning of life*. New York: Simon and Schuster.

Devitt, M. (2008). Resurrecting biological essentialism. *Philosophy of Science*, 75(3), 344–382.

---

* Though the degree of similarity is very variable. Recall, for example, the pangenome of *E. coli* (note 14).

Dupré, J. (2001). In defence of classification. *Studies in History and Philosophy of Biological and Biomedical Sciences*, 32(2), 203–219.

Dupré, J. (2017). The metaphysics of evolution. *Interface Focus.* https://doi.org/10.1098/rsfs .2016.0148.

Dupré, J. (2020). Life as process. *Epistemology & Philosophy of Science*, 57(2), 96–113.

Dupré, J., & O'Malley, M. A. (2009). Varieties of living things: Life at the intersection of lineage and metabolism. *Philosophy and Theory in Biology.* https://quod.lib.umich .edu/cgi/t/text/text-idx?cc=ptb;c=ptb;c=ptpbio;idno=6959004.0001.003;view=text;rgn =main;xc=1;g=ptpbiog.

Ghiselin, M. T. (1966). On psychologism in the logic of taxonomic controversies. *Systematic Zoology*, 26, 207–215.

Ghiselin, M. T. (1974). A radical solution to the species problem. *Systematic Biology*, 23, 536–544.

Hull, D. L. (1976). Are species really individuals? *Systematic Zoology*, 25, 174–191.

Kripke, S. (1980). *Naming and necessity.* Cambridge, MA: Harvard University Press.

Lewis, D. K. (2002). Tensing the copula. *Mind*, 111, 1–14.

Meincke, A. S. (2019). The disappearance of change: Towards a process account of persistence. *International Journal of Philosophical Studies*, 27(1), 12–30.

Nicholson, D. J. (2018). Reconceptualizing the organism: From complex machine to flowing stream. In D. J. Nicholson and J. Dupré (eds.), *Everything flows: Towards a processual philosophy of biology.* Oxford: Oxford University Press.

Nicholson, D. J., & Dupré, J. (2018). *Everything flows: Towards a processual philosophy of biology.* Oxford: Oxford University Press.

Pace, N. R. (2006). Time for a change. *Nature*, 441(7091), 289.

Rasko, D. A., Rosovitz, M. J., Myers, G. S., Mongodin, E. F., Fricke, W. F., Gajer, P., … & Ravel, J. (2008). The pangenome structure of *Escherichia coli*: Comparative genomic analysis of *E. coli* commensal and pathogenic isolates. *Journal of Bacteriology*, 190(20), 6881–6893.

Rieppel, O. (2009). Species as a process. *Acta Biotheoretica*, 57(1), 33–49.

Steward, H. (2015, June). I – What is a continuant? *Aristotelian Society Supplementary Volume*, 89, 109–123.

Stout, R. (2016). The category of occurrent continuants. *Mind*, 125(497), 41–62.

Wilkins, J. S. (2007). The concept and causes of microbial species. *Studies in History and Philosophy of the Life Sciences*, 28(3), 389–408.

14 Metaphysical Presuppositions about Species Stability
Problematic and Unavoidable

Catherine Kendig

CONTENTS

14.1 WHY STABILITY?

Treating species as stable cohesive units presupposes what the Biological Species Concept and other conceptions of species try to explain – the nature of this stability, cohesion, or unity. The claim that they are stable cohesive units suggests that there is some mechanism of stability or some underlying relationship or substance responsible for its maintenance. As such, this means that accurate attributions of species-hood ultimately depend on accurate metaphysical presuppositions, which are

DOI: 10.1201/9780367855604-18

themselves tacit in the putative explanations of stability that are adopted. But stability is liable to be misattributed if the mechanisms of stability are erroneously determined. Stability cannot be used to determine whether or not the clusters themselves are natural. This is because any species attribution that relies on these clusters is anchored to a particular metaphysical picture of the world that it takes to be natural without being able to arbitrate between different candidate pictures of the world. Several new accounts purport to offer metaphysically neutral alternatives to Richard Boyd's Homeostatic Property Cluster Theory. This chapter provides an argument for why – even if we adopt a metaphysically neutral account—we still need to consider cases where views about the metaphysical grounds of stable property clustering have resulted in errors of species attribution.

Species, understood in terms of what Richard Boyd (1991, 1999, 2000, 2010) has called 'Homeostatic Property Cluster Kinds' (or HPC kinds), has received broad support. Boyd explains what his epistemological account of species as HPC kinds entails:

> The homeostatic property cluster that serves to define t is not individuated extensionally. Instead, the property cluster is individuated like a historical object or process: certain changes over time (or in space) in the property cluster or in the underlying homeostatic mechanisms preserve the identity of the defining cluster. In consequence, the properties that determine the conditions for falling under t may vary over time (or space), while t continues to have the same definition. The historicity of the individuation conditions for the definitional property cluster reflects the explanatory or inductive significance of the historical development of the property cluster and of the causal factors that produce it, and considerations of explanatory and inductive significance determine the appropriate standards of individuation for the property cluster itself. The historicity of the individuation conditions for the property cluster is thus essential for the naturalness of the kind to which t refers.

> **(Boyd 1999: 144)**

Criticism of Boyd's epistemic account of HPC kinds has come from those who suggest that his reliance on there being 'underlying homeostatic mechanisms' that preserve the identity of HPC kinds as well as his metaphysical commitment to 'historicity as an essential feature of naturalness' overreaches. They argue that a metaphysically neutral approach to explain the nature of kinds is both preferable and possible (cf. Khalidi 2013; Magnus 2014; Ereshefsky and Reydon 2015; Slater 2015). These epistemology-only or epistemology-first accounts share the assumption that the epistemologically valuable features of Boyd's HPC account can be retained while discharging the unnecessary encumbrance of its metaphysical commitments. For Matthew Slater (2015), it is the 'cliquishness' (or stickiness) of homeostatic property clusters and the putative mechanisms of stability that are really doing the explanatory work and not the metaphysical grounds causing these properties to be stable. Stability need not be metaphysically grounded to be epistemologically valuable.

Stability has also been used in wider discussions concerning the nature of property clusters instantiated by natural kinds, counterfactual invariance, and accounts of natural laws. Slater develops his notion of cohesive stability by analogy with Marc Lange's (2009) account of stability of natural laws in terms of their invariance under counterfactual perturbations. Employing Lange's counterfactual invariance, Slater defines cliquishness:

> for any individual x, the fact that x instantiates any sub-cluster of properties in Ø makes it probable that x instantiates all of them, and defines 'stability' as the cliquishness of Ø is invariant under all relevant counterfactual perturbations.

(Slater 2015: 400)

Epistemology-only accounts have not been without their critics (cf. Martinez 2017; Lemeire 2018). John Grey and I (Kendig and Grey 2021) have argued that metaphysical commitments are ineliminable to any account of natural kinds, at least if we want to safeguard against epistemically justified but metaphysically nefarious accounts. This is because what investigators know depends on what they think is out there in the world and how this informs and shapes their research endeavors.

14.2   SPECIES CONCEPTS THAT TREAT SPECIES AS STABLE COHESIVE UNITS

But are species really the quintessential natural kinds that philosophy has often assumed they are? Answering this question seems to rely on an understanding of what the nature of kindhood is for species as well as what the practice of classifying and kind-making or 'kinding' entails when discussing species concepts (Kendig 2016: 1–3). For Ernst Mayr's Biological Species Concept (BSC), Chung-I Wu's genic species concept, Leigh Van Valen's ecological species concept, and many others, this means treating species as naturally stable cohesive units.

Species concepts provide properties, relationships, patterns of inheritance, or other conditions that must be met for something to belong to the category of species. For Mayr, species are understood as populations of organisms that are genetically and reproductively isolated from members of other populations that are sharply separated from one another. They are natural groupings insofar as they are an interbreeding population that shares 'protected gene pools' (Mayr 1970: 12). Mayr's conception of species is shaped by a perspective that systematically privileges a notion of stability based on linear genetic causes: 'species have reality and an internal genetic cohesion owing to the historically evolved genetic program that is shared by all members of the species' (Mayr 1970: 13). A species is a real, natural grouping, the result of an ancestor–descendant lineage that ensures the stable linear genetic inheritance from parent to offspring. The reality of a species and its stable genetic cohesion are maintained in virtue of a shared 'historically evolved genetic program' (Mayr 1970: 13). For holders of the BSC, treating species as stable cohesive units means privileging this view of species over others.

Advocates of not only Mayr's BSC but also the genealogical species concept (Baum and Shaw 1995), the evolutionary species concept (Simpson 1951), and the mate recognition species concept (Paterson 1985) explicitly rely on the importance of potential or actual physical relationships of sexual reproduction and the exchange of genes between conspecifics it facilitates as justification for reliance on notions of internal species genetic cohesion, genetic isolation from members of other species, and species stability. Later accounts of the evolutionary species concept (Wiley and Mayden 2000) and the genic species concept (Wu 2001) accommodate both sexual as well as asexual taxa. Widely accepted among all of these is that stability serves as a constraint to membership in the category of species. A group of organisms is an evolutionary unit that is 'held together by cohesive forces' (Williams 1970: 357). Following Williams, advocates of the genealogical, evolutionary, mate recognition, genic, and biological species concepts understand these cohesive forces in terms of gene flow, genetic homeostasis, and common selective pressure (cf. Ereshefsky 1992: 385–387). Reproductive relationships and genetic cohesion are also implicitly relied upon as necessary for the neatly bifurcating pattern of unbroken and unreticulated lineages central to both cladistic concepts (Ridley 1989) and those phylogenetic species concepts that rely on monophyly (Panchen 1992; Staley 2006). Genetic cohesion and genetic isolation are assumed by these species concepts in their definition of species as an unbroken lineage – a lineage maintained by a sequence of reproductive relationships restricted to conspecifics and ensured by exclusive vertical transmission of genetic material from one generation to the next. Advocates of these diverse lineage concepts conceive of species in terms of the bifurcating branching tree pattern of evolution, whereas the early form of the Evolutionary Species Concept conceived of a species as 'a lineage (an ancestral–descendant sequence of populations) evolving separately from others and with its own unitary evolutionary role and tendencies' (Simpson 1961: 153). In its stronger form, it relies on identifying species in terms of a group of organisms that share a monophyletic lineage (McKitrick and Zink 1988). Many of these species concepts cash out the stable, unifying, or cohesive forces that make a species a unit of evolution in terms of genetic homeostasis and common selective factors, but others argue that cohesiveness can be ensured by means other than genetic. Building on the original ecological species concept (Van Valen 1976), Frederick Cohan identifies how ecological species lineages are 'bound together' as well as their source of stability. He writes: 'evolutionary lineages are bound together by ecotypé-specific periodic selection', where the ecotype is the cohesive unit defined in terms of a 'set of strains that employ the same ecological resources' (Cohan 2002: 467). As I've argued elsewhere (Kendig 2014), not only can lineages be conceived of in terms of different levels of organisation and identified in terms of the level of their cohesiveness (e.g., mode of extragenetic inheritance, mode of gene flow), but their cohesiveness may also be temporally defined.* That is to say, the causally significant factor or explanation of species cohesion at one point in time may not be the same as at another.

---

* In Kendig (2014), I extended Kevin De Queiroz's (2005, 2007) argument that all specific species concepts talk about species as segments of a lineage by arguing that if we take ecological-evolutionary development seriously, then we should extend the definition of lineage and ancestor–descendant relationships as those that also result from extragenetic inheritance, horizontal inheritance, ecological inheritance, and the acquisition of microbes via symbiotic relationships.

14.3 ASSUMPTIONS OF STABILITY AND LINEAGE

The focus of disagreement between those holding different species concepts has been directed at the claim of species stability itself or the privileging of some properties, mechanisms, or relationships considered to be more significant to ensuring species stability. Two problems seem to follow. The first has already been alluded to – the mechanisms of cohesion may not be the same over time. Second, the assumption of stability may artificially predispose a certain line of inquiry into the nature of species that is both ontologically and epistemologically question-begging. It is ontologically question-begging in its assumption that the category of species is that which is homogeneously realised in all those entities belonging to it, and epistemologically question-begging in its assumption that looking for stable unity is the best way to reveal their nature as species.

At least sometimes, the environment is not a stable set of background causes that either facilitate or frustrate the primary genetic cause and stable unidirectional flow of inheritance facilitated by the vertical transmission of genes from one generation to the next that ensures species cohesion. Susan Oyama argues, 'there is no vehicle of constancy, unless the organism and its niche, as they move along time's arrow are so conceived' (Oyama 2000: 27). The causal priority of any objectively construed mechanism seems to be underpinned by a (possibly mistaken) assumption that there is a discrete vehicle of species constancy.

Understanding the stability of a species relies on implicit assumptions about what constitutes a natural lineage or natural grouping. Natural lineages may be conceived of as lineages of genetic, epigenetic, behavioral, symbolic, cultural, and ecological routes of inheritance and natural groups conceived of as including populations sharing genetic as well as extragenetic resources. All of these resources can be conceived of as potential causes of stability or evolutionarily significant change and developmental organisation depending on which ones are used, how they are used, and when, in what order, and in what combinations they are used during the course of an organism's life cycle (Kendig 2014). Natural lineage could be traced genetically, epigenetically, behaviorally, culturally, or ecologically.

14.4 BACKGROUND COMMITMENTS: WHY PRAGMATISM-ALL-THE-WAY-DOWN ISN'T ENOUGH

One reason why biologists and philosophers of biology discuss the theoretical roles played by the *species category*, as well as a whole range of other concepts, is that they believe that something significant hinges on knowing what is included in the conception of species. Without understanding the theoretical scope of the concept, it is thought that we will be unable to know things like how to measure, model, track, generalise, classify, or compare entities *qua* species. If this isn't bad enough, without understanding the concept of, say, *species*, we won't even be able to do simple things like count them.*

---

* The identification of species as stable cohesive units may be pragmatically useful; e.g., it helps us count how many species there are, but it is not purely for pragmatic reasons that species are treated as stable cohesive units.

Investigators often rely on some underlying commitments to circumscribe the boundaries of those objects of potential interest, even prior to knowing to what ontological category that object of interest really belongs. Scientists either explicitly or implicitly rely on a set of background beliefs that help them assign the putative status of species to objects under investigation even when it is the status of the object itself that is being examined. The reliance on these implicit notions to define entities has been highlighted by Ellen Clarke (2012) and Peter Godfrey-Smith (2015).

Clarke shines a light on some implicitly held concepts botanists rely on that allow them to count plants. In order to count plants, they first need to be able to identify which structures belong to the plant and which do not. This is no easy task, as the way in which plants grow and reproduce means that it is often difficult to determine what qualifies as a new individual. Plants develop and grow in a modular fashion (Clarke 2012: 3–4). This means that botanists need to be careful to count a new organism rather than confuse it with a new part of an existing organism. In some cases, they can only distinguish new organisms from parts of existing organisms once they are able to determine the difference between reproduction and growth.

Even if they rely on methods of identification that are widely used, like John Harper's (1977) distinction between genets and ramets, it is still sometimes unclear what counts as the birth of a new organism or the growth of a new part (Clarke 2012: 323–324). In order to count, the botanist must first decide *what* to count as well as what is the purpose of counting. If the botanist chooses to use a demographic method, he makes a decision to treat plants as populations. It is then the parts of populations that are the objects of counting. Explaining this decision-making process, Clarke writes:

> demographers take ramets or trees to be the countable units. With respect to a species like quaking aspen (*Populus tremuloides*), the demographer would mark off an area of forest. He would then return to the area at regular intervals and record the numbers of births and deaths of trees within the area.

> (Clarke 2012: 331)

Counting offspring is important in this context because it provides a way of measuring reproductive fitness. But the importance of it relies on the demographer first deciding what his, her, or their purpose is for counting. This purpose directly informs their decision about how to measure the fitness of individuals that are of interest to them by choosing to treat plants, in this instance, as populations. By relying on this purpose and background commitment, they secure the individuation criteria required to count them. Measuring how much a tree has genet* grown is used as a proxy for the reproductive fitness that is easy to measure. The use of this proxy is sanctioned because genet size and fitness are correlated with one another. As such, the purpose for counting informs both the ontological commitments scientists make as well as the methods they employ given those commitments.

---

* The trees that one finds in a stand of quaking aspen are genetically identical clones called 'ramets', while the clonal colony itself is called a 'genet'.

Clarke and Godfrey-Smith each show how accounts of biological individuation depend on pre-circumscription activities that employ background commitments. Clarke's analysis shows how concepts of plant individuality are often assumed even though they may be unstated when botanists make decisions about what to measure or which to count, whereas Godfrey-Smith draws attention to three forms of biological reproduction: simple, collective, and scaffolded.* For Godfrey-Smith, what qualifies as a 'Darwinian individual' is determined by the ability of the entity or system to reproduce and be the sort of thing that could be the object of natural selection. If what is desired by scientists is to assess whether or not the entity being investigated is a Darwinian† individual, they might choose to employ their understanding of concepts such as reproduction. Depending on the nature of the entity or system being investigated, they might instead decide that tracking some assembly as an evolutionary individual might be done not in terms of explicit consideration of concepts of reproduction or lineage but in virtue of choosing how to count the assembly, following Clarke's account.

However, if what is of interest is the investigation of a metabolic process, it might be in terms of that particular process that scientists use to individuate processual parts. Deciding whether some putative organic entity is a non-Darwinian individual or a metabolic individual might first require deciding which processes count as instances of reproduction rather than growth and which form lineages. This would also require articulation of different conceptions of inheritance, which might rely on different transmission relationships – e.g., horizontal transfer of genetic material, horizontal or vertical acquisition of microbial symbionts, inherited macro- or microbial environments – or other biological entities that have been transferred or acquired from one organism to another that may form lineages. The investigator may also need to be able to identify different stages in the process of development or patterns of growth being tracked over time. These *individuating activities* inform their choice of measurements and methods they use. But each of these assessments rests on understanding what information about the putative organic entity they can use as evidence in support of that assessment. Scientists must first individuate *some* processes before even beginning to address questions about whether or not the entity, thus individuated by these processes, is a biological individual. Assessments of biological individuality are not gleaned from gathering bare, concept-free facts from conceptually free observations, nor are they the result of operationalising a conception whose understanding follows analytically from its definition. The same is true for assessments of species stability and cohesion. These assessments rely on a series of exploratory investigations and background commitments that not only precede the

---

* Simple, collective, and scaffolded all count as instances of reproduction in virtue of the generative relationship between parent and offspring and can be distinguished from one another in terms of the kind of relationship necessary for production of biological mechanisms and its role in evolution (Godfrey-Smith 2015: 10124).

† 'Darwinian individuals' are members of Darwinian populations, defined as those where evolution by natural selection occurs through variation, heredity, and differences in reproductive success (Godfrey-Smith 2009).

discovery of mechanisms of stability but also serve as the standards and guidelines for how one should proceed to study stability.

Some might object, arguing that scientists are just interested in what works and rely only on pragmatic reasons in defense of their epistemic claims (cf. Barker and Wilson 2010; Slater and Barker forthcoming). I agree that it is often the case that the explanation for their decisions is based on pragmatic reasons. But, I'd argue, their justification for believing this explanation to be a better one than another isn't always an instance of pragmatic reasons that are themselves pragmatically justified, or what I'll call 'pragmatism-all-the-way-down'. Instead, sometimes, what is relied upon is a conceptual framework that seems to underpin these pragmatic choices. What causes these choices to seem to make the most sense of the empirical data – or to be conceived of as most pragmatically useful – is the underlying conceptual commitments through which these data are understood.

14.5   WHAT ALLOWS US TO MAKE GOOD INFERENCES?

Evaluating the stability of a species is not possible without first assessing the metaphysical picture that is used to account for its stability. Even in cases where it seems that stability is only attributed to a species for purely pragmatic or epistemic reasons, the pull of these reasons *as* pragmatic or epistemically justified lies in the implicit commitment to a particular metaphysical picture tacitly held. The suggestion outlined in the last few sections – that metaphysical commitments and perspectives are what make sense of notions of species stability – goes against the current trend of epistemology-only accounts of natural kinds that eschew their metaphysical components.

Examining the literature on natural kinds, which often relies on species as exemplars, shows that what is required is to be able to explain why some property clusters allow us to make good inferences from the existence of some properties in the cluster to the probable existence of the rest of them. In a nutshell, what allows us to make good inferences is that these clusters of properties are stable. If what is relevant to our explanation is that these properties stably cohere together, then it is the stability that licenses our inferences. What is required is 'only that these properties be sufficiently stably co-instantiated to accommodate the inferential and explanatory uses to which particular sciences put [natural kinds]' (Slater 2015: 396). When we observe a green organism with what appears to be flat leaflike structures, stemlike organs, and a tangle of roots, for instance, it is the stability of the coinstantiation of properties like *possession of leaves* and *being capable of photosynthesis* that underwrites our inference that this organism probably photosynthesises.

While some argue that it is possible to figure out the epistemological details of what exactly qualifies as a 'sufficiently stable' coinstantiation of properties (Slater 2015: 396), I argue that doing so is not possible without talking about what grounds that stability. Or to put it another way, while epistemology-only accounts take the epistemic value of natural kinds to be independent of their metaphysical grounds, I argue that the metaphysical grounds are what generate their epistemic value. Metaphysical neutrality is neither possible nor desirable with regard to a stance on natural kinds. We are not justified in accepting the explanations that follow from a commitment to a particular natural kind without examining the underlying metaphysical foundation on which our commitment to it rests (Kendig and Grey 2021: 369–372). The reason for this is that we need to be able to decide whether the putatively defined property clusters actually obtain. We need to be able to answer the question: are we successful in attributing kindhood or specieshood to a particular natural grouping? Tacitly or explicitly, scientists and anyone who is talking about species or kinds are involved in the metaphysical theorising about them. That scientists say they are using concepts such as *individual* or *species* 'just for pragmatic reasons' and do not call what they are doing 'metaphysics' doesn't mean they aren't doing it. This is because when evaluating the stability of a species or kind, some properties and relationships are marked as significant, while others are marked as insignificant. Some counterfactuals are deemed irrelevant and others relevant. Scientists are making decisions as to which property clusters are stable and natural. Epistemology-only accounts suggest we can avoid the problem of scientists' interests by reframing them as judgements about disciplinary 'relevance' and therefore tie stability to a particular scientific context or to a particular set of norms (Slater 2015; Magnus 2018). Context-dependence seems to furnish us with a discipline-specific approach to natural kinds, one that relies on relevant properties and side-steps discussions into metaphysics. According to these, what counts as stability is only determinable once the domain of inquiry is selected.

If natural specieshood is construed as stable property clusterings, but we have no account of the mechanism underlying the clustering, it isn't clear how we might decide which property clusters we should be tracking as potentially stable in the first place. Presumption of stability cannot be used to determine whether or not clusters themselves are natural. Species and kind attribution relies on clusters, but the justification for the clustering is in the metaphysical picture of the world that someone takes to be natural. This means that at least tacitly, what makes some clusters the sorts that are afforded species status and license our epistemic inferences is an assumption that they actually (or are probably) not mistaken attributions. But of course, we can be mistaken in at least two different ways: either we attribute stability to a property cluster that actually lacks stability or we can deny stability to a cluster when it is stable. For convenience, we might refer to these as *false positives* and *false negatives*, respectively (for further elaboration, see Kendig and Grey 2021: 363–368).*

14.5.1 Lichen Stability: A Short History

So, what are these tacitly held metaphysical presuppositions in putative explanations of stability? The history of lichen research, a history interleaved with metaphysical and epistemological debates, provides numerous examples. At first thought to be a single organism, lichens were later discovered to be composed of two organisms: a photobiont (an alga or cyanobacterium) and a mycobiont (a fungus) (Brodo et al. 2001).

---

* For instance, if scientists neglect or ignore some relevant counterfactual perturbations that would make the cluster unstable, then we would have an instance of a false positive.

The Swiss botanist Simon Schwendener first described the dual nature of the lichen in 1869.[*] As a symbiotic system that relies on multiple organisms for survival, reproduction, and growth, there has been intense discussion over the nature of lichen composition and its metaphysical status as an individual ever since Schwendener's initial 'dual hypothesis'. These debates over lichen composition have run in parallel to disputes over lichen species naming conventions and continue in the leading-edge research of A. Elizabeth Arnold (2009), Trevor Goward (2008b), and Toby Spribille (2016).

Researchers have long questioned the mode of lichen growth, how to distinguish it from the means of reproduction, how evolution might occur by means of a symbiotic association, and what, if anything, might be understood as a lichen lineage. The history of lichen naming reveals naming practices that carry metaphysical precommitments, e.g., about the conditions of identity, stability, change, continuity, part–whole relationships, and compositionality, which enable re-identification and renaming and have been used in attempts to settle whether or not lichens were individual species,[†] symbionts, or multi-organism consortia.

What constitutes lichen stability is something that has changed depending upon what underlying assumptions were used. Early lichenologists took lichens to be single organisms rather than a composite system of bionts. There was intense disagreement about the seemingly incongruent morphological and physiological features of lichens. One issue over which there was sustained speculation was why there were green cells within the lichen thallus cortex[‡] that appeared to be very different in origin from those of the rest of the structure of the lichen. Another questioned the mode of lichen generation, given the apparent absence of any obvious seed from which new lichens might be seen to germinate. Some lichenologists attempted to resolve the no-seed problem by conceiving of lichens as the product of air and moisture, reflecting their damp environment (Luyken 1809). Others believed that lichens arose by way of a vegetable *infusorium* that grew in the presence of moisture, air, algae, and mosses (Hornschuch 1819). In addition to these, some argued that lichens were a liminal substance capable of sublimating from dry lichen to slippery algae in the presence of water (Raab 1819), while still others suggested that the curious green cells observed in the lichen thallus, causing so much speculation, were actually imperfectly developed gonidia or 'unfortunate brood cells' (Wallroth 1825).

It was in response to these competing accounts and speculation surrounding lichen germination and composition that Schwendener (1829–1919) introduced his

---

[*] Schwendener's dual hypothesis was published eight years before the forest pathologist, Albert Bernhard Frank (1877), introduced the term 'symbiosis' in his research on crustose lichens.

[†] The questions 'what are lichens?' and 'what kind of species are lichens?' have been answered in ways that often confound more traditional conceptions of species. What lichens *really* are – ontologically speaking – continues to be a topic of ongoing debate and one that is informed by different metaphysical commitments to stability, identity over time, and homeostatic mechanisms. For instance, lichen species have been considered to be 'culture chambers for photobiont cells'; 'alga-induced galls'; 'emergent entities'; or 'evolutionary/ecosystem consortia' (see Section 14.6.1) for further discussion of these ontological categories and their accompanying metaphysical commitments).

[‡] The lichen's body is called a 'thallus'. A typical thallus has an upper and lower cortex made up of fungal hyphae, inner layers that include an algal zone and fungal hyphae, and rhizines that allow it to anchor to rocks, trees, or other substrates. [Note to typesetter: do not change the spelling of any of these terms]

dual hypothesis of lichens. According to this, lichens were composite systems made up of two organismal parts, a fungus and an alga. Although Schwendener's dual theory answered several questions, it also generated far more, such as: what exactly is the nature of the stable relationship between algae and fungi? Schwendener's dual hypothesis was adopted by several botanists, including Heinrich Anton de Bary (1831–1888), Albert Bernhard Frank (1839–1900), and Melchoir Treub (1851–1910). But, prominent lichenologists, such as James Mascall Morrison Crombie (1831–1906), resisted Schwendener's dual characterisation. Outrightly dismissing the dual hypothesis, Crombie argued that it amounted to

> a degradation [of lichens] from the position they have so long held as an indepen-
> dent class to [nothing but a fictitious] romance of lichenology, or the unnatural union
> between a captive algal damsel and a tyrant fungal master.
>
> **(Crombie 1874: 262)**

Crombie was not alone in his rejection of Schwendener's dual hypothesis; his views were shared by William Lauder Lindsay (1829–1880), Wilhelm Nylander (1822–1899), and many other lichenologists who were unwilling to demote lichen from a distinctive class of plants to fungi that parasitise algae.

At the center of these disagreements was resistance to what appeared to be a change in lichen metaphysics: from lichens being conceived of as one species of organism to lichens being conceived of as two (from different kingdoms!). This biological debate surrounding lichen composition made particular trouble for naming lichens. Crombie, Nylander, and other detractors rejected what they saw as a new fashion in lichen naming based not on the whole lichen but on only part of it. For Crombie and Nylander, the use of whole lichen species names had been common practice and one that had been gaining strength. Support for whole lichen naming was not limited to Crombie and Nylander. Seeking to ensure that the convention for whole lichen names would continue, American lichenologist Bruce Fink (1911) went so far as to campaign, writing to 75 American and 75 European botanists and lichenologists to ask: 'Should the Lichen be maintained as a distinct class of plants or should they be distributed among the Fungi?' Charles Plitt reports,

> Of the 115 replies [Fink] received, 19, or about 17%, favored distribution; 14, or about
> 12%, thought that Lichens might be distributed, but for one reason or other prefer
> that they should remain a distinct group. In other words, 83% of the 115 believe that
> Lichens should be maintained as a distinct group ... [and the] lichenists [were] nearly
> unanimous in favor of maintaining the group Lichens.
>
> **(Plitt 1919: 84)**

Fink was clearly not alone. Others, such as Eugen Thomas (1912–1986), suggested that lichen symbionts* be assigned a specific nomenclature that was distinct from

---

* Thomas preferred the terms to describe a lichen as a whole rather than as a 'symbiont', which implies that the lichen is a union of two partners that one can (and should) name individually.

the nomenclature of their fungal or algal partners. Thomas' specific recommendation was:

> use the lichen name only as a name for the lichens. The lichen fungi are to be named by the genus name of the lichen with the ending -*myces* and the species name of the lichen in the genitive. Due to ambiguity, the term '*Gonidia*' is used to denote the lichen algae.
>
> **(Thomas 1939: 200)**

Naming practices focusing on the whole lichen, like Thomas', were ultimately rejected, despite early support, and the convention for naming lichens as separate from their fungal or algal partners halted. A new convention, set out in an amendment to the International Code of Botanical Nomenclature (ICBN), anchored the nomenclature of the lichen symbiont exclusively to its fungal partner (Ahlner 1950). Article 76 states: 'For nomenclatural purposes names given to Lichens shall be considered as applying to their fungal components ...' (Ahlner 1950: 809). The 1950 Amendment meant that for the purposes of naming, lichens should be considered as if they were fungi.

Formal names, in order to be universally accepted, must comply with the 1950 Amendment (Barron et al. 2015: 5). The new naming convention was initially strongly resisted by those who argued that some fungi were present in more than just one lichen species and so were not suitable to anchor lichen names (Ciferri and Tomaselli 1955: 190–192), while others argued that some lichen-specific properties, like the presence of secondary metabolites, were not the result of fungal traits but arose only out of the symbiotic state (Culberson 1961: 161–163).

14.5.2   How Is Lichen Stability Understood (or Rather, How Should It Be)?

As this short history shows, how lichen stability and lichen composition were understood informed scientists' decisions about what they took to be a well-justified naming practice. How a lichen was conceived of as stable, what was considered to be the cause of their stability over time, their growth, and their reproduction were also used to direct decisions about how lichens should be studied.

On the classical binary view of lichen composition, a lichen is conceived of as a fungus capable of lichenisation given a particular environment where the right algae or cyanobacteria are available for it to develop into a lichen.* On this view, the explanation of the stability of this cluster of properties was the presence of the same fungus over time. Although the lichen is a multi-organismal system, it is the dominant fungus (referred to as the 'lichenising fungus') that is still considered to be the necessary cause of the lichen and underpins classificatory practices that rely on a single fungus (not the algae or cyanobacteria or any secondary metabolites) to name and track lichens as species.

---

* The classical binary view of lichen composition privileges the role of the fungus, which is typically an ascomycete, over that of the algae or cyanobacteria.

In terms of lichen stability, the single lichenising fungus together with the photo-biont has typically been thought to ground the stability of the properties characteristic of the lichen symbiont (cf. Kendig and Grey 2021: 366). However, recent research suggests that the one-lichen, one-fungus picture ignores relevant counterfactual alternatives; namely, the existence of a lichen may require many more organisms than just two (Henskens, Green, and Wilkins 2012; Spribille et al. 2016; Chagnon et al 2016). Mounting evidence shows that lichens do not *just* include the mycobiont and photobiont but also other fungi, epibionts, and other organisms that are known to be parts present within the lichen symbiont (cf. Arnold et al. 2009; Spribille et al. 2016). *Pace* Schwendener, the dual theory of lichens as a symbiotic system comprised of an algal partner and a fungal partner is an over-simplification of the real lichen relationship of what might be best described as a lichen multi-organism consortium or holobiont.*

Over the past decade, A. Elizabeth Arnold and colleagues extensively researched the role played by other fungi present within the lichen system that were not the mycobiont. These endolichenic fungi were found to be part of healthy lichens and associate closely with the algal partner within the lichen thalli and contribute to their stability (Arnold et al. 2009: 283). These endolichenic fungi play a significant role in lichen evolution and speciation: 'endolichenism appears to have served as an evolutionary source for transitions to parasitic/pathogenic, saprotrophic, and especially endophytic states' (Arnold et al. 2009: 293). These findings reveal the evolutionary impact of non-mycobiont endolichenic fungi on diverse lichen phenotypes – findings that help explain diverse mechanisms of stability. Following Arnold, more researchers (Aschenbrenner et al. 2016; Tuovinen et al. 2019) have revealed a host of functionally and evolutionarily significant relationships and mechanisms that are significant in ensuring lichen stability. For instance, the stability of symbiotic unions has recently been found to be influenced by ecological specialisation and physiological responses of the algal (or cyanobacterial) partners in association with the fungal partner (Muggia et al. 2020: 2–3).

This research shows that fungal diversity and ancestry are not the only things that matter. Algal diversity and phylogeny are also necessary in order to understand the causes of lichen stability as well as the evolution of the lichen holobiont. But, it is not just these that matter; endolichenic fungi, endolichenic algae, cyanobacteria, or epibionts also play significant evolutionary roles in maintaining lichen stability (cf. Arnold et al. 2009; Aschenbrenner et al. 2016; Tuovinen et al. 2019). In some species of lichen, a basidiomycete yeast as well as a host of lichen-specific heterotrophic bacteria has been found to be potentially responsible for morphological diversity in the lichen thallus cortex by maintaining the lichen's shape and structure, appearing to play essential roles in the stability of the lichen system (Aschenbrenner et al. 2016; Spribille et al. 2016).

---

* I'd argue that the lichen symbiont is the most *holobionty* holobiont there is. But it is, of course, not the only one. For further philosophical research on the notion of the holobiont in areas beyond lichenology, see Gilbert, Sapp, and Tauber 2012: Pradeu 2012: and Skillings 2016.

Previous studies ignored many of the counterfactual perturbations relevant for lichenisation by ignoring many constituents of the cluster of properties of lichen symbionts that have later turned out to be responsible for its being stably instantiated (such as the presence of endolichenic fungi, endolichenic algae, and lichen-specific heterotrophic bacteria just discussed). Stability was mistakenly thought to be due to one fungus and one photobiont. As such, it is an instance of a false positive because the stable properties are not explained by its one-lichen, one-fungus, bipartite composition. In many lichens, stability relies on the presence of not two but many more symbiotic microbial species.

14.6   GOOD KIND ATTRIBUTIONS DEPEND ON GOOD METAPHYSICAL PRESUPPOSITIONS

If scientists give weight to irrelevant counterfactuals, then they miss the property cluster that is actually stable because they view it wrongly as being unstable. This occurs because the counterfactuals that would show that the cluster is stable are incorrectly considered to be irrelevant in assessing it as a natural kind. For Slater's account and other epistemology-only accounts of stability, the epistemic value of natural kinds depends on our judgements about the relevance or irrelevance of such possibilities (Kendig and Grey 2021: 364–370). The lichen cases illustrate a series of problems that occur when relying on characterisations of clusters alone when attempting to make kind attributions. Good kind attributions depend on good metaphysical presuppositions, which are tacitly held in the putative explanations of stability that we adopt.

Questioning the basis of natural kind attributions requires treating the activities and assumptions involved in individuating and kinding within scientific practice as significant in the articulation of them as kinds. Doing so enables critical examination of how these individuating and kinding activities affect the acquisition of knowledge, the context of discovery, and the categorisation of the contents of the world. Those holding metaphysically neutral alternatives still need to consider cases where views about the metaphysical grounds of property clustering have resulted in errors of natural kind attribution. This is because stability can be misattributed if the mechanism of stability is erroneously determined. As the example of lichens illustrates, errors in stability attribution arise if one has a mistaken explanation of the stability of the cluster of properties as a result of misidentification of the underlying mechanisms responsible for that stability that explains why the clusters hang together. I contend that Slater's, Barker's, Magnus' and other epistemology-only accounts that seek good explanations of what it is to be a natural kind rely not just on the tendency of properties to cluster in a way that is observable and statistically measurable as properties of clusters, but also 'on the hypothetical claim that these clusters' clustering is and will be stably maintained under various counterfactual changes' (Kendig and Grey 2021: 363). However, if scientists use a faulty perspective or erroneous view when looking at property clusters, they are vulnerable to considering the wrong counterfactual changes. This is because judgements of relevance are themselves informed by their metaphysical assumptions and perspectives, and so these – as well as their empirical data and practices – need to be attended to in order to avoid false positives and false negatives.

14.6.1 FOUR KINDS OF STABILITY ASSUMPTIONS

Metaphysics is usually thought of as something that should be avoided in scientific investigations: a spooky set of *a priori* assumptions without any empirical basis that are best expunged from one's scientific toolbox. Because of this, it is very likely that the claim that scientists rely on metaphysical commitments in their investigations will be seen as highly controversial. In order to alleviate some of this understandable wariness, it is necessary to be clear about what is meant by the phrase 'metaphysical commitments' and what sorts of things qualify as metaphysical commitments. A metaphysical commitment is an extra-empirical assumption about the kind of category something belongs to. This categorisation partially informs and directs scientists' investigative practices, as it furnishes the frame within which the objects of study are and should be understood. Metaphysical commitments, like those that inform one's understanding of part–whole relationships, causation, and composition, shape how researchers conceive of stability and identity in numerous ways. Take, for example, the different views of lichen stability. As mentioned in the previous section, the view of lichen stability – as the result of a two-organism symbiont made up of a fungal partner and algal partner – is not the only view. The stability of the lichen symbiont has also been attributed to:

1. The role of the fungi that are using the photosynthetic properties of algae or cyanobacteria as a different dietary strategy thus ensuring lichen stability. On this view, lichens are a 'very sophisticated culture chamber for photobiont cells' (Honegger 1998: 197);
2. The role of the alga, acting as the architect, which constructs the lichen as an 'alga-induced gall' (Goward 2008a: 2). On this view, the formation of stable lichens ensures long-term algae propagation;
3. The lichen as a self-sufficient and self-replicating organism. Lichens are visible to evolution as an organism in roughly (but not precisely) the same ways as their fungal and algal partners. On this view, 'the lichen thallus is an emergent entity arising from biont interactions' (Goward 2008a: 6);
4. The evolutionary complex or consortia of the lichen, which is the locus of many parts that remain constant from one thallus to the next and from one generation to the next (Goward 2008b: 153–162). On this view, a lichen is 'a cosy mixture of reasonably compatible, self-supporting species of algae and fungi' (Fox 1996: 11).

Each of these four perspectives on stability qualifies as (or at least functions as) a metaphysical presupposition. On the basis of each of these, different entities are treated as having causal priority, and different processes are conceived of as essential to determining continuity over the symbiont's existence or across generations. For instance, causal priority is assigned to the fungi (on option 1); the algae (on option

2); the emergent organism (on option 3); or the evolutionary consortia (on option 4). Processes and mechanisms conceived of as essential to determining continuity over time include fungal control over the photosynthetic metabolism of algae (option 1); algal control over the structural development of fungi (option 2); selective pressures on the unit of selection construed as an autopoietic system (option 3); or the set of homeostatic causal mechanisms acting within symbiogenesis ensuring evolutionary continuity (option 4). Commitment to a particular metaphysical presupposition provides a different filter through which the entities and processes being investigated are either seen as essential to stability or are not. What is considered relevant in determining species stability, evolutionary or developmental significance, or taxonomic salience depends on one's prior metaphysical commitments rather than being purely a consequence of one's metaphysically neutral investigations.* Commitment to each of these very different notions of stability results in different investigative paths that may be taken and leads to different decisions about the lines of research that are pursued by the botanist, mycologist, lichenologist, or naturalist holding them. For instance, if option 1 is taken, the stability of the lichen symbiont lies in the fungus exploiting a different dietary/ metabolic strategy. It will be the fungal partner, rather than the algal partner, that is considered to be the object of investigation, whose reproductive processes are attended to, whose phylogeny is considered taxonomically significant, and whose name grounds that of the lichen symbiont.

14.6.2 How Different Notions of Stability Inform Different Assessments of Identity

The lichen commonly called 'Panther Pelt' is usually referred to with the Code-compliant name *Peltigera britannica*, following the name of its fungal partner. However, *Peltigera britannica* is part of two very different lichens – Panther Pelt and Deciduous Pelt. In Panther Pelt, the photobiont is a *Nostoc* cyanobacterium, whereas for Deciduous Pelt, it is a *Coccomyxa* alga.† According to most lichenologists, they are two forms or *morphs* of the same lichen – Panther Pelt, described as the *cyanomorph*, and Deciduous Pelt, as the *chloromorph* (Goward 2008b: 159).

Whether one understands Panther Pelt and Deciduous Pelt to be *the same* lichen depends on which of the above conceptions of lichen stability (1–4 listed earlier) one adopts. That is, how one understands lichen stability has a direct effect on what one considers to be the appropriate criteria of identity to assess the status of Panther Pelt and Deciduous Pelt.

Depending on which conception of lichen stability one holds will determine (or at least inform) whether or not they consider Panther Pelt and Deciduous Pelt to be two forms of the same lichen. If a group of researchers conceive of lichen stability as being

---

* These commitments may, of course, be found to be deficient on the basis of empirical observations. This does not mean that these commitments are then exclusively empirically based, only that a distorted metaphysical picture of species stability or relatedness may be revealed to be erroneous on the basis of empirical results.

† Although *Coccomyxa* is the dominant photobiont, it is actually not the only one present in Deciduous Pelt (Goward 2008b: 159).

the result of the metabolic solution for fungi (option 1), then it is the fungus that is given priority in the lichen symbiont. It is the fungal partner that is considered causally and evolutionarily relevant to lichen stability, insofar as it is the one that utilises the photobiont partner in order to solve its metabolic problem (e.g., by relying on the photobiont to make its food). On this view, the means by which it does so is of lesser evolutionary or of taxonomic relevance. That the fungal partner uses *Nostoc* instead of *Coccomyxa* is not significant for the assessment of either lichen stability or identity.

As the preceding example illustrates, one's conception of lichen stability directs (or at least informs) one's assessment of lichen identity. The four options listed are only a sample of the possible ways lichen stability has been assessed. If, instead of the fungi-focused conception of option 1, a group of investigators viewed lichen stability as primarily the result of the role of the algae or cyanobacteria (option 2), then an assessment of whether Panther Pelt and Deciduous Pelt are the same lichen would turn out very differently. Holding option 2, the photobiont rather than the mycobiont is given priority in causal explanations of lichen stability. And so, if lichen stability is ensured by the presence of *Nostoc* cyanobacteria in one and *Coccomyxa* algae in the other, Panther Pelt and Deciduous Pelt would not be treated as being the same lichen.*

Which conception of stability one holds does not only have consequences for how many different lichens one identifies when one looks at *Peltigera britannica*; it also affects what sorts of stable relationships are anticipated or expected to be significant in future research. For instance, the recent discovery that some lichens are composed of three or more rather than two symbiotic organisms relies on holding option 4, a consortium view of the lichen symbiont, which corrected (at least in some species) the use of an erroneous metaphysical assumption that lichens were always bipartite. On this view, the symbiotic lichen system of mycobiont and photobiont may possess other microbial species; it may have a third fungal partner, which helps maintain the shape and structure of the lichen thallus (U'Ren et al. 2016); or there may be endolichenic microbial species within the lichen thallus that play a significant role in the way lichens assemble, in the causes of their formation, and in making the instantiation of the associated properties stable over time (Chagnon et al. 2016). As these examples reveal, commitment to a particular metaphysical picture of the causes of lichen stability plays a role in determining what entities, processes, and mechanisms are considered significant in assessing stable clusters of properties and shapes how inferences about both stability and identity are justified.

14.7 METAPHYSICAL COMMITMENTS SUPPORT EPISTEMOLOGICAL CLAIMS

As both the earlier discussion of false positives and negatives and the use of different assessments of stability in lichen research show, stability cannot be used to determine whether or not the clusters themselves are natural. This is because any kind

---

* Taking a phylogenetic perspective, the fungus *P. britannica* belongs to one fungal species taxonomic rank, whereas with a system perspective, taking the lichen species names 'Deciduous Pelt and Panther Pelt' results in two subspecies at the taxonomic lichen rank.

attribution that relies on these clusters is anchored to a particular metaphysical picture of the world that it takes to be natural. This means that the metaphysical presumption of stability cannot be used to arbitrate between different candidate pictures of the world. Attributions of stability are often made on the basis of assumptions about which counterfactual perturbations are relevant. As such, the epistemic value of a natural kind and the claim of its stability are contingent upon certain tacit metaphysical commitments underpinning judgements of relevance. These commitments inform the choices scientists make about what they take to be stability in the first place and guide them in their evaluation of a purported kind. But of course, scientists (like the rest of us) can be fallible with respect to their metaphysical theorising. This means that justifying our inferences based on putative species because of their relative stability means we need to have reliable identification of what qualifies as a mechanism of stability. Purely epistemological accounts that attempt to determine what considerations are relevant to the identification of species without consideration of the underlying metaphysical commitments implicit in the investigator's view assume not only that the investigators' goals fully determine which mechanisms of stability are sought, but also that these are not open to criticism (Kendig and Grey 2021: 373).

The aim of this chapter was to show that because these metaphysical commitments are what cause these inferences to make sense, investigating the role they play in explanations of species stability seems warranted. In order for the epistemological claims – or even epistemology-only accounts – to be justified, some metaphysical view needs to be presupposed that makes a group of properties cluster together (in a cliquish way) *in the first place.*[*] The examples in the foregoing show why metaphysical commitments are necessary to support at least some epistemological claims and illustrate how different commitments lead to different research choices.

14.8   ACKNOWLEDGEMENTS

Generous thanks go to Frank Zachos, John Wilkins, and Igor Pavlinov for their helpful critical feedback and comments on an earlier draft.

REFERENCES

Ahlner, S. 1950. Some aspects of nomenclature and taxonomy of lichens. In *Proceedings of the 7th International Botanical Congress*, eds. H. Osvald, and E. Åberg. Stockholm: Almquist & Wiksell.

Arnold, A., Miadlikowska, J., Higgins, K., Sarvate, S., Gugger, P., Way, A., Hofstetter, V., Kauff, F., & Lutzoni, F. 2009. A phylogenetic estimation of trophic transition networks for ascomycetous fungi: Are lichens cradles of symbiotrophic fungal diversification? *Systematic Biology* 58(3): 283–97.

Aschenbrenner, I., T. Cernava, G. Berg, and M. Grube. 2016. Understanding microbial multi-species symbioses. *Frontiers in Microbiology* 7: 180.

---

[*] I argue elsewhere (Kendig, forthcoming) that this is a ubiquitous problem that arises whenever natural kinds are identified in terms of properties that aim to be kind-constitutive.

Baum, D., and K. Shaw. 1995. Genealogical perspectives on the species problem. In *Experimental and molecular approaches to plant biosystematics*, eds. P. Hoch and A. Stephenson. St. Louis, MO: Missouri Botanical Garden.

Barker, M. and Wilson, R. 2010. Cohesion, gene flow, and the nature of species. *The Journal of Philosophy* CVII(2): 61–79.

Barron, E., C. Sthultz, D. Hurley, and A. Pringle. 2015. Names matter: Interdisciplinary research on taxonomy and nomenclature for ecosystem management. *Progress in Physical Geography: Earth and Environment* 39(5): 640–60.

Boyd, R. N. 1991. Realism, anti-foundationalism and the enthusiasm for natural kinds. *Philosophical Studies*, 61(1): 127–48.

Boyd, R. 1999. Homeostasis, species and higher taxa. In *Species: New interdisciplinary essays*, ed. R. A. Wilson, 141–86. Cambridge, MA: MIT Press.

Boyd, R. 2000. Kinds as the 'workmanship of men': Realism, constructivism, and natural kinds. In *Rationalität, realismus, revision: Vorträge des 3. Internationalen Kongresses der Gesellschaft für Analytische Philosophie*, ed. J. Nida-Rümelin, 52–89. Berlin: De Gruyter.

Boyd, R. 2010. Realism, natural kinds and philosophical methods. In *The semantics and metaphysics of natural kinds*, eds. H. Beebee and N. Sabbarton-Leary, 212–34. New York: Routledge.

Brodo, I., S. D. Sharnoff, and S. Sharnoff. 2001. *Lichens of North America*. New Haven: Yale University Press.

Chagnon, P.-L., J. M. U'Ren, J. Miadlikowska, F. Lutzoni, and A. E. Arnold. 2016. Interaction type influences ecological network structure more than local abiotic conditions: Evidence from endophytic and endolichenic fungi at a continental scale. *Oecologia* 180(1): 181–91.

Ciferri, R., and R. Tomaselli. 1955. The symbiotic fungi of lichens and their nomenclature. *Taxon* 4: 190–92.

Clarke, E. 2012. Plant individuality: A solution to the demographer's dilemma. *Biology & Philosophy* 27(3): 321–61.

Cohan, F. 2002. What are bacterial species? *Annual Review of Microbiology* 56: 457–87.

Crombie, J. 1874. On the lichen-gonidia question. *Popular Science Review* 13: 260–77.

Culberson, W. 1961. Proposed changes in the international code governing the nomenclature of lichens. *Taxon* 10: 161–65.

De Querioz, K. 2005. Ernst Mayr and the modern concept of species. *Proceedings of the National Academy of Sciences* CII(1): 6600–607.

Ereshefsky, M. 1992. Species, higher taxa, and the units of evolution. In *The units of evolution*, ed. M. Ereshefsky, 227–246. Cambridge, MA: MIT Press.

Ereshefsky, M., and T. Reydon. 2015. Scientific kinds. *Philosophical Studies* 172: 969–86.

Fox, B. 1996. Presidential address 1995. *British Lichen Society Bulletin* 78: 11–122.

Frank, A. 1877. Ueber die biologischen Verhiltnisse des Thallus einiger Krustenflechten. *Cohn's Beitrige zur Biologie Pflanzen* 2: 123–200.

Gilbert, S., J. Sapp, and A. Tauber 2012. A symbiotic view of life: We have never been individuals. *The Quarterly Review of Biology* 87(4): 325–41.

Godfrey-Smith, P. 2009. *Darwinian populations and natural selection*. Oxford: Oxford University Press.

Godfrey-Smith, P. 2015. Reproduction, symbiosis, and the eukaryotic cell. *Proceedings of the National Association of Sciences* 112(33): 10120–25.

Goward, T. 2008a. Twelve readings on the lichen thallus IV. Re-emergence. *Evansia* 26(1): 1–6.

Goward, T. 2008b. Twelve readings on the lichen thallus. VII. Species. *Evansia* 26 (4): 153–62.

Harper, J. 1977. *Population biology of plants*. London: Academic Press.

Henskens, F. L., T. G. Green, and A. Wilkins. 2012. Cyanolichens can have both cyanobacteria and green algae in a common layer as major contributors to photosynthesis. *Annals of Botany* 110(3): 555–63.

Honegger, R. 1998. The lichen symbiosis – what is so spectacular about it? *Lichenologist* 30(3): 193–212.

Hornschuch, F. 1819. *Einige Beobachtungen über das Entstehen der Algen, Flechten und Laubmoose*. Flora ii.140–144.

Kendig, C. 2014. Towards a multidimensional metaconception of species. *Ratio* 27(2): 155–72.

Kendig, C. 2016. Activities of *kinding* in scientific practice. In *Natural kinds and classification in scientific practice*, ed. C. Kendig, 1–13. Abingdon and New York: Routledge.

Kendig, C. forthcoming. *Philosophy of synthetic biology*. Cambridge: Cambridge University Press.

Kendig, C., and J. Grey. 2021. Can the epistemic value of natural kinds be explained independently of their metaphysics? *British Journal for the Philosophy of Science* 72(2): 359–76.

Khalidi, M. 2013. *Natural categories and human kinds*. Cambridge: Cambridge University Press.

Lange, M. 2009. *Laws and lawmakers*. Oxford: Oxford University Press.

Lemeire, O. 2021. No purely epistemic theory can account for the naturalness of kinds. *Synthese* 198: 2907–2925.

Luyken, J. 1809. *Tentamen Historiae Lichenum in genere cui accedunt primae lineae distributionis novae*. Göttingen: Henry Dieterich.

Magnus, P. D. 2014: NK ≠ HPC. *The Philosophical Quarterly* 64: 471–77.

Magnus, P. D. 2018. Taxonomy, ontology, and natural kinds. *Synthese* 195(4): 1427–39.

Martinez, E. 2017. Stable property clusters and their grounds. *Philosophy of Science* 84(5): 944–55.

Mayr, E. 1970. *Populations, species, and evolution*. Cambridge, MA: Harvard University Press.

McKitrick, M. C., and R. M. Zink.1988. Species concepts in ornithology. *Condor* 90: 1–14.

Muggia, L., Muggia, Nelsen, M., Kirika, P., Barreno, E., Beck, A., Lindgren, H., Lumbsch, H., Leavitt, S. 2020. Formally described species woefully underrepresent phylogenetic diversity in the common lichen photobiont genus *Trebouxia* (Trebouxiophyceae, Chlorophyta): An impetus for developing an integrated taxonomy. *Molecular Phylogenetics and Evolution* 149: 106821.

Oyama, S. 2000. *The ontogeny of information*. Durham, NC: Duke University Press.

Panchen, A. 1992. *Classification, evolution, and the nature of biology*. Cambridge: Cambridge University Press.

Paterson, H. 1985. The recognition concept of species. In *Species and speciation*, ed. E. Vrba. Pretoria: Transvaal Museum. 21–29.

Plitt, C. 1919. A short history of lichenology. *The Bryologist* 22(6): 77–85+xii.

Pradeu, T. 2012. *The limits of the self: Immunology and biological identity*. Oxford: Oxford University Press.

Raab, J. 1819. Excursion in die Gegend von Muggendorf. *Flora* ii: 289–304.

Ridley, M. 1989. The cladistic solution to the species problem. *Biology & Philosophy* 4: 1–16.

Rieppel, O. 2007. Species: Kinds of individuals or individuals of a kind. *Cladistics* 23(4): 373–84.

Schwendener, S. 1869. *Die algentypen der Flechtengonidien*. Basel: Schultze.

Simpson, G. 1951. The species concept. *Evolution* 5(4): 285–98.

Simpson, G. 1961. *Principles of animal taxonomy*. New York: Columbia University Press.

Skillings, D. 2016. Holobionts and the ecology of organisms: Multi-species communities or integrated individuals? *Biology and Philosophy* 31(6): 875–92.

Slater, M. H. 2015. Natural kindness. *The British Journal for the Philosophy of Science* 66(2): 375–411.

Slater, M. and M. Barker. forthcoming. *Norms of Scientific Classification*. Oxford: Oxford University Press.

Spribille, T, V. Tuovinen, P. Resl, D. Vanderpool, H. Wolinski, C. Aime, K. Schneider, E. Stabentheiner, M. Toome-Heller, G. Thor, H. Mayrhofer, H. Johannesson, and J. McCutcheon. 2016. Basidiomycete yeasts in the cortex of ascomycete macrolichens *Science* 353(6298): 488–92.

Staley, J. 2006. The bacterial species dilemma and the genomic-phylogenetic species concept. *Philosophical Transactions of the Royal Society London B* 361: 1899–909.

Thomas, E. 1939. *Über die Biologie von Flechtenbildnern*, 200. Bern: Buchdruckerei Büchler & Co.

Tuovinen, V., S. Ekman, G. Thor, D. Vanderpool, T. Spribille, and H. Johannesson. 2019. Two basidiomycete fungi in the cortex of Wolf lichens. *Current Biology*. 29(3): 476–483.

U'Ren, J., J. Miadlikowska, N. Zimmerman, F. Lutzoni, J. Stajich, and A. Arnold. 2016. Contributions of North American endophytes to the phylogeny, ecology, and taxonomy of Xylariaceae (Sordariomycetes, Ascomycota). *Molecular Phylogenetics and Evolution* 98: 210–32.

Van Valen, L. 1976. Ecological species, multispecies, and oaks. *Taxon* 25(2–3): 233–39.

Wallroth, F. W. 1825–1827. *Naturgeschichte der Flechten*. 2 vols. F. Wilman: Frankfurt am Main.

Wiley, O., and R. Mayden. 2000. The evolutionary species concept. In *Species concepts and phylogenetic theory*, eds. Q. Wheeler and R. Meier. Columbia, OH: Columbia University Press.

Williams, M. 1970. Deducing the consequences of evolution: A mathematical model. *Journal of Theoretical Biology* 29: 343–85.

15 Critique of Taxonomic Reason(ing)

Nature's Joints in Light of an 'Honest' Species Concept and Kurt Hübner's Historistic Philosophy of Science

Frank E. Zachos

CONTENTS

15.1 INTRODUCTION

When Charles Darwin cautiously discussed the species issue with friends and colleagues in the 1840s, well before the eventual publication of his evolutionary thinking in 1858/1859, Joseph Dalton Hooker famously told him in a letter that 'no one has hardly a right to examine the question of species who has not minutely described many' (DCP, letter 915). Hooker's claim has sometimes been repeated in the often-heated debates about the species problem, particularly when practicing taxonomists or those who used to be practicing taxonomists tried to disparage opponents who came from a more theoretical or even philosophical background. It has also been claimed that this comment of Hooker's was the main culprit for Darwin waiting so long with his evolutionary coming-out and immersing himself

DOI: 10.1201/9780367855604-19

in a decade-long study of barnacles to meet Hooker's 'condition'. Of course, things are much more complicated than that (see Desmond and Moore 1991), and I have always found Hooker's comment rather strange. Are taxonomists really supposed to describe and delimit species based on some kind of intuitive 'gut feeling' and only then ask themselves what they have been doing all the time? Of course, Hooker may not have meant exactly that, but I still think the opposite of his famous claim makes a lot more sense: only with a sound theoretical underpinning and an awareness of potential principal limitations of taxonomy should we embark upon the formidable task of describing and classifying biodiversity. Taxonomy is one of the most important and most fundamental of all biological disciplines, as most, if not all, other disciplines depend on it in one way or another. From this, it follows, and this is an important conclusion, that taxonomy and its theoretical underpinnings can be ignored but not escaped.

The present chapter is, of course, not intended to be a complete treatise on the nature of taxonomy, and in that sense, the Kantian pun in the title may seem a bit bold, but it will touch on two related fundamental issues or problems of theoretical taxonomy – species concepts, on the one hand, and the theoretical evaluation of taxonomy as a scientific discipline, on the other. I will briefly highlight the importance of taxonomy and then give a description of the discipline in terms of its inherently twofold character. From this, I deduce what I call here the Honest Species Concept, which really only describes (rather than being a real concept) what taxonomists do in practice despite all occasional theoretical claims to the contrary. Based on these insights into the discipline of taxonomy, I will then briefly discuss taxonomy from the viewpoint of philosophy of science. Taxonomists, wary of being dismissed as not doing 'proper' science, often emphasise that the description of a species is a hypothesis. Mostly, this is viewed in a Popperian framework, and indeed, Popper reigns supreme among the philosophers of science when it comes to biological systematics, but maybe he shouldn't. A philosophy of science in much better accordance with the nature of taxonomy as presented here is Kurt Hübner's historistic philosophy of science, which I will briefly discuss in the context of its applicability to biological systematics and taxonomy in particular.

15.2   THE NATURE OF TAXONOMY

15.2.1   WHY TAXONOMY MATTERS: THE POWER OF NAMES

Despite its image as a rather dull and purely descriptive enterprise,[*] the significance of taxonomy for the life sciences and beyond can hardly be overestimated. Names do matter, most of all whether there is a name for a taxon or not. In 1975, a by now famous article was published in *Nature* in which the Loch Ness monster was described as a species (*Nessiteras rhombopteryx*) by Sir Peter Scott and Robert

---

[*] Ernest Rutherford's famous line that all science is either physics or stamp collecting seems particularly applicable to taxonomy if one wants to belittle it.

Rines (Scott and Rines 1975). Whether serious or not – the scientific name happens to be an anagram of 'monster hoax by Sir Peter S' – the reasoning behind this taxonomic act was perfectly valid: if an organism or a population of organisms is not acknowledged taxonomically, it does not have legal standing and will not be protected. And that *Nessiteras rhombopteryx*, if it exists, is so rare that it deserves protection, nobody will deny. All joking aside, there are many very real examples where recognition as a species or at least a subspecies makes a critical difference to the fate of the population or whether a region is considered a biodiversity hotspot or not. Apart from implications for potentially endangered populations, there is also a societal and financial dimension to the issue – whether real estate developments are legal, for example, may hinge on the taxonomic evaluation of unique populations, as in the notorious case of the California gnatcatcher, a little songbird of the southwestern United States that has become iconic for the 'real-world' ramifications of taxonomy. For more details on this, from Nessie and gnatcatchers to reptiles and rhinos, see chapter 7 in Zachos (2016). Suffice it to say here that Robert May's quip of 'taxonomy as destiny' (May 1990) really hits the nail on the head.

It doesn't stop there, though. The taxonomic rank of the species is the most commonly used currency for biodiversity not only in conservation biology but also in comparative biology, macroecology, and evolutionary biology. It has repeatedly been pointed out and shown that, for example, diversification analyses, which usually use species numbers and a dated phylogenetic tree as raw data to arrive at comparative speciation rates in different lineages, critically depend on taxonomic decisions (Faurby et al. 2016; Willis 2017) – decisions that are fundamentally underdetermined (see later) and thus, not completely scientific or 'objective'. This introduces a serious bias into purportedly quantitative analyses, such that the claim that 'diversification rates are X times higher in lineage A than in lineage B' almost by necessity could just as well be 'diversification rates are Y times higher in lineage A than in lineage B'. I am not doubting that *qualitatively*, most such analyses will yield correct and relevant results, but their quantitative claim is skewed by the fuzziness of taxonomic decisions at and around what we perceive as the species level. This is a fundamental problem of taxonomy that, as will be argued, cannot be solved. It is also the reason why there have been (and probably will continue to be) endless, and often inherently futile, debates between 'splitters' and 'lumpers' and why there are different competing lists of species for many groups, especially well-studied taxa such as birds (see, for example, McClure et al. 2020 and Neate-Clegg et al. 2021). Again, this is not of solely academic interest – the same taxonomic names in different lists (and thus, also in agreements or evaluations like the Convention on International Trade in Endangered Species of Wild Fauna and Flora (CITES) or the International Union for Conservation of Nature (IUCN) Red List) may refer to different groups of organisms, or the same populations may be listed under different names. This is why there have recently been approaches to create ways of governing such lists with the aim of standardised and transparent procedures to solve conflicts and arrive at a single authoritative global list accepted not only by taxonomists themselves but importantly, also by the various stakeholders and users of taxonomic lists (Garnett et al. 2020; see also Conix et al., this volume).

The species problem has (at least) two different dimensions: a philosophical and a biological one. To philosophers, its main issues are the theoretical significance of species, the ontology of species taxa, and perhaps, communication about species. To biologists, however, the species problem is most real when it comes to the use of species as units of evolution or biodiversity, i.e., in conservation biology and comparative analyses in ecology and evolution (see earlier). Biologists may think that using species as *the* fundamental currency works well, but just because such analyses are being published (including in high-impact journals), and algorithms are becoming ever more sophisticated (e.g., for species delimitation, see Smith and Carstens, this volume), that does not mean that there is no fundamental problem with this approach. Species may well have some theoretical meaning (see Reydon, this volume), but while philosophically speaking, that may save them (at least partly), it is not enough to justify the widespread and uncritical use of the species category as the gold standard in conservation, ecology, and evolutionary biology.

15.2.2 Taxonomy as a Twofold Enterprise akin to Map-Making

Why is there unavoidable taxonomic fuzziness around the species level? The short answer is because *taxonomy is a discrete system imposed on (the results of) a continuous process*. That process, of course, is the process of evolution. This friction between discrete names and a continuous process of the sundering of populations creates a grey area where two populations can be considered either conspecific or two different species, because essentially all criteria used for species delimitation vary continuously, such as morphological or genetic distances or distinctness (how distant, how distinct …?), ecological differences (how different is different enough to count as a distinct niche …?), gene flow, interbreeding and interfertility, and so on. I have dealt with this in more detail elsewhere (Zachos 2016, particularly chapter 6) and will not do so again here. This grey area between clearly a single species (e.g., in the case of a single panmictic population) and undoubtedly at least two different species (as in the case of 'good species'; see later and the contributions by Amitani and Wilkins in this volume) comprises the early or earlier stages of population lineage sundering. Some of the criteria used by the 30+ species concepts published so far will be met at any given time horizon within the grey area, but others will not (see figures 5.4, 5.3, and 5.2 in de Queiroz 1998, de Queiroz 2005, and Zachos 2016, respectively). Many, if not most, real-world cases of closely related (meta-)populations, allopatric lineages, etc. will fall somewhere within this grey area. This is the underlying reason why it has been repeatedly claimed that unequivocal species delimitation can only be done in hindsight (e.g., Sober 1984), i.e., once two or more diverging population lineages have left the grey area and diverged irreversibly.* It is also the background against which Mayden (1997) and de Queiroz (1998, 1999, 2005, 2007) made their by now well-known distinction between primary or ontological species concepts that tell us what a species is – basically, separately evolving population-level lineages (formalised

---

* As long as the two lineages are in the grey area, it is open to debate whether they are nascent species or nascent 'non-species', as it were (Freudenstein et al. 2017, p. 648).

as the Evolutionary Species Concept *sensu* Simpson (1951, 1961) and Wiley (1978) in Mayden and the General Lineage or Unified Species Concepts in de Queiroz) – and secondary or operational species concepts. The latter, on this view, are rather identification criteria that tell us when and if the condition of the ontological species concept is met. For example, if two populations exhibit different niches (Ecological Species Concept), they are different species not because of that but because the very fact that they are so different ecologically is evidence of their being on separate evolutionary trajectories. In other words: they are not distinct species because they have different niches, but they have different niches because they are different species. While I personally think that this hierarchy of species concepts is a valuable conceptual insight – Richards (2010, chapter 5) calls it the 'division of conceptual labor solution' to the species problem – it does not do much in the way of solving practical issues of the kind described earlier (see Section 15.2.1).

To fully realise the extent of the problem, it is important to grasp what the grey area of taxonomy entails. The evolutionary status of any given population in the grey area can be determined as precisely as our current methodologies allow in terms of genetic and reproductive isolation, morphological and ecological distinctness, etc., but at some point, there is a decision to be made with regard to the taxonomic conclusion that we want to draw from what we know about the population. Because of the continuous nature of population sundering (which for a long time, sometimes millions of years, remains reversible), the discrete decision (one species or two) will always, by necessity, be an executive decision that could go either way.* Taxonomy, therefore, is essentially a twofold enterprise (Zachos et al. 2020). The first part is strictly scientific and deals with the description and quantitative analysis of biodiversity in space and time (including state-of-the-art molecular, morphological, ecological, phylogenetic, etc. analyses). The second, however, is an executive decision about how the results of the first part are translated into names. This will always entail some level of qualitative and subjective judgement (cf., among others, Mishler and Theriot 2000a; Sites and Marshall 2004; chapter 6 in Zachos 2016). This means that while species *taxa* are truly discovered in nature, the act of describing and classifying them as *species* taxa is not a discovery at all. What is discovered when 'a new species is described' (according to the usual phrasing) is a hitherto unknown portion of biodiversity, and that is a true increase in our knowledge of the structuring of the living world. This portion of biodiversity is then described *as* a new species (Zachos et al. 2020), i.e., it is *deemed worthy of being classified at species level*. It is important to realise that this is not the same as saying that a species is described. Also, the often-heard statement that many, if not most, species are still unknown to science ('undescribed' – the so-called Linnean shortfall), at least partly (unless hitherto properly unknown organisms are discovered, e.g., in the deep sea), means that what is really unknown so far is the structuring of the living world, i.e., knowledge about the status of populations with respect to the grey area of lineage sundering.

---

* It is important to note that this subjective element of the executive decision only applies in the grey area, not above or below, where taxonomic decisions are comparatively easy and straightforward (at least in theory). Above the species level, for example, monophyly is an objective guide for the delimitation of taxa.

Recently, Thiele et al. (2021, p. 617) have argued that taxonomy is akin to map-making and that this analogy is 'free from the tensions caused when it [taxonomy] is forced into a hypothetico-deductive straitjacket with an over-emphasis on apparent objectivity'. They say that

> Just as cartographers examine patterns in landscapes and from these draw and label simplified maps to represent them, so taxonomists examine patterns in nature and from these delimit and label (name) taxa, thus contributing to a 'map' that others use to navigate biodiversity. A map is a conceptual model of a geographic area, and a taxon is a conceptual model of an enduring evolutionary lineage. The patterns of variation that taxonomists study are bewilderingly complex, and it is the task of taxonomists to abstract this complexity into a more or less simple system that others can follow and use without becoming themselves bewildered.

Importantly, the map is not the same as the territory that is being mapped. Between the level of the territory and that of the map, there is an act of abstraction and a number of decisions on what to include and what not with respect to the level of resolution aimed for and the kind of information considered important as opposed to information that is considered irrelevant. The map-making analogy fits in seamlessly with the conception of taxonomy as a twofold enterprise, as explained earlier. Think of two cartographers who are supposed to map a mountain range.* They will use equipment to measure elevation, and these altitudinal data will be entered into grid cells with coordinates. This is equivalent to the first part of taxonomy, the description and quantification of biodiversity. As long as the two cartographers use the same kind of measuring techniques, they will get the same results. Importantly, if they don't, they can settle disagreements objectively; i.e., they can find out who is right and who is wrong (within limits, of course). But, now comes the second part: if you ask the two how many mountains there are in the mountain range, you are very likely to get two different answers, and these answers will be reflected in the two maps the cartographers produce. One will consider a group of peaks to belong to one and the same mountain, while the other one may think that the depression between two of the peaks should count as a valley separating two mountains in their own right. Translated into taxonomy, the same biological raw data will yield a single species ('mountain') with several subspecies ('peaks') in the first case and two different species in the second. Importantly, that difference is a difference in subjective judgement and cannot be settled based on the data alone.† It is here that different taxonomic 'philosophies' like lumping and splitting come into play, along

---

* Wilkins (2018, p. 342f.) also likens species to mountains as they are 'real phenomena [...] they are things observed that call for explanation, they are explicanda' (p. 342, italics removed).

† According to the Food and Agriculture Organization of the United Nations, the area of the world's land surface classified as covered by mountains has recently increased by about 10% from 29 million km² to 32 million km². The reason for this increase is improved resolution and accuracy of remote sensing models (FAO 2015, p. 9). Upon reading this, a colleague of mine quipped, 'the pen is mightier than plate tectonics', although strictly speaking, it was changes in the raw data that went into the classification scheme that caused the difference.

with the various species concepts that are favored by different taxonomists or taxonomic communities.

The two parts of taxonomy as described here refer to the two well-known distinct taxonomic activities of grouping and ranking. Grouping refers to what since Plato has metaphorically been called 'carving nature at its joints', i.e., finding discontinuities in the spatiotemporal distribution of biodiversity (by means of morphological, genetic, ecological, phylogenetic, etc. analyses). This part is strictly scientific and produces testable hypotheses, e.g., that two groups of populations are reciprocally monophyletic. Discontinuities, however, come at different levels, and the Tree of Life is made up of lineages within lineages in a hierarchical, or enkaptic, structure. Which level of inclusiveness among these various levels is considered the level of species (and something analogous applies to subspecies), however, is the issue of ranking. It is mostly agreed that the supraspecific ranks in the Linnean hierarchy (genus, family, order, etc.) are artificial and incommensurable (see Zachos 2011 and references therein; but see Sigwart, this volume for a different view), but in fact, it very much looks as if the same applies to the species rank as well. Ranking groups – natural groups if they represent 'nature's joints' – at species level (rather than at subspecies level, for instance) is an executive decision that is decidedly not a testable hypothesis, which is why taxonomists often disagree on whether two populations are conspecific or not and why they equally often engage in futile discussions about who is right and who is wrong. These different taxonomic views almost exclusively refer to populations in the grey area, and hence, the issue cannot be solved objectively.* In the grey area, then, there will always be cases where 'A and B are still conspecific' and 'A and B are already distinct species' are equally right or wrong in that neither of the two can empirically be shown to be superior to the other. In line with the Kantian title of this chapter, and in analogy with Kant's antinomies of pure reason, one could call this dilemma a fundamental taxonomic antinomy. But perhaps scientifically, it doesn't matter all that much anyway (leaving aside the legal dimension of taxonomic categories, e.g., in environmental law). If the biological facts are known, and we know that evolution produces fuzzy and vague boundaries between taxa, conceptually, it may not make a difference if we call two populations subspecies or species. An example close to home is that of three hominid taxa including our own. There is evidence beyond reasonable doubt that *Homo sapiens*, Neanderthals, and a third taxon called the Denisovans interbred at some time, which left observable traces in the three respective genomes. Does that make all three conspecific? Particularly, Neanderthals have been classified as *Homo neanderthalensis* or as *H. sapiens neanderthalensis*, and they still could fit either category (species or subspecies) depending on where one chooses to draw the line. Svante Pääbo, one of the leading human palaeogeneticists, says (in Gibbons 2011), 'I think discussion of what is a species and what is a subspecies is a sterile academic endeavor' and 'Why take a

---

* One might argue that the conditions of the various species concepts (criteria *sensu* Mayden and de Queiroz) are objectively testable. That is true, but the decision on which of them should be the yardstick is not, and a delimitation criterion that is objectively testable is not the same as an objective delimitation criterion.

stand on it when it will only lead to discussion and no one will have the final word?'
One does not have to be as pessimistic as Pääbo to acknowledge that he is making a
good point here.

Nature's boundaries are inherently fuzzy, because nature's entities, including species, are often rather processes than discrete objects (Rieppel 2009; Dupré 2017 and
this volume; Nicholson and Dupré 2018), and a discrete classification system will
always experience friction. But while that is a problem for taxonomic studies in the
grey area, it is hardly taxonomy's fault. Just as you cannot blame physicists for the
uncertainty principle and the resulting impossibility of determining impulse and
location at the same time in the microcosmos, you equally cannot blame taxonomists for not being able to objectively settle questions in the grey area. In both cases,
uncertainty or vagueness is the very nature of the beast.

Still, as sophisticated as taxonomists are, it is surprising how little awareness
there often is of the fundamental difference between grouping and ranking and their
respective level of objectivity or lack thereof. It may be due to an urge to over-
compensate for being looked down upon (remember Rutherford's stamp collecting),
but taxonomists frequently overemphasise that their discipline is hypothesis-driven
even where it is not (ranking). It is probably not an exaggeration to say that among
taxonomists, there is a widespread denial of, or ignorance about, the grey area and
the necessary executive decisions involved in ranking. A typical recent example is
Padial and De la Riva (2021). They claim that everything is fine as long as taxonomic
hypotheses are open to falsification (which holds for grouping, but not for ranking),
and they 'circumvent' the grey area problem by referring to Hennig's famous tok-
ogeny/phylogeny divide, i.e., the biological level at which anastomosing reproductive ('horizontal') relationships among organisms give way to strictly hierarchical
('vertical') phylogenetic relationships. However, this divide is an idealisation that
comes with a large grey area of 'some gene flow' and many exceptions (chapter 6.2
in Zachos 2016; see also Vrana and Wheeler 1992). The tokogeny/phylogeny divide,
therefore, is not a clear break but depends on thresholds – thresholds that are ultimately arbitrary in at least some way. And, this does not even take into account the
case of allopatry, where by definition, there are no tokogenetic relationships among
populations. Back to thresholds and executive decisions again ….

Nothing in the Padial and De la Riva paper directly deals with the grey area,
let alone refutes it. They ignore the many cases in which their simplified framework fails, and they implicitly base their arguments mainly on 'good species' (see
Wilkins, this volume). The result is something like a simplistic but 'feel-good' theoretical description of taxonomy that ignores the necessary executive decisions and
thus, a fundamental issue of the discipline.

15.2.3   SPECIES DELIMITATION: BONA FIDE OR FIAT BOUNDARIES?

What follows from this for the issue of species delimitation (see also Quinn, and
Smith and Carstens in this volume)? Do species have bona fide or fiat boundaries?
Species boundaries are, in a way, both. Species delimitation comprises both grouping
(species *delimitation*) and ranking (*species* delimitation). Grouping, if real natural

patterns (discontinuities in biodiversity space) are represented, uncovers bona fide boundaries. However, due to the enkaptic nature of the living world (groups nested within groups of increasingly higher inclusiveness), a decision is needed on which of the various levels of bona fide groups is the species level, and that decision is by fiat.[*] This does not mean that this decision is completely random, but one has to justify (implicitly or explicitly) why a certain level is ranked as a species. Some level of biological or evolutionary relevance will be the guideline here (see Mishler 'and Theriot 2000b; Freudenstein et al. 2017, and Mishler and Wilkins 2018 for a similar view), but there will be many cases where more than one species classification can be defended.[†] Of course, there can be conflict about grouping as well, e.g., conflict among different datasets (morphology vs. genetics, gene trees vs. organism trees, and more), but this kind of conflict is based on real data and can be evaluated accordingly; i.e., it can be addressed by means of an integrative approach or arbitrated based on consilience, etc. The ranking decision is of a different nature.

This has important consequences for all disciplines using species numbers as raw data (among them conservation, macroecology, and evolutionary biology; see earlier). The often-raised question 'how many species of X are there?', intended to be a quantitative and biologically meaningful question,[‡] is essentially at least partly a pseudoquestion. One can meaningfully ask how many mammal species have been described or how many are considered valid at the moment, but the answer to that question tells us as much about taxonomists and their theoretical predilections as it tells us about mammals and the 'objective world out there'. The first sentence in Larsen et al. (2017, p. 230, italics removed) is 'The number of species on Earth is one of the most fundamental numbers in science'. In light of the arguments about the nature of taxonomy presented in this chapter, I am not sure biologists (or anyone else) should uncritically subscribe to such a view. It also means that species counts, for whatever purpose they are done, should always be viewed with caution. Although biologists still seem to cling to species counts and be unaware of their pitfalls, this is

---

[*] This is particularly obvious in the case of monophyly, which is key to a number of different species concepts (Zachos 2016, chapter 4), most notably the monophyly version of the Phylogenetic Species Concept. There is unequivocal connection among the organisms of a monophylum (as they and only they have a certain common ancestor), but delimitation is not possible in an unambiguous way because of the fractal nature of the Tree of Life, which consists of monophyla nested within more inclusive monophyla (cf. Kunz 2012, p. 133f.). It is here that the fiat decision comes in that 'defines' the *species* level of monophyly.

[†] This is analogous to phylogeographic studies where intraspecific lineages are named based on phylogenetic trees or networks. Some level of clade or haplogroup is singled out and named based on certain traits (e.g., geographical distribution), but one could always choose different, more or less inclusive groups instead.

[‡] In particular, the question of how many species there are in total across the whole Tree of Life has received considerable attention (e.g., Mora et al. 2011; Larsen et al. 2017). However, no matter how sophisticated these estimates are, none of them solves the problem of the grey area in which a single correct number cannot be given (Larsen et al. 2017 at least mention the possibility of species incommensurability across the Tree of Life). It is, therefore, maybe not too surprising that results often differ by several orders of magnitude; Locey and Lennon (2016) even hypothesise the existence of ca. one trillion (10^{12}!) species in the microbial realm alone. That would be six orders of magnitude more than the number of all species described to date.

not a new insight. Heywood (1998, p. 211) more than 20 years ago said that 'species are equivalent by designation, only not in terms of their state of evolutionary, genetic or ecological differentiation or divergence', and similarly, Hey (2001a, p. 187) said that the 'genuine problem with species counts, even repeatable ones that are arrived at with a consensus on methods, is that we don't know just what they are counts of'.

The question of how many species there are is equivalent to the question of how many mountains there are in a mountain range (see earlier) or how many families live in a certain community or city. One will count parents and their children, but if two siblings each have children, does that count as one or two families? And, what about cousins? Second cousins, third cousins, etc.? Even if one has all the raw data (elevational data on a grid or complete pedigrees of all people living in an area), the question has no single objective answer. The species problem in biological practice is thus mainly one of incommensurability. We think we use the same currency for our analyses, but what we are really doing is mixing euros, yen, dollars, pounds, and others in the same balance. Put a bit differently:

Species lists are lists of apples and oranges. Perhaps they could – through the adoption of a consistent but necessarily arbitrary threshold – be turned into a list of apples only, but then they could also always be turned into an equally justifiable list of oranges, simply by agreeing on a different threshold or criterion.[*]

15.3 AN 'HONEST' SPECIES CONCEPT

Where does all this leave us in taxonomic practice? What I am arguing for is a long-over-due acknowledgement by the biological community at large that the species *category* is no different from the so-called higher categories, which are known – at least, this is the majority consensus – to be incommensurable artefacts. Just as the *taxon* Mammalia is a real entity, but the *class* Mammalia is an artificial category, so *Homo sapiens* is a real entity, but there is no reason to believe that it is the same kind of entity (and treat it as such) as, say, *Drosophila melanogaster* or *Escherichia coli*.[†] A truly 'honest' species concept must take this into consideration. Importantly, the term 'honest' denotes a theoretical commitment on my part and does not entail that whoever holds opposing views is being dishonest. It highlights the limitations of the concept of a species concept itself. Species concepts, after all, are definitions of the species category, i.e., attempts at defining what all species taxa have in common that allows them to be ranked at species level. If there is no such thing as an objective level or rank of species, then an honest species concept is very different from other species concepts. Note that the denial of

---

[*] The incommensurability problem says nothing about the reality of species *taxa*. Every single species taxon may capture a real entity (in the sense of extramental reality), but these entities are not the same across the board. It is the species *category* that seems to lack theoretical justification. The taxa that comprise lions and dandelions are both real entities, but there is no reason to believe that they are *the same kind* of entity. This is the essence of the species problem in biological practice.

[†] That species taxa are real is also suggested by the fact that they are recognised not only by humans but also by other living beings: lions recognise lions as potential mates or competitors, parasites recognise their specific hosts, and pollinators (like certain wasps) recognise specific plants (figs in the case of the wasps), and so forth. This is convincing evidence of the reality of lions and figs, but nothing about it requires lions and figs to be the same kind of real entity.

an objective rank of species does not mean that the species category has no theoretical or heuristic meaning at all (see Reydon, this volume). It does mean, however, that it is sufficiently vague or fuzzy not to be able to function as a fundamental unit or currency in conservation or comparative biology. Pigliucci (2003), for example, has argued that the species category is a case of Wittgensteinian family resemblance, i.e., there are a couple of characteristics of the species category (basically, the content of the 30+ species concepts: interfertility, morphological diagnosability, ecological distinctness, etc.), but that not all species taxa ever exhibit all of them.[*] This is an interesting point of view conceptually, but it also confirms that simple species counts lump very different kinds of entities together (reproductively isolated species with those that are not, ecologically distinct species with those that are not, etc.). Comparative studies, particularly those aiming at quantitative results like the diversification analyses mentioned earlier, cannot and should not be carried out on such a basis.

As stated earlier, taxonomy's fundamental problem can be described as the friction arising when a discrete system is imposed on a continuous process. This results in the grey area mentioned earlier. Everyone can agree (and most biologists and philosophers of biology do) that species taxa are separately evolving population-level lineages that show cohesion or homeostasis at any point in time of their existence. This view is a combination of the Evolutionary, General Lineage, and Unified Species Concepts that emphasise the aspect of separate lineages with Templeton's (1989) Cohesion Species Concept. Separation among species taxa and cohesion or homeostasis within them are two sides of the same coin. And both of these sides, and here lies the rub, come in degrees – being separate just as much as being coherent. 'Abandon all hope ye who enter here', one might be inclined to say, when it comes to saving the species category as a class of commensurable species taxa, and an honest species concept must not turn a blind eye to this. Unlike many other species concepts, it is not only normative (be it ontologically or operationally) but also descriptive, in that it includes what taxonomists in real life are actually doing, in addition to and often despite all theoretical claims. Somewhat surprisingly, taxonomists quite often do not make explicit reference to one or several concrete species concepts when they describe new species (Luckow 1995), but given the need for an executive decision in the grey area, this is perhaps not so unusual. Data are presented, the distinctness of a taxon is hypothesised based on these data, and then, the often-silent implication is that 'this is different enough to warrant species status'. Given what has been said of the nature of taxonomy and its limitations so far, this is ultimately indeed the only feasible way, and probably, this is the reason why it is so common.

---

[*] Note that this is a family resemblance at the level of the species category. Usually, family resemblance concepts have been invoked in discussions on the metaphysical status of species taxa (Boyd's homeostatic property clusters being the most famous; see Boyd 1991, 1999). Family resemblance is a kind of polythetic similarity: membership in polythetic classes, unlike common classes with essential necessary and sufficient characters, is based on the more-than-random probability that some of a group of characters are present (usually in the form of a minimum quorum, which, once again, depends on a subjective threshold).

Honestly facing and expressing these limitations is not so common, unfortunately.*
Putting all this together, an honest definition or circumscription of species (taxa) that
also takes taxonomic practice into account could be something like this:

> Species taxa are populations or metapopulations/groups of populations that show
> genetic and/or ecological and demographic cohesion among them and at the same time
> are separated from other such groups in such a way that they are on their own evolu-
> tionary trajectory and have a good probability of staying separate in the future (i.e.,
> do not merge again with related groups). All the criteria that establish cohesion, on
> the one hand, and separation, on the other, vary continuously; delimitation decisions
> pertaining to the species/subspecies rank therefore entail executive threshold decisions
> that need to be justified and are ideally made as consistently as possible across as many
> taxa as possible. Importantly, in the grey area of lineage sundering, two or more taxo-
> nomic conclusions are equally justifiable.

Obviously, this is not nearly as pithy as most species concepts, but it is not intended to be
pithy, just honest. It contains an element of skepticism, because ranking decisions (and
every *species* delimitation is also a ranking decision) are necessarily partly arbitrary.

An important term that has been mentioned is 'good species' (see earlier and the
chapters by Amitani and Wilkins in this volume). Amitani (2015) links it to what
is called prototypical reasoning in cognitive psychology. Prototypes are essentially
highly exemplary instances of a concept – in the case of species, those that meet many
different identification criteria, such as reproductive isolation, morphological and ecological distinctness, etc., and thus, those where there is no argument over their status as
a separate species taxon. In other words, good species are those outside the taxonomic
grey area, and it is these examples that will come to mind when we think of species,
e.g., the platypus, *Homo sapiens*, and others. The Honest Species Concept highlights
that many taxonomic decisions, however, must be made on populations that are within
that area and hence, not 'good' species in the sense given here. It should also be mentioned that while there is agreement that good species are not non-species, according
to certain species concepts that are prone to splitting, they are often classified as more
than just a single species. Particularly, the diagnosability version of the Phylogenetic
Species Concept, according to which any two consistently diagnosably different populations are also different species, leads to the splitting of many good species into several, simply because different populations within the single good species are in the
grey area of separation (e.g., allopatric populations that have been sundered for some
time). A point in case is the tiger (*Panthera tigris*), which is often used as an example
of a good species but which has been split into two species by some authors based on a
few diagnostic base pair positions in the mitochondrial DNA (Cracraft et al. 1998). The
application of the same species concept has also led to a 'doubling' in species numbers

---

* But see Seifert (2020), who in his recently published GAGE (Gene and Gene Expression) Species
Concept makes explicit reference to thresholds of evolutionary divergence, which 'should be fixed by
consensus among the experts under the principle of avoiding oversplitting or lumping' (pp. 1033 and
1036). This would be something like a democratic or majority-rule version of the Taxonomic (Regan
1926; Blackwelder 1967) or Cynical Species Concept (Kitcher 1984), according to which a species is
what an expert taxonomist says it is.

for the Bovidae (cattle, wild sheep, antelopes, and relatives) (Groves and Grubb 2011) – a decision that has been severely criticised (Heller et al. 2013). So, maybe good species are not so 'good' after all; but at least, it seems safe to say that good species are almost never merged with another taxon and that when they are split, the resulting species are much less 'good' than the more inclusive species was. That also holds in cases where there is general agreement that a formerly 'good species' must be split into two or more, as is the case in sibling species.

An honest species concept must place emphasis on thresholds and decisions made by taxonomists – decisions that could be different and thus, are somewhat contingent. This view of taxonomy fits well with a philosophy of science as advanced by Kurt Hübner, to which I will now turn in the last part of this chapter.

15.4 BIOLOGICAL SYSTEMATICS AND KURT HÜBNER'S HISTORISTIC PHILOSOPHY OF SCIENCE

The discipline of taxonomy is part of biological systematics. The terms *systematics*, *taxonomy*, and *phylogenetics* are often used with different definitions, but here, I will treat systematics as a higher-level discipline comprising phylogenetics and taxonomy. Taxonomy and phylogenetics are different disciplines but clearly related ones that depend on one another. Any phylogenetic reconstruction depends on some a priori information or decisions relating to taxonomy. One needs at least some preliminary 'operational taxonomic units' (OTUs) among which phylogenetic relationships can be established, and without an a priori idea on these units, adequate sampling for morphological, genetic, etc. analyses is impossible. Conversely, alpha taxonomy, the description, delimitation, and naming of species, if done properly, will take into account phylogenetic and phylogeographic information on the organisms under scrutiny. The classification* of higher taxa (sometimes called beta taxonomy) is primarily based on phylogenetic results anyway (strictly so in a phylogenetic systematic or cladistic framework, not quite as much in evolutionary systematics).

It is probably fair to say that physics is the model discipline used in philosophy of science – its *Drosophila melanogaster* or *Escherichia coli*, as it were. However, biology is arguably no less philosophically relevant than physics (see Simpson 1963 for an early plea not to let physics rule supreme here). And within biology, the philosophical foundations of systematics have probably been more hotly debated than most other topics in the life sciences.[†] These often fierce debates, sometimes known

---

* The term *classification* is also ambiguous. Philosophically, it suggests an ordering scheme based on classes (entities defined by essential properties as opposed to, for example, philosophical individuals). To avoid this ontological connotation, the term *systematisation* is sometimes used instead. I will not pursue this further. By classification in the context of biological systematics, I mean the representation of the hierarchical pattern that, for instance, the species *Homo sapiens* is part of *Homo*, which, together with other genera like *Pan*, *Gorilla*, *Australopithecus*, etc., is part of Hominidae, which in turn is a part of Primates, Mammalia, Vertebrata, etc. None of these taxa are classes in the ontological sense, but I use classification nonetheless because it is the most commonly used term.
† Especially cladists have been described as being 'obsessed with the philosophical underpinnings of science' (Perkins 2011, p. 895).

as the 'systematics wars', started with the cladistic revolution in the wake of Willi Hennig's seminal publication *Phylogenetic Systematics* (Hennig 1966). The gold standard for many when it came to transforming systematics into a rigorously scientific ('objective') discipline was Karl Popper's falsificationism, where falsifiability serves as the demarcation criterion between empirical science, on the one hand, and pseudoscience (but also metaphysics and mathematics), on the other. Popper was arguably the most influential philosopher of science in the twentieth century, and that is certainly true for biology and particularly, evolutionary biology and systematics. More recently, the application of Popperian philosophy of science to systematics has been heavily criticised (e.g., Vogt 2014a, b and de Santis 2020 and references therein; for a general discussion of the topic, see Rieppel 2003). One of the contentious issues has been whether and in what way cladograms are falsifiable (Vogt 2008 against, for example, Platnick 1977). However, strict falsificationism is, rather, a framework in which to evaluate theories. In practice, it is not as straightforward as the logical all-or-nothing examples might lead one to think[*]: a single white raven refutes the universal statement that 'all ravens are black', but especially in biology, the science of exceptions, historical contingency, and fuzzy boundaries, things are seldom as simple as that. Here, the 'ravens' are often grey or black with white spots, and then, the question arises how big a white spot or how many white spots are allowed for the raven still to be called black. As a result, and not only in biology, falsifiability comes in degrees and depends on conventions (as does corroboration; see Rieppel 2003), and this certainly also holds for taxonomy and its twofold nature with a grey area and the need for executive decisions based on qualitative judgements. Rather than discuss Popper yet another time in more detail, I would like to present a few thoughts on how the theoretical foundations of biological systematics may be framed in terms of the ideas of another philosopher of science, less well known than Popper, namely, those of Kurt Hübner (1921–2013), who would have celebrated his 100th birthday at the time of writing.

Hübner was interested in many different aspects of philosophy, but particularly in philosophy of science and in building bridges between philosophy and natural sciences.[†] His major publication in the philosophy of science is the *Critique of Scientific Reason* (Hübner 1983,[‡] henceforth abbreviated as CSR), obviously a reference to Kant (and another inspiration for the title of the present chapter), in which, building on Pierre Duhem (1861–1916), he develops a historistic philosophy of science that he contrasts with logical positivism, Popper's critical rationalism, and other views that he considers fundamentally ahistorical. For Hübner, theory and history of science

---

[*] All methods to reconstruct phylogenetic trees, for example, allow for some level of homoplasy, and thus, no character distribution in the dataset will be able to *logically* falsify a certain tree topology.

[†] For a comprehensive obituary (in German) including an outline of his life and work, see Deppert (2015). For a complete bibliography, see Deppert et al. (2015). Although Hübner's philosophy of science contains elements of an evolutionary process of gaining knowledge (evolutionary epistemology), he seems to have been strangely critical of the theory of evolution and even biological evolution itself, potentially because of his religious upbringing (Deppert 2015).

[‡] I am quoting from the 1985 English paperback edition; the German original was published in 1978 (Hübner 1978).

are connected in a fundamental way: 'A theory of science without a history of science is empty, while a history of science without a theory of science is blind' (CSR, p. 71).* I will not give a full summary of Hübner's ideas, let alone an in-depth treatment. Rather, I will focus on some central ideas of his philosophy of science insofar as they are relevant for a theoretical evaluation of biological systematics. Perhaps most importantly, Hübner thinks that science (he uses physics as a model but later applies his ideas to the study of history as well), or more precisely, the construction and judgement of a theory, depends on a number of a priori categories that he calls precepts (*Festsetzungen* in German; some of them he sometimes also calls *Grundsätze*, or principles). This he already found in the writings of Duhem, but he presents an emendation of Duhem's theory in that in his view, these a priori precepts 'can only be grounded and understood historically. They are contingent precepts. And because the historical element of a theory resides in these precepts, a historistic theory of science will have to begin by establishing a guiding thread to be used in their systematic disclosure' (CSR, p. 42). Hübner then distinguishes five (groupings of) such precepts (CSR, pp. 42f.), but he explicitly states that he makes no claim to being exhaustive:

1. *Instrumental precepts*: 'precepts that lead to the procurement of measurement results (precepts concerned with the validity and function of employed instruments and means, etc.)';
2. *Functional precepts*: 'precepts that are used in the formulation of functions or natural laws on the basis of measurement results and observations (for example, limitations placed upon the selection of measurement data, margin of error theories, etc.)';
3. *Axiomatic precepts*: 'precepts that serve as axioms introduced for the deduction of natural laws and by means of which experimental predictions are made with the aid of boundary conditions';
4. *Judicative precepts*: 'precepts that govern the acceptance or rejection of theories on the basis of experiments. (The following belong to this group: (*a*) precepts that serve as a basis for deciding whether the theoretically deduced predictions agree with the given results of measurements or observations; (*b*) precepts that indicate whether the theory in question is to be rejected or retained in the event of a failure of agreement and, if retained, whether it is to be altered, as well as where such alterations are to be made)';
5. *Normative precepts*: 'precepts that serve prescriptively in the determination of those characteristics which a theory should generally possess (for example, simplicity, a high degree of falsifiability, self-evidence [...], the satisfaction of certain causal principles or empirical criteria of meaning, and other similar traits)'.

---

* The German term for theory of science is *Wissenschaftstheorie*, for which there is no standard English term. It is the part of philosophy of science dealing with the structure, preconditions, evaluation, etc. of science and scientific theories. As noted by Hübner's translators, this quote is a reworking of Kant's famous quip 'thoughts without content are empty, intuitions without concepts are blind' (*Critique of Pure Reason*, B 75).

Hübner considers these precepts as absolutely necessary for science (physics in this case):

> These five concepts describe those kinds of precepts which are indispensable for the formation, examination and judgment of theories in physics [...] It is therefore a condition of the possibility [another very Kantian thought] of a theory of physics that there be particular precepts pertaining to it in such a manner that each of these precepts must fall into at least one of the above-mentioned five concepts, and thus that none of these remains empty. For this reason we can then call these concepts *scientific-theoretical categories* (*wissenschaftstheoretische Kategorien*).
>
> **(CSR, pp. 43f.)**

He then gives an example for each of the five categories for physical theories (CSR, pp. 44f.): rigid bodies follow the laws of Euclidean geometry (instrumental precept); a function can be derived from measurement data by means of the Newtonian interpolation formula (functional precept); all systems of inertia should be viewed as equal (axiomatic precept); theories must be abandoned if a theoretically predicted result is not obtained (judicative precept); and, all theories must be in agreement with a deterministic and therefore unlimited principle of causality (normative precept).

These examples, with the exception of the radical version of the principle of falsification as a judicative precept, are, of course, tailored to physics, not to biology, but I think that Hübner's approach can be very fruitfully applied to biological systematics, because it has the potential to yield a more detailed and therefore, more appropriate picture of the theoretical underpinning of the discipline than the often rather one-dimensional Popperian falsificationism. This is not to deny that empirical scientific claims should be falsifiable but rather, to add additional layers of theoretical considerations to this very general normative guideline. For example, in the grey area, as explained earlier, taxonomic decisions regarding species status (or the lack thereof) are not falsifiable but depend on generally agreed-upon conventions or as is often the case, on the taxonomic 'taste' of the single taxonomist making that decision. Maureen Kearney closes a chapter on *Philosophy and Phylogenetics* with the thought that '[f]ertile ground for future discussion between systematists and philosophers lies in the critical examination of what it means to be objective *and* scientific within an evolutionary worldview' (Kearney 2007, p. 231). I believe that this is where Hübner's philosophy of science can play an important role.

A complete analysis of Hübner's five categories or precepts as applied to biological systematics would be beyond the scope of this chapter, but in the following, I will highlight a few aspects of systematics whose re-interpretation in a Hübnerian framework will hopefully show the potential of this approach. *Instrumental and functional precepts* in biological systematics would refer to the collection of data relevant to taxonomy and phylogenetics, for example DNA sequencing, three-dimensional geometric morphometrics, measuring ecological variables, and the like, and how and why these data are suitable for systematic purposes, i.e., the choice of relevant characters. To give just one example, Richard Owen (1804–1892) published a classification of mammals into four subgroups based on brain anatomy (Owen 1858): Lyencephala, Lissencephala, Gyrencephala, and Archencephala, with humans the only representatives of Archencephala. This choice of character is almost certainly anthropocentrically motivated, even if Owen states that the cerebral traits he used show 'association with concurrent modifications in other important systems of organs' (p. 13).

Axiomatic precepts in biological systematics are perhaps somewhat ambiguous. When comparing physics and historical sciences, Hübner says that these axiomatic precepts or principles

> are understood as those which make up the core of a theory. In the natural sciences these are assumptions about the fundamental laws of a natural system (for example, the Schrödinger equations). In the historical sciences, however, these are assumptions about the fundamental basic rules of a historical system.

(CSR, p. 184)

The rub then lies in whether biological systematics is a historical discipline (as is most of evolutionary biology) or not. This mainly hinges on how one views the relationship between systematics and evolutionary history. Hennig and the later school of process cladists see phylogenetics as a discipline concerned with reconstructing the results of the process of evolution, particularly grouping based on the recency of common ancestry, and they will allow for evolutionary arguments in the methodology to arrive at phylogenetic trees. Pattern cladists, on the other hand, argue that the patterns of character distribution have logical primacy and that they must be arrived at without evolutionary premises. In a second step, these patterns can then be interpreted in the light of evolutionary history. This is why pattern cladists often present topologies as sets within sets or boxes within boxes based on character trait distributions rather than as trees, because trees have evolutionary connotations, while simple schemes of boxes or sets do not. If presented as trees, the resulting cladograms are not to be seen as phylogenetic trees but primarily as 'synapomorphy schemes' (Nelson and Platnick 1981, p. 141). While synapomorphies can be (and are) secondarily interpreted as being inherited from a common ancestor, in systematics s. str., they are just the result of arranging groups into a hierarchical scheme with a minimum of contradictions with respect to character distribution. In short, process cladists explicitly aim at reconstructing history, while pattern cladists simply build parsimonious topologies and thus, want to divorce systematics and the classification based on its results from evolutionary theory. Whether this is really possible[*] or even desirable is a different issue and need not bother us here (for a very readable short overview of the history of cladistics, see Rieppel 2014).

---

[*] All cladistic analyses are based on data matrices, and while it is possible to analyse such a matrix without evolutionary assumptions, the very construction of the matrix requires a priori decisions on which traits of a taxon should be compared with which traits of the other taxa in the matrix. This, however, is an a priori assumption on homology at a higher level and therefore, an a priori commitment that might be interpreted as historical or evolutionary. At the very least, homology is not simply deductive. For a similar view, see Wägele (2001, chapter 6.1.10).

The line between axiomatic and normative precepts is sometimes perhaps rather fine. Process cladists certainly take evolution as an a priori fact for their systematic reconstruction; for them, evolution is a 'basic rule of the historical system' they are studying and in that sense, an axiomatic precept. Evolutionary assumptions are also part of many present algorithms of tree reconstruction. While the classical mode of inference in cladistics was maximum parsimony, which indeed is pattern-based, many modern methods of tree reconstruction for molecular datasets such as maximum-likelihood and Bayesian approaches explicitly make use of models of DNA sequence evolution (so-called substitution models), i.e., are inherently historical. Even classical distance-based methods like neighbor-joining can implement these substitution models to arrive at more sophisticated values of genetic distance among taxa.

Another candidate for an *axiomatic precept* is the dichotomous branching of phylogenetic trees. This clearly is an a priori precept, as no results from empirical evolutionary research prohibit the fragmentation of a population into more than two entities. The dichotomy requirement, however, is the consequence of a paradigm shift in systematics/phylogenetics. Hennig made systematics rigorously scientific by means of the requirement that taxa be justified on the basis of derived character states (apomorphies); closest relatives could be inferred by the presence of shared derived character states (synapomorphies). This put an end to the free choice of characters (including plesiomorphies) with which to circumscribe and justify taxa. However, ancestors ('stem species') lack apomorphies beyond those of the whole group they gave rise to, which means that they are in principle 'invisible'. The search for ancestors thus was replaced by the search for the sister group, and modern phylogenetics can indeed be encapsulated as the search for sister group relationships. Hennig himself was a zoologist and entomologist, but the paradigm shift from ancestors to sister groups was particularly relevant to paleontology, and it is probably no coincidence that among the early influential generation of cladists, some paleontologists took the lead. In fact, the rejection of the possibility of identifying ancestors in the fossil record was a precondition of the adoption of an ahistorical approach (pattern cladistics) in the first place.

Historically, there have been a number of interesting, although by now obsolete, axiomatic precepts in biological systematics in the form of certain a priori concepts that forced the way we conceive of biodiversity into a metaphysical straitjacket. *Scala naturae* thinking, the idea that there is a continuously ascending 'great chain of being' (Lovejoy 1936; for remnants of it, see Rigato and Minelli 2013), is one of those concepts, as is the strange nineteenth-century quinarian system by William Sharp Macleay (1792–1865), William John Swainson (1789–1855), and others, according to which all taxa were subdivided into exactly five subgroups (e.g., Pietsch 2012, pp. 52ff.). Relationships or affinities (both what we call homologies and analogies today) were depicted in diagrams based on five circles, and the whole system of numerical mysticism is somewhat reminiscent of the ancient Pythagoreans. Both the *scala naturae* and the quinarian system are powerful reminders of how much our conception of taxonomy has changed through time and how a priori conceptions of the discipline impact what we are looking for and also what we find. This is the essence of Hübner's historistic approach.

Systematics depends on a number of *judicative precepts* as well. A very obvious (and much debated) such precept is the alpha value or significance threshold in statistical analyses, which, of course, is also relevant to many tests based on taxonomic data, for example when it comes to population differentiation. The usual 5% is an arbitrary threshold, as would be any other value. This may seem like a rather minor issue, but the decision on a significance threshold is also a decision on whether scientists accept that there is an effect of something or not. I have no doubt that the contents of our science textbooks would look very different if we had historically chosen different alpha values.

Similar thresholds, whether explicit or implicit, are also needed when the resolution of phylogenetic trees is evaluated. How high a value of bootstrap support or Bayesian posterior probability is high enough? These thresholds are important and unavoidable in alpha taxonomy as well. How different must two populations be to count as different subspecies or even species? Seifert (2020, pp. 1033 and 1036, see earlier) is unusually open when he admits that thresholds of evolutionary divergence 'should be fixed by consensus among the experts under the principle of avoiding oversplitting or lumping'. The guideline he gives for these thresholds, namely, to avoid both lumping and oversplitting, is an additional, normative precept, and one that is again open to interpretation and in need of a consensus on what constitutes excessive lumping and splitting.

Numerical thresholds are not new in taxonomy, of course. Bacterial species have operationally often been based on a 70% homology threshold in DNA-DNA hybridisation assays and 5 °C or less difference in melting temperature between homologous and heterologous hybrids. Using direct DNA sequencing techniques, this threshold was 'translated' into 97% (or slightly higher values, depending on the authors) sequence identity at the 16s rRNA gene. Similarly, for subspecies designation, there is a '75% rule' (Amadon 1949) according to which 'subspecies A is recognised taxonomically if, and only if, $\geq 75\%$ of the individuals in group A lie outside 99% of the range of variation of group B for the character or set of characters under consideration' (Patten 2015, p. 482). It is easy to see that the specific values chosen in all these cases are, within limits, arbitrary and take their authority simply from repeated use and consensus within the taxonomic community. However, it follows from the twofold nature of taxonomy that *some* threshold value cannot be escaped, and if taxonomists are not happy with one, all they can do is choose another.

Finally, it is rather easy to also identify a number of a priori *normative precepts*. For both taxonomy and phylogenetics, probably all biologists would argue that they should be at least compatible with evolutionary theory; non-pattern-cladists would even go further and say they should reflect the process of evolution. That is clearly a normative decision, and one that is very different from past ideas of what taxonomy was supposed to represent. So is the principle of parsimony or 'Ockham's razor', according to which, of two hypotheses or theories with equal explanatory power, the simpler is to be preferred. In phylogenetics, this means that tree topologies that require a minimum of character state changes and a minimum of homoplasies in the dataset are preferred over competing trees *by virtue of* their implying fewer (evolutionary) steps. Parsimony is not seen in nature; it is a priori imposed on it, based on a normative idea of how a theory should be. We thus impose a kind of process economy on evolution itself, one that is in fact not easy to justify intrinsically.

The requirement for monophyly (which is only rejected by a minority of systematists) is another normative precept. Being raised scientifically in the spirit of Hennigian phylogenetic systematics and process cladistics, I have no problem arguing for monophyly, but it remains a fact that it entails an a priori decision on what kind of biological entities systematics should aim for or allow, and it is at least possible to make alternative decisions as well (see Sigwart, this volume). Monophyly is not only relevant to supraspecific phylogenetics; it is also often discussed in a taxonomic context at and even below the species level. Some species concepts explicitly require species to be monophyletic (Mishler and Theriot 2000b; Mishler and Wilkins 2018), and there are some definitions of so-called Evolutionarily Significant Units (ESUs) for conservation at the infraspecific level that are based on monophyly as well (Moritz 1994). It does not matter here that the applicability of monophyly at that taxonomic level (due to processes like hybridisation that break up strictly hierarchical, 'vertical' patterns) is contentious (Rieppel 2010; Zachos 2016, chapter 5.6.1); what counts in the present context is that monophyly is viewed, at least by some, as an a priori normative principle about what taxonomically recognised entities should be like. Monophyly is also interesting in a different respect, because it entails certain judicative precepts. Monophyly is a historical relationship (not considering pattern cladistics now) and as such, not observable; i.e., it has to be inferred. That, in turn, necessitates a consensus about when, given our current knowledge, a taxon is considered monophyletic beyond reasonable doubt. I have already mentioned the issue of tree resolution, and it is here that this becomes vital: certain levels of node support will be considered sufficient, while others won't (this also holds for pattern cladistics). That will depend on some quantification based on bootstrapping or Bayesian posterior probabilities etc. Also, a certain tree topology will probably not be considered well supported if it is only minimally more parsimonious (say, one additional step) than an alternative topology, but how much more parsimonious is enough, then?

One more example of a normative precept may suffice. It is, however, an important and particularly fuzzy one: biological relevance. When looking at the Tree of Life, it immediately becomes clear that not every monophyletic group, or clade, is named (despite the view of some that this should be done). That is not very surprising, given that the number of all enkaptic clades is astronomical. Biologists, therefore, make a choice of which clades to name, and the guideline is relevance, as only relevant group names are needed for communication. For example, the clade Mammalia is highly relevant for a number of reasons, and therefore, naming it is indispensable. On the other hand, the clade comprising, say, red deer and their two closest living relatives (wapitis and sika deer) is only relevant in a comparatively limited number of contexts and can easily be circumscribed by naming the three or four species. There is an equivalent of this with respect to species concepts, if monophyly is considered a defining trait of the species category. The living world has a fractal structure of clades within more inclusive clades within even more inclusive clades, and so on, until one reaches the complete Tree of Life – the most inclusive clade comprising all life on Earth. Monophyla, therefore, come at many different levels, and a decision must be made which of these levels is supposed to be the species level. Mishler and Theriot, for example, proponents of the monophyly version of the Phylogenetic Species Concept, are very much aware of that when they say that '[t]axa are ranked as species rather than at some higher level because they are *the smallest monophyletic groups deemed worthy of formal recognition*, because of the amount of support for their monophyly and/or because of their importance in biological processes operating on the lineage in question' (Mishler and Theriot 2000b, p. 46f., italics added; see Mishler and Wilkins 2018 for a similar view; Freudenstein et al. 2017 also highlight the need for biological relevance in the decision of ranking at species level).

More generally, David Hull formulated a basic dilemma of species concepts with respect to biological relevance (which he calls theoretical significance) and applicability: 'if a species concept is theoretically significant, it is hard to apply, and if it is easily applicable, too often it is theoretically trivial' (Hull 1997, p. 358). In analogy to Heisenberg's uncertainty principle in quantum physics, Adams (2001) dubbed this the 'species delimitation uncertainty principle', and he gives a very convincing example:

> Species of helminths (parasitic worms) may be defined tentatively as a group of organisms, the lipid free antigen of which, when diluted to 1:4000 or more, yields a positive precipitin test within 1 h with a rabbit antiserum produced by injecting 40 mg of dry weight liquid free antigenic material and withdrawn ten to 12 days after the last of four intravenous injections administered every third day.

(Adams 2001, p. 156, quoted from Wilhelmi 1940)

This is probably the most operational definition of species taxa I have ever seen, but it is also without a doubt the least theoretically significant or biologically relevant one. Still, what really constitutes biological relevance – what is 'deemed worthy of formal recognition' in the words of Mishler and Theriot – is open to debate, and it seems obvious that views can differ considerably here. The diagnosability version of the Phylogenetic Species Concept has been accused of rendering the species category biologically trivial (the case mentioned earlier of two different tiger species based on a few diagnostic base pairs in the mitochondrial genome would be a point in case here), but its adherents argue that even the smallest distinct groups are real and deserve formal recognition as species. One may disagree (in fact, I do), but one should also be aware that all one can do is make a different normative decision as to what species taxa are supposed to represent. The argument here is not about empirical taxonomic content, but about a priori taxonomic philosophy. This is the reason why these arguments go on and on – neither side can ever be proved wrong.

I hope to have shown that Kurt Hübner's historistic approach to a philosophy of science as applied to biological systematics (both phylogenetics and taxonomy) is a potentially fruitful endeavor and that it is well suited to raise awareness among biologists of necessary a priori assumptions or decisions – assumptions and decisions that can change with time and have done so repeatedly in the past! Taxonomy is not alone in that (it applies to all scientific disciplines), but taxonomy is special in that one of its main parts, the ranking of taxa at species level, is particularly fuzzy and dependent upon an executive decision. Viewing taxonomy and biological systematics in general in a simplified Popperian framework does not do justice to the complexity of the discipline, but Hübner's approach might offer a promising alternative.

15.5 FINAL THOUGHTS

As stated earlier, taxonomy is one of biology's fundamental disciplines. The fact that many taxonomic results are the topic of heated debates is due to the twofold nature of taxonomy, which includes a partly arbitrary ranking decision and which can be more systematically and formally explicated in the framework of Hübner's scientific-theoretical categories: the five kinds of a priori precepts that are the condition of the possibility of a scientific theory. For taxonomy, the deeper root of its vague boundaries is the fact that it is a discrete ordering system imposed on a continuous historical process (evolution). We certainly discover order in nature, but we also impose order on the natural world. Jody Hey (2001a, b) has put forward an epistemological hypothesis on the origin of the species problem. Inspired by W.V.O. Quine, his starting point is the acknowledgement of friction between our discrete language and a largely continuous nature. Our minds, he argues, are wired in such a way that they think in terms of categories arising from recurrent patterns that we perceive. Our scientific insights into the evolutionary history of life have compromised our categorical thinking, but our minds keep pigeon-holing:

> In brief, modern biologists suffer two imperatives. The first is the ancient one of all people and that is to devise categories and invent just as many kinds of organisms as we want or need to give voice to our thoughts about that diversity. The second is to understand the causes of that diversity. Indeed, our pursuit of that second imperative has been so successful that it has given us a species problem.
>
> **(Hey 2001a, p. 108)**

Our evolutionary insights have shown us that our 'innate' view of species is often flawed, and that also applies to a number of 'good species' when morphologically very similar organisms turn out to be different entities after all (sibling species, cryptic species). Species as we perceive them have thus two causes: '(1) the evolutionary processes that have caused biological diversity; and (2) the human mental apparatus that recognises and gives names to patterns of recurrence' (Hey 2001b, p. 328). Hey never mentions Kant, but it is difficult to not see a parallel here to Kantian epistemology, where the objectively real thing (the 'thing-in-itself' or *Ding an sich*) is inaccessible to us and is always mediated by our cognitive powers with their a priori categories and pure intuition, making all empirical knowledge ultimately a hybrid of both. Categorical thinking may be deeply engrained in us, but on top of that, we can choose from more than 30 species concepts for these categories. Our taxonomy, then, is a hybrid of real structure in biodiversity, our way of reasoning, and our taxonomic taste. Viewed from this angle, taxonomic debates are hardly surprising – neither is the fact that they are often settled not on the basis of empirical tests but by tradition or majority rule, etc.

I have rejected a simple falsificationist approach to the scientific nature of taxonomy. However, another of Popper's concepts is well worth considering in the debate about species: methodological nominalism. Rather than getting hung up on the somewhat essentialist question 'What is a species?', we should focus on the empirical data and ask ourselves what we should call a biological entity with certain traits like genetic distinctness, ecological uniqueness, etc. This would sensitise biologists to the arbitrariness necessarily involved in assigning the species rank, and it would reduce the danger of arguing about names (semantics) rather than biological phenomena (Zachos 2018). Coming back to the issue of species status of modern humans, Neanderthals, and Denisovans, does it really matter if we classify them at species or subspecies level? The biological facts are untouched by that, and it is those that count and that shed light on our evolutionary history. The species category is not a single objectively real level in the hierarchy of life, and it is time we realised that and acted accordingly. Taxonomy must be as rigorously scientific as possible. We cannot afford to let it be caricatured as 'stamp collecting', but we also need to acknowledge its inherent limitations due to the fuzzy boundaries of its very study objects. Only then can taxonomy play the role we need it to play: contributing to the overall reference system to which all other biology refers.

ACKNOWLEDGEMENTS

I would like to express my gratitude to the people at the Konrad Lorenz Institute for Evolution and Cognition Research in Klosterneuburg, Austria, for hosting me as a visiting fellow during my work on and for this book. I would also like to thank my co-editors, John S. Wilkins and Igor Pavlinov, for many insightful exchanges about species and more, as well as for helpful comments on an earlier version of this chapter.

REFERENCES

Adams, B. J. 2001. The species delimitation uncertainty principle. *Journal of Nematology* 33:153–60.
Amadon, D. 1949. The seventy-five percent rule for subspecies. *Condor* 51:250–58.
Amitani, Y. 2015. Prototypical reasoning about species and the species problem. *Biological Theory* 10:289–300.
Blackwelder, R. E. 1967. *Taxonomy: A text and reference book*. New York: Wiley.
Boyd, R. 1991. Realism, anti-foundationalism and the enthusiasm for natural kinds. *Philosophical Studies* 61:127–48.
Boyd, R. 1999. Homeostasis, species, and higher taxa. In *Species. New interdisciplinary essays*, ed. R. A. Wilson, 141–85. Cambridge, MA: MIT Press.
Cracraft, J., Feinstein, J., Vaughn, J., and K. Helm-Bychowski. 1998. Sorting out tigers (*Panthera tigris*): Mitochondrial sequences, nuclear inserts, systematics, and conservation genetics. *Animal Conservation* 1:139–50.
DCP, Letter 915: Darwin Correspondence Project, 'Letter no. 915.' https://www.darwinproject.ac.uk/letter/DCP-LETT-915.xml (accessed April 5, 2021).
de Queiroz, K. 1998. The general lineage concept of species, species criteria, and the process of speciation. In *Endless forms: Species and speciation*, eds. D. Howard, and S. H. Berlocher, 57–75. Oxford: Oxford University Press.
de Queiroz, K. 1999. The general lineage concept of species and the defining properties of the species category. In *Species. New interdisciplinary essays*, ed. R. A. Wilson, 49–89. Cambridge, MA: MIT Press.

de Queiroz, K. 2005. A unified concept of species and its consequences for the future of taxonomy. *Proceedings of the California Academy of Sciences* 56:196–215.

de Queiroz, K. 2007. Species concepts and species delimitation. *Systematic Biology* 56:879–86.

de Santis, M. D. 2020. Scientific explanation and systematics. *Systematics and Biodiversity* 19(3):312–21. DOI: 10.1080/14772000.2020.1844339.

Deppert, W. 2015. Ein großer Philosoph: Nachruf auf Kurt Hübner und Aufruf zu seinem Philosophieren. *Journal for General Philosophy of Science* 46:251–68.

Deppert, W., Görg, E., and M. Sojka. 2015. Kurt Hübner – Bibliographie. *Journal for General Philosophy of Science* 46:269–77.

Desmond, A., and J. Moore. 1991. *Darwin*. New York: Warner Books.

Dupré, J. 2017. The metaphysics of evolution. *Interface Focus* 7:20160148.

FAO (Food and Agriculture Organization of the United Nations). 2015. Mapping the vulnerability of mountain peoples to food insecurity. Romeo, R., Vita, A., Testolin, R., and Rome, H. T.

Faurby, S., Eiserhardt, W. L., and J.-C. Svenning. 2016. Strong effect of variation in taxonomic opinion on diversification analyses. *Methods in Ecology and Evolution* 7:4–13.

Freudenstein, J. V., Broe, M. B., Folk R. A., and B. T. Sinn. 2017. Biodiversity and the species concept – lineages are not enough. *Systematic Biology* 66:644–56.

Garnett, S. T., Christidis, L., Conix, S., et al. 2020. Principles for creating a single authoritative list of the world's taxa. *PLoS Biology* 18(7):e3000736.

Gibbons, A. 2011. The species problem. *Nature* 331:394.

Groves, C., and P. Grubb. 2011. *Ungulate taxonomy*. Baltimore, MD: The Johns Hopkins University Press.

Heller, R., Frandsen, P., Lorenzen, E. D., and H. R. Siegismund. 2013. Are there really twice as many bovid species as we thought? *Systematic Biology* 62:490–93.

Hennig, W. 1966. *Phylogenetic systematics*. Urbana, IL: University of Illinois Press.

Hey, J. 2001a. *Genes, categories, and species. The evolutionary and cognitive causes of the species problem*. Oxford: Oxford University Press.

Hey, J. 2001b. The mind of the species problem. *Trends in Ecology and Evolution* 16:326–29.

Heywood, V. H. 1998. The species concept as a socio-cultural phenomenon – a source of the scientific dilemma. *Theory in Biosciences* 117:203–12.

Hübner, K. 1978. *Kritik der wissenschaftlichen Vernunft*. Freiburg, München: Karl Alber.

Hübner, K. 1985 (1983). *Critique of scientific reason*. Translated by P. R. Dixon, Jr., and H. M. Dixon. Chicago, IL: The University of Chicago Press.

Hull, D. L. 1997. The ideal species concept – and why we can't get it. In *Species: The units of biodiversity*, eds. M. F. Claridge, H. A. Dawah, and M. R. Wilson, 357–80. London: Chapman & Hall.

Kearney, M. 2007. Philosophy and phylogenetics. Historical and current connections. In *The Cambridge companion to the philosophy of biology*, eds. D. L. Hull, and M. Ruse, 211–32. Cambridge: Cambridge University Press.

Kitcher, P. 1984. Species. *Philosophy of Science* 51:308–33.

Kunz, W. 2012. *Do species exist? Principles of taxonomic classification*. Weinheim: Wiley-Blackwell.

Larsen, B. B., Miller, E. C., Rhodes, M. K., and J. J. Wiens. 2017. Inordinate fondness multiplied and redistributed: The number of species on Earth and the new pie of life. *The Quarterly Review of Biology* 92:229–65.

Locey, K. J., and J. T. Lennon. 2016. Scaling laws predict global microbial diversity. *Proceedings of the National Academy of Sciences of the United States of America* 113:5970–75.

Lovejoy, A. O. 1936. *The great chain of being: A study of the history of an idea*. Cambridge, MA: Harvard University Press.

Luckow, M. 1995. Species concepts: Assumptions, methods, and applications. *Systematic Botany* 20:589–605.

May, R. M. 1990. Taxonomy as destiny. *Nature* 347:129–130.

Mayden, R. L. 1997. A hierarchy of species concepts: The denouement in the saga of the species problem. In *Species: The units of biodiversity*, eds. M. F. Claridge, H. A. Dawah, and M. R. Wilson, 381–424. London: Chapman & Hall.

McClure, C. J., Lepage, D., Dunn, L., et al. 2020. Towards reconciliation of the four world bird lists: Hotspots of disagreement in taxonomy of raptors. *Proceedings of the Royal Society B* 287(1929):20200683.

Mishler, B. D., and E. C. Theriot. 2000a. A critique from the Mishler and Theriot phylogenetic species concept perspective: Monophyly, apomorphy, and phylogenetic species concepts. In *Species concepts and phylogenetic theory – A debate*, ed. Q. D. Wheeler, and R. Meier, 119–32. New York: Columbia University Press.

Mishler, B. D., and E. C. Theriot. 2000b. The phylogenetic species concept (*sensu* Mishler and Theriot): Monophyly, apomorphy, and phylogenetic species concepts. In *Species concepts and phylogenetic theory – A debate*, eds. Q. D. Wheeler, and R. Meier, 44–55. New York: Columbia University Press.

Mishler, B. D., and J. S. Wilkins. 2018. The hunting of the SNaRC: A snarky solution to the species problem. *Philosophy, Theory, and Practice in Biology* 10:1.

Mora, C., Tittensor, D. P., Adl, S., Simpson, A. G. B., and B. Worm. 2011. How many species are there on Earth and in the ocean? *PLoS Biology* 9(8):e1001127.

Moritz, C. 1994. Defining 'evolutionary significant units' for conservation. *Trends in Ecology and Evolution* 9:373–75.

Neate-Clegg, M. H. C., Blount, J. D., and C. H. Sekercioglu. 2021. Ecological and biogeographical predictors of taxonomic discord across the world's birds. *Global Ecology and Biogeography* 30:1258–70.

Nelson, G., and N. I. Platnick. 1981. *Systematics and biogeography: Cladistics and vicariance*. New York: Columbia University Press.

Nicholson, D. J., and J. Dupré. 2018. *Everything flows: Towards a processual philosophy of biology*. Oxford: Oxford University Press.

Owen, R. 1858. On the characters, principles of division, and primary groups of the class mammalia. *Journal of the Proceedings of the Linnean Society. Zoology* II:1–37.

Padial, J. M., and I. De la Riva. 2021. A paradigm shift in our view of species drives current trends in biological classification. *Biological Reviews* 96:731–51.

Patten, M. A. 2015. Subspecies and the philosophy of science. *The Auk* 132:481–85.

Perkins, S. L. 2011. Beyond cladistics: The branching of a paradigm. *Systematic Biology* 60:895–97. (book review)

Pietsch, T. W. 2012. *Trees of life. A visual history of evolution*. Baltimore, MD: The Johns Hopkins University Press.

Pigliucci, M. 2003. Species as family resemblance concepts: The (dis-)solution of the species problem. *BioEssays* 25:596–602.

Platnick, N. I. 1977. Cladograms, phylogenetic trees, and hypothesis testing. *Systematic Zoology* 26:438–42.

Regan, C. T. 1926. Organic evolution. *Report of the British Association for the Advancement of Science* 1925:75–86.

Richards, R. A. 2010. *The species problem – A philosophical analysis*. Cambridge: Cambridge University Press.

Rieppel, O. 2003. Popper and systematics. *Systematic Biology* 52:259–71.

Rieppel, O. 2009. Species as a process. *Acta Biotheoretica* 57:33–49.

Rieppel, O. 2010. Species monophyly. *Journal of Zoological Systematics and Evolutionary Research* 48:1–8.

Rieppel, O. 2014. The early cladogenesis of cladistics. In *The evolution of phylogenetic systematics*, ed. A. Hamilton, 117–37. Berkeley, CA: University of California Press.

Rigato, E., and A. Minelli. 2013. The great chain of being is still here. *Evolution: Education and Outreach* 6:18.

Scott, P., and R. Rines. 1975. Naming the Loch Ness monster. *Nature* 258:466–468.

Seifert, B. 2020. The Gene and Gene Expression (GAGE) species concept: An universal approach for all eukaryotic organisms. *Systematic Biology* 69:1033–38.

Simpson, G. G. 1951. The species concept. *Evolution* 5:285–98.

Simpson, G. G. 1961. *Principles of animal taxonomy*. New York: Columbia University Press.

Simpson, G. G. 1963. Biology and the nature of science. *Science* 139:81–88.

Sites, J. W. Jr., and J. C. Marshall. 2004. Operational criteria for delimiting species. *Annual Review of Ecology, Evolution, and Systematics* 35:199–227.

Sober, E. 1984. Sets, species, and evolution: Comments on Philip Kitcher's 'species'. *Philosophy of Science* 51:334–41.

Templeton, A. R. 1989. The meaning of species and speciation: A genetic perspective. In *Speciation and its consequences*, eds. D. Otte, and J. A. Endler, 3–27. Sunderland, MA: Sinauer Associates.

Thiele, K. R., Conix, S., Pyle, R. L., et al. 2021. Towards a global list of accepted species I. Why taxonomists sometimes disagree, and why this matters. *Organisms Diversity & Evolution* 21:615–22.

Vogt, L. 2008. The unfalsifiability of cladograms and its consequences. *Cladistics* 24:62–73.

Vogt, L. 2014a. Why phylogeneticists should care less about Popper's falsificationism. *Cladistics* 30:1–4.

Vogt, L. 2014b. Popper and phylogenetics, a misguided rendezvous. *Australian Systematic Botany* 27:85–94.

Vrana, P., and W. Wheeler. 1992. Individual organisms as terminal entities: Laying the species problem to rest. *Cladistics* 8:67–72.

Wägele, J.-W. 2001. *Grundlagen der Phylogenetischen Systematik*. München: Verlag Dr. Friedrich Pfeil.

Wiley, E. O. 1978. The evolutionary species concept reconsidered. *Systematic Zoology* 27:17–26.

Wilhelmi, R. W. 1940. Serological reactions and species specificity of some helminths. *The Biological Bulletin* 79:64–90.

Willis, S. C. 2017. One species or four? Yes!... and, no. Or, arbitrary assignment of lineages to species obscures the diversification processes of Neotropical fishes. *PLoS ONE* 12(2):e0172349.

Wilkins, J. S. 2018. *Species. The evolution of the idea*. 2nd ed. Boca Raton, FL: CRC Press/ Taylor & Francis.

Zachos, F. E. 2011. Linnean ranks, temporal banding, and time-clipping: Why not slaughter the sacred cow? *Biological Journal of the Linnean Society* 103:732–34.

Zachos, F. E. 2016. *Species concepts in biology. Historical development, theoretical foundations and practical relevance*. Switzerland: Springer International Publishing.

Zachos, F. E. 2018. (New) Species concepts, species delimitation and the inherent limitations of taxonomy. *Journal of Genetics* 97:811–15.

Zachos, F. E., Christidis, L., and S. T. Garnett. 2020. Mammalian species and the twofold nature of taxonomy: A comment on Taylor et al. 2019. *Mammalia* 84:1–5.

Afterword

16 Continuing after Species

Robert A. Wilson[*]

As the current volume attests, biologists who play philosopher, as well as philosophers who play biologist, continue after species. For the most part, they do so in the shadow of 'the species problem', poking a stick at it while making some interesting observations about species taxa in one or more globs of biospace. Shamelessly refusing to leave the comfort of the armchair for now, I would hazard the guess that philosophers and biologists alike who have responded to the species problem for the past 50 years with more than a poke have done so in three opposed ways: mostly offering solutions to the species problem (Hausdorf 2011; Richards 2010), occasionally declaring that a solution is impossible (Hull 1997; Reydon 2005), and elsewhere, arguing that there is in fact no problem to be solved (Pavlinov 2013).

The tendency to adopt one of these responses hasn't shown signs of decline in recent years, with all three tendencies clearly manifest in the current volume. Before saying something about 'the species problem' beyond the 'It's solved!', 'It can't be solved!', and 'It's not a problem!' responses to it, some initial stock-taking on the concept of species itself.

At least ideologically, as a card-carrying enthusiast for naturalistic philosophy of science, I remain attracted to utilising the latest technical modes of empirically informed philosophical methodology despite my confessed bodily affection for the laziness of the armchair, a professional hazard for all (but not only) philosophers. I was somewhat disappointed, however, with Siri's refusal to answer my simple query 'Are species real?' Undaunted and ever-resourceful, I went directly to my methodological next stop: a Google Scholar search for 'species'. While not as common as either 'cells' or 'organisms' – and even less so once we add the singulars 'cell' and 'organism' – nonetheless, 'species' delivers well over 6 million hits on Google Scholar. GS hits for the terms cell(s), gene(s), and species occur in roughly a 3:2:1 ratio, shifting roughly to 3:1:2 for searches since 2020. (Interestingly, 'gene(s)' shows roughly the same usage level and trajectory as 'organism(s)'.)

One thing that can be safely concluded is that talk about species has a robust academic history, one that shows little signs of dissipating to become merely history. In addition, with over 500 million hits on Google itself, and with roughly the same ratio of hits to 'cell(s)' there as it has on Google Scholar, 'species' prima facie earns its keep in more vernacular contexts. The tendency to make use of or appeal to the

[*] This paper is dedicated to the memory of Richard N. Boyd and his infectious enthusiasm for scientific realism, naturalism, and natural kinds over the past 50 years. I am grateful to Dick for modelling how to think systematically and deeply about species, amongst other things. Thanks to John Wilkins for the invitation to write these reflections, and to John and to Frank Zachos for reading and commenting on an earlier draft.

DOI: 10.1201/9780367855604-21

concept of species has been, and remains, strong across academic and non-academic contexts.

One reason for the robustness of the concept of species within both common sense and scientific thinking is a *clumpiness* to biological nature at the level of populations that is hard to ignore. As Kim Sterelny has put it in his under-appreciated 'Species as Ecological Mosaics':

> The mechanisms of evolution have produced on Earth an astounding variety of life forms. Together with adaptive design, the evolution of that diversity is the central explanatory target of evolutionary biology. Though great, however, the diversity of life on Earth is limited in important ways. Diversity is bunched or clumped. ... Life's mechanisms have produced *phenomenological species*: recognizable, reidentifiable clusters of organisms. This fact makes possible the production of bird and butterfly field guides, identification keys for invertebrates and regional floras, and the like.
>
> **(Sterelny 1999, p. 119)**

Population-level clumpiness is typically conceptualised as cohesiveness, and *species cohesion* is a primary phenomenon to be explained (for example, by 'gene flow'; see Barker and Wilson 2010). As Sterelny goes on to note, the reality of phenomenological species is the beginning, rather than the end, of reflection on the nature of species and their relationships to the mechanisms of evolution and the diversity of life forms (see also Minelli, this volume). For some, the existence of phenomenological species serves as a foundation for realism not only about species *taxa* but also about the species *category*, even if this proves to be a form of pluralistic or promiscuous realism (Dupré 1981, 1999; see also Wilson 1996; Barker, this volume). For others, attempting to make more of the concept of species beyond phenomenological species is to invite potential confusions of various kinds. They accordingly opt for a deflationary understanding of species, whereby species are little more than phenomenological species, if that (Ereshefsky 1992a, 1999; see also Mishler 1999; Wilkins, this volume).

What of 'the species problem', a focus for much of the ongoing philosophical discussion of species? The species problem is really a cluster of problems. That cluster both stems from and fuels at least four kinds of ongoing discussions in the philosophical deep end of the biological pool. These discussions arise from and involve:

1. The diverse characterisations that have been given of species by biologists over the past 100 years;
2. Questions about the standing of species as a distinctive rank in the Linnaean hierarchy, which has remained the dominant scheme of taxonomic classification of populations of organisms since it established itself as such over 250 years ago;
3. Debates over the ontological status of species, such as whether species are natural kinds, individuals, processes, feedback systems, or even, as John Locke might have put it, creatures of the understanding;
4. Emerging technologies for species delimitation (such as coalescent models) and the corresponding issue of species discordance.

If we continue to locate species as one of the Big Four concepts applied in describing and explaining the world of living things – gene, cell, organism, and species – perhaps the confluence of the preceding four dimensions partially explains why there is a species problem but no corresponding gene, cell, or organism problems.

The other part to such an explanation, however, is surely historical, concerning two distinct spotlights thrown on species in the history of biological thought. First, species were cast on stage by the title of Darwin's *On the Origin of Species by Means of Natural Selection* (1859). Darwin's own reflections on and hesitations about species continue to attract scrutiny and discussion, even in recent years (de Queiroz 2011; Mallett 2010, 2013; Stamos 2007). Second, the term 'biological species concept' was both introduced and promoted by the ornithologist Ernst Mayr (Mayr 1942) and others (Wright 1940; Dobzhansky 1950) as labelling a settled view of the nature of species. That view, according to which species are reproductively isolated populations stabilised by gene flow resulting from interbreeding, emerged from the so-called Modern Synthesis of the 1930s–1940s that integrated the theory of natural selection and genetics, or to put it more crudely, Darwin and Mendel.

In the introduction to *Species: New Interdisciplinary Essays* (1999), I noted that 'the last decade has seen something of a publication boom on the topic' (Wilson 1999, p. ix), going on to chiefly cite collections with dominant contributions from biologists: 'Otte and Endler 1989; Ereshefsky 1992b; Paterson 1994; Lambert and Spence 1995; Claridge, Dawah, and Wilson 1997; Wheeler and Meier 1999 [sic]; Howard and Berlocher 1998' (p. ix). Much of that publication boom was in response to the growing recognition of the breakdown of whatever consensus there was around Mayr's 'biological species concept' and how that did (and didn't) affect the continuation of thinking about processes of speciation and the goals of conservation and the preservation of biological diversity.

One editorial directive given to me in the invitation to write this Afterword was to reflect especially on the development of discussions of species and the species problem over the past 25 years. The trading zone between philosophers of biology and biologists of a greater variety of stripes has expanded, even if the bulk of recent work on core topics, such as speciation and phylogenetic constraints on taxonomy, continues to operate on the territory of biologists beyond that zone. Both the expanded trading zone and the tendencies to live beyond it can be seen in *Species: New Interdisciplinary Essays*, particularly in two essays it contains that develop ideas which have been especially influential during this time, the first more so amongst biologists, the second more so amongst philosophers.

Kevin de Queiroz's 'The General Lineage Concept of Species and the Defining Properties of the Species Category' provided one of his earliest and most extended discussions of the general lineage concept of species, while Richard Boyd's 'Homeostasis, Species, and Higher Taxa' remains the most discussed paper on the homeostatic property cluster view of natural kinds that has an application to species. Each of these views might be best thought of as providing an overarching framework for thinking about species, one completed by further biological details that might well vary from context to context.

De Queiroz's (1998, 1999) general lineage conception of species (later, *metapopu-lation lineage* conception, de Queiroz 2005, 2007) has come to be viewed by many as the successor to Mayr's Biological Species Concept, for now at least being the closest thing we have to an adequate, monistic conception of species. For de Queiroz, species are 'segments of population-level lineages' (de Queiroz 1999: 53), with dif-ferent extant species conceptions providing different criteria for the segmentation, and 'populations' covering both sexual and asexually reproducing organisms. While its generality in subsuming more specific views of the nature of species is no doubt one source of its appeal for biologists, philosophers are more likely to view *segments of population-level lineages* as taking us little way to identifying what species are. In the morass of biological entities, many things – from nuclear families to multi-Order clades – are such segments: a parent and its offspring form a segment of a popula-tion-level lineage; a pair of these form a metapopulational lineage segment. Since the ancestor–descendant relationship holds of many biological entities, segments of population-level lineages are ubiquitous in the living world, as are metapopulations. The heavy lifting, conceptually speaking, lies elsewhere in delineating species as a particular type of metapopulational lineage.

As influential amongst philosophers of biology as the metapopulational lineage conception has been amongst biologists has been Boyd's homeostatic property cluster view of species and of natural kinds more generally (Boyd 1999; see also Griffiths 1999; Wilson 1999). The HPC view, originally developed in defending moral realism (Boyd 1988), has been articulated as part of a broader naturalistic form of scientific realism (Boyd 1991, 2010, 2019). For Boyd, species are phe-nomena that cohere due to a variety of homeostatic mechanisms, such as gene exchange, reproductive isolation, coadapted gene complexes, developmental con-straints, and niche construction (e.g., Boyd 1999: 164–165), and these phenomena are mostly plausibly viewed as natural kinds (pp. 167–169). Here, the general idea has been to reconceptualise what natural kinds are, showing how thinking of spe-cies as such reconceptualised kinds both allows one to address many issues about species and serves as a paradigm for other applications (see also Wilson, Barker, and Brigandt 2007; Wilson 2005: ch. 3–6).

In commissioning and assembling the essays in *Species*, I took its subtitle, *New Interdisciplinary Essays*, not only to refer to the interface between biologists and philosophers but also to signal an invited expansion in the nature of the interdis-ciplinary contributions to the discussion. In addition to innovative and influential essays from evolutionary theorists (such as de Queiroz) and philosophers of biology (such as Boyd), the volume also included contributions from a pair of developmental psychologists (Keil and Richardson 1999) and a cultural anthropologist (Atran 1999), each discussing folk biological knowledge and its development and universality, and a ciliatologist specialising in eukaryotic protists (Nanney 1999) using the case of *Tetrahymena* to call attention to ways in which microbial taxonomy and the Modern Synthesis (including Mayr's Biological Species Concept) had bypassed each other.

Research since then on folk biology in both developmental psychology and cul-tural anthropology has blossomed, with perhaps its best-known application being in the sophistication of discussions of Indigenous classification and taxonomy of the living world (Ludwig and El-Hani 2020; Ludwig and Poliseli 2018; Ludwig and Weiskopf 2019). Yet, this work has had only a limited impact on core discussions of species and the species problem amongst philosophers and biologists (see Kendig 2020 and Kendig, this volume, for exceptions).

By contrast, the corresponding discussions, not so much of eukaryotic protists in particular but of the microbial world more generally, have informed those discussions. In this regard, it is interesting to note that Wilkins devotes to microbial species roughly one-third of a chapter newly added to the second edition of his *Species: The Evolution of the Idea*, 'The Development of the Philosophy of Species' (esp. pp. 317–330; see also Wilkins 2006). Insights gained from this attention to the microbial world have informed more general views of the nature of species. For example, explorations of the significance of lateral or horizontal gene transfer, more prevalent in the microbial than in the macrobial world, have reinforced one idea fuelling pluralistic views of the species category: that whatever 'species' are in bacteria, they are really something different in kind from species in plants and animals (O'Malley 2014: ch. 2; see also Franklin 2007; Franklin-Hall 2010).

The microbial geneticist Ford Doolittle has developed perhaps the most sophisticated forms of this view over a number of years (e.g., Doolittle 2009, 2013, 2019; Doolittle and Zhazybayeva 2009; Novack and Doolittle 2020). Other and contrasting recent discussions have defended the application of Mayr's biological species concepts to bacteria (and even viruses) by assimilating homologous recombination to interbreeding (Bobay and Ochman 2017, 2018), building on earlier work by Dykhuizen and Green (1991) on *Escherichia coli* and extending this assimilative exercise to encompass the *pangenome*, the complete set of genes in all strains of a given microbial species (Bobay 2020).

The associated development of metagenomics alongside techniques for large-scale sampling of genomic elements, both of which have been pioneered in the microbial world, forms part of this shift informing a species literature that has been skewed historically by an overwhelming focus on the macrobial world. The growth of computationally driven delimitation methods, the most popular of which are multispecies coalescent models (Carstens et al. 2013), has been one important development in the species literature. Viewed as integrating population genetics with phylogenetic analysis in order to more accurately construct species trees, multispecies coalescent models now provide a large number of algorithmic species discovery procedures that have been taken to build on de Queiroz's general lineage conception of species (see Quinn, this volume; Smith and Carstens, this volume).

Excitement about the microbial world has also become manifest in the emergent interdisciplinary field of multispecies ethnography (Haraway 2008; Kirksey and Helmreich 2010; Kirksey 2014). In contrast with earlier ethnobiological interest in the human universality of species-like categories across cultures, this ethnographic perspective on species shifts to focus on the entangled relationships between a variety of human and nonhuman species, including plants, nonhuman animals, fungi, and microbes such as chytrids and *Wolbachia* (Kirksey 2015). Based in environmental, animal, and science and technology studies, multispecies ethnography views species relations as primary, with species emerging as enactive and intertwined entities.

Multispecies ethnography reflects a broader performative turn in cultural anthropology and practices of care (Hartigan Jr. 2017) and as such, relishes the destabilisation of what are viewed as ossified and problematic ways of thinking of the natural world. (For some anthropologists (Ingold 2013), this includes species and so, multispecies! Echoes of debate over eliminativism about species, to be sure.) To date, however, the enrichment of discussions by philosophers and biologists working on the species problem of this recent provocative work has been limited, despite its clear relevance both to lively topics (e.g., species and holobionts) and to methodological shifts (e.g., viewing species through the science as practice lens).

In light of these developments, it is easy to forget that until the turn of the twenty-first century, microbes – bacteria and the eukaryotic protists, for example – were largely ignored in the species literature. Furthermore, attempts to shed that ignorance were often met with ridicule and hostility. Bacteria were typically simply bracketed out from the rest of the living world in discussions of species; in an informal interview, Ernst Mayr went so far as to call eukaryotic protists a 'sort of garbage can group' (Mayr 2004). Microbial biologists, whether working on bacteria, eukaryotic protists, or other microbial organisms and clonelines, mostly just got on with the job of describing the diversity they found in the microbial world (Nanney 1999; Warren et al. 2016). They chiefly opted for phenetic views of microbial groups, including of those referred to by Linnaean binomials and so regarded as species.

Although both the labels 'pheneticism' and 'numerical taxonomy' (Sokal and Crovello 1970; Sneath and Sokal 1973) have made a quiet exit from discussions of species over the past 25 years, it is perhaps worth reflecting on the relationship between multispecies coalescent models and pheneticism. Both ultimately rely on computational strategies that require judgements of similarity, revealing a conventional dimension to taxonomic decisions, including with respect to species. The reliance of pheneticism about species on similarity judgements of this kind served as a red flag for those skeptical of either mind-dependent or theoretically neutral conceptions of species (e.g., Hull 1997, 1999; Mayden 1997). Whether the same holds true of multispecies coalescent models, or whether they are, by contrast, taken to manifest a kind of pluralistic realism about species (Nathan 2019), remains to be seen.

This niche in the species literature promises to become further sophisticated by the continuing integration of focused discussions of genealogical discordance (Haber 2019; Velasco 2019). Here, the lineages of component entities (such as genomes) diverge from those of which they are components (such as organisms and species). More general discussion of the phenomenon of discordance (Haber and Molter 2019) has already informed views of multispecies biofilms (Pedroso 2019) and broader evaluations of methodologies within phylogenetic reconstructions (Quinn 2019).

Both de Queiroz's metapopulational lineage conception of species and Boyd's homeostatic property cluster view of species have suggested ways to develop kinds of pluralistic realism about species: 'pluralism' because of the multiplicity of criteria for lineage segmentation (de Queiroz) and of homeostatic mechanisms (Boyd), and 'realism' as acknowledgment that these criteria and mechanisms constitute the joints at which nature itself is carved. The attraction of an eliminativist view of species has proved strong for some who have weighted the pluralistic dimension to such views more heavily (Mishler and Donoghue 1982; Ereshefsky 1992a), leading to the idea that, as Mishler and Wilkins (2018: 1) have noted, 'the species rank should disappear as part of a general move to rankless taxonomy (Ereshefsky 1999; Mishler 1999; Pleijel 1999)'.

Barker (2019b) has recently argued that a dilemma facing Ereshefsky's (1992a) arguments for eliminative pluralism holds more generally for all extant forms of eliminative pluralism. Call the categories putatively replacing the species concept the *successor categories*. Barker's dilemma turns on the features and relationships shared by such successor categories. If these are shared across successor categories, then they are the basis for forming a superordinate category – such as species – thus undermining the *eliminativist* part of eliminativist pluralism. If these are not shared across successor categories, however, then this undermines the *pluralism* of the view, since without these, each successor category inherits whatever doubts there are about the scientific interest of the species category (see Barker 2019b: 672–673).

*****

Given that this Afterword began with some faux-data about 'species', it seems appropriate that it should end with a perhaps apocryphal story about philosophy. In the darkness before naturalistic light illuminated philosophy of biology as a new field within the philosophy of science just over 50 years ago, the end of philosophy was being predicted by many in the Anglo-American traditions dominated by 'linguistic philosophy'. What had been figured out was that philosophical problems were puzzles to be dissolved, matters simply of the language games we chose to play or not to play. And, all that was needed was a record of what philosophy was, a record to be created as the *Encyclopedia of Philosophy*.

Contained within the eight large volumes of the original *Encyclopedia* was barely a whiff of anything about the philosophy of biology, let alone about species, except insofar as Aristotle seemed to have had views about them. Wind forward to the contemporary online treasure-trove, the *Stanford Encyclopedia of Philosophy*, and things couldn't look more different. Not only does 'species', of course, have its own substantive entry, but discussions of species can be found in articles on conservation biology, biodiversity, Darwinism, human nature, philosophy of macroevolution, and biological individuals. Rather than being records of a moribund past, these discussions are very much part of ongoing interchanges between philosophers and biologists, often drawing as much on articles in journals in the evolutionary, ecological, and other biological sciences as on those in *Philosophy of Science* or *Biology and Philosophy*. If they are a record of anything, they are a record of the healthy future that the species problem has for philosophers and biologists alike.

REFERENCES

Atran, S., 1999. The Universal Primacy of Generic Species in Folkbiological Taxonomy: Implications for Human Biological, Cultural, and Scientific Evolution. In R.A. Wilson (ed.). *Species, New Interdisciplinary Essays*, Cambridge, MA: Bradford/MIT Press, pp.231–261.

Barker, M. J., 2019a. Evolving Lineages, Including Many Species, as Evolutionary Feedback Systems, *Philosophy, Theory, and Practice in Biology* 11 (13). doi:10.3998/ptpbio.16039257.0011.013.

Barker, M.J., 2019b. Eliminative Pluralism and Integrative Alternatives: The Case of Species, *British Journal for the Philosophy of Science* 70: 657–681.

Barker, M.J., this volume. We Are Nearly Ready to Begin the Species Problem.

Barker, M.J. and R.A. Wilson, 2010. Cohesion, Gene Flow, and the Nature of Species, *The Journal of Philosophy* 107: 59–77.

Bobay, L.M., 2020. The Prokaryotic Species Concept and Challenges. In H. Tettlin and D. Medini (eds.), *The Pangenome: Diversity, Dynamics and Evolution of Genomes.* Cham, Switzerland: Springer Nature, pp.21–49.

Bobay, L.M., and H. Ochman, 2017. Biological Species are Universal Across Life's Domains, *Genome Biology and Evolution* 9(3): 491–501.

Bobay, L.M., and H. Ochman, 2018. Biological Species in the Viral World, *Proceedings of the National Academy of Sciences 115*(23): 6040–6045.

Boyd, R.N., 1988. How to Be a Moral Realist. In G. Sayre-McCord, (ed.), *Essays in Moral Realism,* Ithaca, NY: Cornell University Press, pp. 181–228.

Boyd, R.N., 1991. Realism, Anti-Foundationalism and the Enthusiasm for Natural Kinds, *Philosophical Studies* 61: 127–148.

Boyd, R.N., 1999. Homeostasis, Species, and Higher Taxa. In R. A. Wilson (ed.), *Species: New Interdisciplinary Essays,* Cambridge: Cambridge University Press, pp. 141–85.

Boyd, R.N., 2010. Homeostasis, Higher Taxa and Monophyly, *Philosophy of Science* 77(5): 686–701.

Boyd, R.N., 2019. Rethinking Natural Kinds, Reference and Truth: Towards More Correspondence with Reality, Not Less, *Synthese*, 23 Feb 2018: 1–41. https://doi-org.ezproxy.library.uwa.edu.au/10.1007/s11229-019-02138-4

Carstens, B.C., T.A., Pelletier, N.M., Reid, and J.D., Satler, 2013. How to Fail at Species Delimitation, *Molecular Ecology,* 22(17): 4369–4383.

Claridge, M.F., A.H. Dawah, and M.R. Wilson, 1997. *Species: The Units of Biodiversity.* London: Chapman and Hall.

Darwin, C., 1859. On the Origin of Species by Means of Natural Selection. London: John Murray.

de Queiroz, K., 1998. The General Lineage Concept of Species, Species Criteria, and the process of Speciation: A Conceptual Unification and Terminological Recommendations. In D.J. Howard and S.H. Berlocher (eds.), *Endless forms: Species and Speciation.* Oxford: Oxford University Press, pp.55–75.

de Queiroz, K., 1999. The General Lineage Concept of Species and the Defining Properties of the Species Category In R.A. Wilson (ed.), *Species, New Interdisciplinary Essays.* Cambridge, MA: Bradford/MIT Press, pp.49–89.

de Queiroz, K., 2005. Ernst Mayr and the Modern Concept of Species, *Proceedings of the National Academy of Sciences* 102 (Supp. 1): 6600–6607. doi:10.1073/pnas.0502030102.

de Queiroz, K., 2007. Species Concepts and Species Delimitation, *Systematic Biology* 56(6): 879–886. doi:10.1080/10635150701701083.

de Queiroz, K., 2011. Branches in the Lines of Descent: Charles Darwin and the Evolution of the Species Concept, *Biological Journal of the Linnean Society 103*(1): 19–35.

Dobzhansky, T., 1950. Mendelian Populations and their Evolution, *American Naturalist* 84: 401–418.

Doolittle, W.F., 2009. The Practice of Classification and the Theory of Evolution, and What the Demise of Charles Darwin's Tree of Life Hypothesis Means for Both of Them, *Philosophical Transactions of the Royal Society B: Biological Sciences 364*(1527): 2221–2228.

Doolittle, W.F., 2013. Microbial Neopleomorphism, *Biology and Philosophy* 28(2): 351–378.

Doolittle, W.F., 2019. Speciation without Species: A Final Word, *Philosophy, Theory, and Practice in Biology* 11(14): 1–16.

Doolittle, W.F., & O. Zhaxybayeva, 2009. On the Origin of Prokaryotic Species, *Genome Research* 19(5): 744–756.

Dupré, J., 1981. Natural Kinds and Biological Taxa, *The Philosophical Review* 90(1): 66–90. doi:10.2307/2184373.

Dupré, J., 1999. On the Impossibility of a Monistic Account of Species. In R.A. Wilson (ed.), *Species: New Interdisciplinary Essays*, Cambridge, MA: MIT Press, pp.3–22.

Dykhuizen, D.E., and L. Green., 1991. Recombination in Escherichia coli and the Definition of Biological Species, *Journal of Bacteriology* 173(22): 7257–7268.

Ereshefsky, M., 1992a. Eliminative Pluralism, *Philosophy of Science* 59: 671–90.

Ereshefsky, M., 1992b (ed.). *The Units of Evolution*. Cambridge, MA: MIT Press.

Ereshefsky, M., 1999. Species and the Linnaean Hierarchy. In R.A. Wilson (ed.), *Species, New Interdisciplinary Essays*. Cambridge, MA: Bradford/MIT Press, pp.285–305.

Franklin, L. R., 2007. Bacteria, Sex, and Systematics, *Philosophy of Science* 74(1): 69–95.

Franklin-Hall, L.R., 2010. Trashing Life's Tree, *Biology and Philosophy* 25(4): 689–709.

Griffiths, P.E., 1999. Squaring the Circle: Natural Kinds with Historical Essences. In R. A. Wilson (ed.), *Species: New Interdisciplinary Essays*, Cambridge, MA: MIT Press, pp. 209–228.

Haber, M.H. 2019. Species in the Age of Discordance, *Philosophy, Theory, and Practice in Biology* 11 (21). doi:10.3998/ptpbio.16039257.0011.021.

Haber, M., and D. Molter, 2019. Species in the Age of Discordance: Meeting Report and Introduction, *Philosophy, Theory, and Practice in Biology* 11(12): 1–8.

Haraway, D., 2008. *When Species Meet*. Minneapolis: University of Minneapolis Press.

Hartigan Jr, J. 2017. *Care of the Species: Races of Corn and the Science of Plant Biodiversity*. Minneapolis: University of Minnesota Press.

Hausdorf, B., 2011. Progress Towards a General Species Concept, *Evolution*: 65(4): 923–931. doi:10.1111/ j.1558-5646.2011.01231.x

Howard, D.J. and S.H. Berlocher (eds.), 1998. *Endless forms: Species and Speciation*. Oxford: Oxford University Press.

Hull, D.L., 1997. The Ideal Species Definition and Why We Can't Get It. In M.F. Claridge, A.H. Dawah, and M.R. Wilson (eds), *Species: The Units of Biodiversity*. London: Chapman and Hall, pp.357–380.

Hull, D.L., 1999. On the Plurality of Species: Questioning the Party Line. In R. A. Wilson (ed.), *Species: New Interdisciplinary Essays*, Cambridge, MA: MIT Press, pp. 23–47.

Ingold, T. 2013. Anthropology Beyond Humanity. *Suomen Antropologi: Journal of the Finnish Anthropological Society*, 38(3): 5–23.

Keil, F.C., and D. Richardson, 1999. Species, Stuff, and Patterns of Causation. In R. A. Wilson (ed.), *Species: New Interdisciplinary Essays*, Cambridge, MA: MIT Press, pp. 263–281.

Kendig, C., 2020. Ontology and Values Anchor Indigenous and Grey Nomenclatures: A Case Study in Lichen Naming Practices Among the Samí, Sherpa, Scots, and Okanagan, *Studies in History and Philosophy of Science Part C: Studies in History and Philosophy of Biological and Biomedical Sciences* 84: 101340.

Kendig, C., this volume. Metaphysical Presuppositions about Species Stability.

Kirksey, E. 2014. (ed.) *The Multispecies Salon*. Durham: Duke University Press.

Kirsksey, E. 2015. Species: A Praxiographic Study. *Journal of the Royal Anthropological Institute* (*N.S.*) 21: 758–780.

Kirksey, S.E. and S. Helmreich, 2010. The Emergence of Multispecies Ethnography. *Cultural Anthropology* 25(4): 545–576.

Lambert, D. and H. Spence, 1995. *Speciation and the Recognition Concept.* Baltimore: Johns Hopkins University Press.

Ludwig, D., and C.N. El-Hani, 2020. Philosophy of Ethnobiology: Understanding Knowledge Integration and Its Limits. *Journal of Ethnobiology 40*(1): 3–20.

Ludwig, D., and L. Poliseli, 2018. Relating Traditional and Academic Ecological Knowledge: Mechanistic and Holistic Epistemologies Across Cultures, *Biology and Philosophy* 33(5–6): 1–19.

Ludwig, D., and D.A. Weiskopf, 2019. Ethnoontology: Ways of World-Building Across Cultures, *Philosophy Compass 14*(9): e12621.

Mallet, J., 2010. Why Was Darwin's View of Species Rejected by 20th Century Biologists? *Biology and Philosophy* 25: 497–527.

Mallet, J., 2013. Darwin and Species. In Michael Ruse (ed.). *The Cambridge Encyclopedia of Darwin and Evolutionary Thought.* New York: Cambridge University Press, pp.109–115.

Mayden, R.L., 1997. A Hierarchy of Species Concepts: The Denouement in the Saga of the Species Problem. In M. F. Claridge, A. H. Dawah, and M. R. Wilson (eds.), *Species. The units of biodiversity.* London: Chapman & Hall, pp.381–424.

Mayr, E., 1942. *Systematics and the Origin of Species.* New York: Columbia University Press.

Mayr, E., 2004. The Evolution of Ernst: Interview with Ernst Mayr. *Scientific American*, July 6th. http://www.scientificamerican.com/article.cfm?id=the-evolution-of-ernst-in

Minelli, A., this volume. Species Before and After Linnaeus.

Mishler, B., 1999. Getting Rid of Species? In R. A. Wilson (ed.), *Species: New Interdisciplinary Essays*, Cambridge, MA: MIT Press, pp. 307–15.

Mishler, B.D., and M.J. Donoghue. 1982. Species Concepts: A Case for Pluralism, *Systematic Zoology* 31: 491–503. doi:10.2307/2413371.

Mishler, B.D., and J.S. Wilkins, The Hunting of the SNaRC: A Snarky Solution to the Species Problem. *Philosophy, Theory, and Practice in Biology*, 10 (001). doi: 10.3998/ptpbio.16039257.0010.001

Nanney, D.L., 1999. When Is a Rose?: The Kinds of *Tetrahymena*. In R. A. Wilson (ed.), *Species: New Interdisciplinary Essays.* Cambridge, MA: MIT Press, pp. 93–118.

Nathan, M., 2019. Pluralism Is the Answer! What Is the Question? *Philosophy, Theory, and Practice in Biology* 11 (15). doi:10.3998/ptpbio.16039257.0011.015.

Novick, A., and W.F. Doolittle, 2020. Horizontal Persistence and the Complexity Hypothesis. *Biology and Philosophy 35*(1): 1–22.

O'Malley, M., 2014. *Philosophy of Microbiology.* New York: Cambridge University Press.

Otte, D. and J. Endler, 1989. *Speciation and its Consequences.* Sunderland, MA: Sinauer.

Paterson, H., 1994. Evolution and the Recognition Concept of Species: Collected Writings. Baltimore: Johns Hopkins University Press.

Pavlinov, I., 2013. The Species Problem: Why Again? In I. Pavlinov (ed.), *The Species Problem: Ongoing Issues.* Intechopen. https://www.intechopen.com/books/the-species -problem-ongoing-issues/the-species-problem-why-again-.

Pedroso, M., 2019. Forming Lineages by Sticking Together, *Philosophy, Theory, and Practice in Biology* 11(16). doi:10.3998/ptpbio.16039257.0011.016.

Pleijel, F., 1999. Phylogenetic Taxonomy, a Farewell to Species, and a Revision of Heteropodarke (Hesionidae, Polychaeta, Annelida), *Systematic Biology* 48(4): 755–789. doi:10.1080 /106351599260003.

Quinn, A., 2019. Diagnosing Discordance: Signal in Data, Conflict in Paradigms, *Philosophy, Theory, and Practice in Biology* 11 (17) doi:10.3998/ptpbio.16039257.0011.017.

Quinn, A., this volume. Species in a Time of Big Data.

Reydon, T.A., 2005. On the Nature of the Species Problem and the Four Meanings of 'Species', *Studies in History and Philosophy of Science Part C: Studies in History and Philosophy of Biological and Biomedical Sciences*, 36(1): 135–158.

Richards, R., 2010. *The Species Problem*. New York: Cambridge University Press.

Smith, M., and B.C. Carstens, this volume. Species Delimitation Using Molecular Data.

Sneath, P.H., and R.R. Sokal, 1973. *Numerical Taxonomy: The Principles and Practice of Numerical Classification*. San Francisco: W.H. Freeman.

Sokal, R.R., and T. Crovello. 1970. The Biological Species Concept: A Critical Evaluation, *American Naturalist* 104: 127–153. doi:10.1086/282646.

Stamos, D.N., 2007. *Darwin and the Nature of Species*. Albany, NY: SUNY Press.

Sterelny, K., 1999. Species as Ecological Mosaics. In R.A. Wilson (ed.), *Species: New Interdisciplinary Essays*, Cambridge, MA: MIT Press, pp. 119–138.

Velasco, J.D., 2019. The Foundations of Concordance Views of Phylogeny, *Philosophy, Theory, and Practice in Biology* 11(20). doi:10.3998/ptpbio.16039257.0011.020.

Warren, A., et al. 2016, Beyond the 'Code': A Guide to the Description and Documentation of Biodiversity in Ciliated Protists (Alveolata, Ciliophora) *The Journal of Eukaryotic Microbiology* 64: 539–554.

Wheeler, Q. and R. Meier, 2000. *Species Concepts and Phylogenetic Theory*, New York: Columbia University Press.

Wilkins, J.S., 2006. The Concept and Causes of Microbial Species, *History and Philosophy of Life Sciences* 28(3): 389–407.

Wilkins, J.S., 2018. *Species: The Evolution of the Idea*. 2nd ed. Boca Raton, FL: CRC Press.

Wilkins, J.S., this volume. The Good Species.

Wilson, R.A., 1996. Promiscuous Realism, *British Journal for the Philosophy of Science* 47(2): 303–316.

Wilson, R.A., 1999a. Realism, Essence, and Kind: Resuscitating Species Essentialism?. In R. A. Wilson (ed.), *Species: New Interdisciplinary Essays*, Cambridge, MA: MIT Press, pp. 187–207.

Wilson, R.A., 2005. *Genes and the Agents of Life: The Individual in the Fragile Sciences: Biology*. New York: Cambridge University Press.

Wilson, R.A., M.J. Barker, and I. Brigandt, 2007. When Traditional Essentialism Fails: Biological Natural Kinds, *Philosophical Topics* 35: 189–215.

Wright, S., 1940. In J. Huxley (ed.), *The New Systematics*. London: Oxford University Press, pp.161–183.

Index

Printed in the United States
by Baker & Taylor Publisher Services